Visual Analytics for Data Scientists

Natalia Andrienko • Gennady Andrienko
Georg Fuchs • Aidan Slingsby • Cagatay Turkay
Stefan Wrobel

Visual Analytics
for Data Scientists

 Springer

Natalia Andrienko
Fraunhofer Institute Intelligent
Analysis and Information Systems IAIS
Schloss Birlinghoven
Sankt Augustin, Germany

Department of Computer Science
City, University of London
Northampton Square, London, UK

Georg Fuchs
Fraunhofer Institute Intelligent
Analysis and Information Systems IAIS
Schloss Birlinghoven
Sankt Augustin, Germany

Cagatay Turkay
Centre for Interdisciplinary Methodologies
University of Warwick
Coventry, UK

Gennady Andrienko
Fraunhofer Institute Intelligent
Analysis and Information Systems IAIS
Schloss Birlinghoven
Sankt Augustin, Germany

Department of Computer Science
City, University of London
Northampton Square, London, UK

Aidan Slingsby
Department of Computer Science
City, University of London
Northampton Square, London, UK

Stefan Wrobel
Fraunhofer Institute Intelligent
Analysis and Information Systems IAIS
Schloss Birlinghoven
Sankt Augustin, Germany

University of Bonn
Bonn, Germany

ISBN 978-3-030-56145-1   ISBN 978-3-030-56146-8   (eBook)
https://doi.org/10.1007/978-3-030-56146-8

This Springer imprint is published by the registered company Springer Nature Switzerland AG
The registered company address is: Gewerbestrasse 11, 6330 Cham, Switzerland

To our families, friends, colleagues and partners

Preface

There are several disciplines concerned with developing computer-oriented methods for data analysis: statistics, machine learning, data mining, as well as disciplines specific to various application domains, such as geographic information science, microbiology, or astronomy. Until recent years, it was customary to believe in numbers and rely in data analysis solely on mathematical and computational techniques. Visualisation appeared after the end of the analysis processes for making illustrations for reports. Currently, the situation is different, and the importance of visualisation as a means to convey information to humans for involving human understanding and reasoning in the processes of analysing complex data and solving complex problems is acknowledged. As a sign of this, data visualisation is included in the list of the skills required for being a data scientist.

Visual analytics is a research discipline that is based on acknowledging the power and the necessity of the human vision, understanding, and reasoning in data analysis and problem solving. It aims at developing methods, analytical workflows, and software systems that can support unique capabilities of humans by providing appropriate visual displays of data and involving as much as possible the capabilities of computers to store, process, analyse, and visualise data. During the time of the existence of visual analytics, the researchers have created a large amount of knowledge on human-computer data analysis that could be used by practitioners, particularly, by data scientists. There is a problem, however: the practicable knowledge is scattered over a great number of scientific papers, which are not only abundant but also not practitioner-oriented. It is very hard for non-researchers to find what can be useful to them in this ocean of publications.

At the same time, the use of visual representations in data analysis becomes a widely accessible and seemingly easy activity due to the appearance of the Python, R and Javascript languages and proliferation of open-access packages with codes for data processing, analysis, modelling, and visualisation. Creation and execution of analytical workflows involving both computations and visualisations is supported by the Jupyter Notebook application, and there are myriads of analytical notebooks created by various people and published on the Web. These notebooks are often taken by other people for being adapted to their needs or as examples of what is possible.

While this is a very positive development, it has a back side. The notebooks are often created or adapted by people having quite little idea of how to choose appropriate visualisation techniques and design correct and effective visualisations of the data they deal with, and also have no good understanding of why, when, and how visualisations need to be used in analysis. Some visualisations occurring in the publicly accessible example notebooks may look impressive and convincing to non-specialists, but, in fact, they may communicate spurious patterns in inadequate ways. Those who view these visualisations and think of doing the same for their data and tasks often lack knowledge that would enable critical assessment and understanding of the suitability of the techniques. Other notebooks include only basic graphics having little analytical value, whereas better ways exist for representing the relevant information.

Besides poor visualisation literacy, detrimental for analysis is a propensity to uncritically trust computers and take the outcome of a single run of an analysis algorithm with default parameter settings, or with settings previously used by someone else, as the final result. Naive analysts may not realise that a slight change in the data or parameters can sometimes significantly change the result; therefore, they may not bother to examine the reaction of the algorithm to such changes and to check results of several runs for consistency. More experienced and critically minded analysts, who usually take the trouble to evaluate and compare what they get from computers, may tend to rely solely on statistical measures rather than trying to gain better understanding with the help of visualisations.

Visual analytics has not only generated the body of knowledge on how to create meaningful visualisations and how to use them effectively in data analysis together with computer operations but also developed a philosophy that should underlie analytical activity. The main principles are the primacy of human understanding and reasoning and awareness of the weaknesses of computers, which cannot see, understand, and think, and thus need to be led and controlled by humans. Both the knowledge and the philosophy should be transferred to practitioners to help them to do better analyses and come to valid conclusions. With this textbook, we make an attempt to do this.

In this book, we do not aim to present the latest results of the visual analytics research and the most innovative and advanced techniques and methods. Instead, we shall present the main principles and describe the techniques and approaches that are ready to be put in practice, which means that they, first, proved their utility and, second, can be reproduced with moderate effort. We put emphasis on describing examples of analyses, in which we explain the need for and the use of visualisations.

Sankt Augustin,   *Natalia Andrienko*
London   *Gennady Andrienko*
June 2020   *Georg Fuchs*
                                                                    Aidan Slingsby
                                                                    Cagatay Turkay
                                                                    Stefan Wrobel

Acknowledgements

This book is a result of our collaboration with many different groups of partners. We are thankful to

- visual analytics researchers who developed methods and analytical workflows presented and discussed in out book;

- students of the master program on data science[1] at the City, University of London, whose feedback on our visual analytics module helped us to understand what and how needs to be taught to them;

- numerous project partners in a series of research projects funded by EU (European Union) and DFG (German Research Foundation), who introduced us to a variety of application domains of visual analytics, helped us to shape and develop our vision, and critically evaluated our approaches;

- our colleagues at Fraunhofer Institute IAIS – Intelligent Analysis and Information Systems[2], especially the excellent team of the KD – Knowledge Discovery department[3], and at City, University of London[4], specifically, the dynamic and vibrant team of GICentre[5];

- our colleagues Linara Adilova and Siming Chen - for careful reading of different versions of our manuscript and providing helpful comments and constructive critiques.

---

[1] https://www.city.ac.uk/study/courses/postgraduate/data-science-msc

[2] www.iais.fraunhofer.de/en.html

[3] https://www.iais.fraunhofer.de/en/institute/departments/knowledge-discovery-en.html

[4] www.city.ac.uk

[5] www.gicentre.net

Writing of the book was financially supported by the German Priority Research Program SPP 1894 on Volunteered Geographic Information, EU projects Track&Know and SoBigData++, EU SESAR project TAPAS, and Fraunhofer Cluster of Excellence on "Cognitive Internet Technologies".

Contents

Acronyms

ACF **AutoCorrelation function**: correlation of a signal with a delayed copy of itself as a function of delay

AI **Artificial Intelligence**

ASW **Average Silhouette Width**: a cluster quality measure

AUC **Area Under the Curve**: a measure of the quality of a classifier; see ROC.

CMV **Coordinated Multiple Views**: two or more visual displays where special techniques support finding corresponding pieces of information.

DTW **Dynamic Time Warping**: a distance function for measuring similarity for time series data.

GPS **Global positioning system**: a satellite-based navigation system that enables a GPS receiver to obtain location and time data without a requirement to transmit any data.

IQR **interquartile range**: the difference between the upper (third) and the lower (first) quartiles of a set of numeric values

KDE **Kernel Density Estimation**: a statistical technique used for data smoothing

LDA **Latent Dirichlet Allocation**: a topic modelling method

LSA **Latent Semantic Analysis**: a topic modelling method

MDS **MultiDimensional Scaling**: a method for data embedding

ML **Machine Learning**

NER **Named Entity Recognition**: a text processing method

NMF **Nonnegative Matrix Factorization**: a topic modelling method

OSM   **OpenStreetMap**: a database of crowdsourced worldwide geographic information and a set of services, including generation and provision of map tiles

PCA   **Principal Component Analysis**: a method for dimensionality reduction

PLSA   **Probabilistic Semantic Analysis**: a topic modelling method

POI   **Point Of Interest**, or **Place Of Interest**: a place in the geographical space.

RMSE   **Root Mean Squared Error**: a measure of model error

ROC   **Receiver Operating Characteristic** Curve: a plot for assessing classifier's performance

SOM   **Self-Organising Map**: a machine learning method that can be used for data embedding

STC   **Space-time cube**: a visual display providing a perspective view of a 3D scene where the dimensions represent 2D space and time.

SVM   **Support-vector machine**: a machine learning algorithm for classification and regression analysis

t-SNE   **T-distributed Stochastic Neighbor Embedding**: a method for data embedding

U-matrix   **unified distance matrix**: the matrix of distances between data items represented by nodes (neurons) of a network resulting from SOM

WGS84   **World Geodetic System** established in 19**84**: a standard reference system for specifying geographic coordinates

XAI   **eXplainable Artificial Intelligence**

Part I

Introduction to Visual Analytics in Data Science

Chapter 1
Introduction to Visual Analytics by an Example

Abstract An illustrated example of problem solving is meant to demonstrate how visual representations of data support human reasoning and deriving knowledge from data. We argue that human reasoning plays a crucial role in solving non-trivial problems. Even when the primary goal of data analysis is to create a predictive model to be executed by computers, this cannot be done without human reasoning and derivation of new knowledge, which includes understanding of the analysis subject and knowledge of the computer model built. Reasoning requires conveying information to the human's mind, and visual representations are best suited for this. Visual analytics focuses on supporting human analytical reasoning and develops approaches combining visualisations, interactive operations, and computational processing. The underlying idea is to enable synergistic joint work of humans and computers, in which each side can effectively utilise its unique capabilities. The ideas and approaches of visual analytics are therefore very relevant to data science.

1.1 What is visual analytics? (A brief summary)

Visual analytics is the science and practice of analytical reasoning by combining **computational processing** with **visualisation**. These are tightly-coupled using **interactive** techniques so that each informs the other. In this book, we will discuss and illustrate the benefits of this approach for doing analysis.

The involvement of computational processing allows computers to do **what computers are good at**: transforming and summarising data and searching for specific pieces of information. Interactive visual interfaces involve the human analyst, allow him or her do **what human analysts are good at**: interpreting and reasoning. There is a long tradition of using visualisation to help interpret results of computing, but visual analysis develops this further, emphasising the benefits of an iterative cycle of doing computations, understanding and evaluating the results to refine the compu-

N. Andrienko et al., *Visual Analytics for Data Scientists*,
https://doi.org/10.1007/978-3-030-56146-8_1

tation analysis or to investigate complementary findings. In visual analytics, there is close-coupling between the use of computational techniques and the interactive visualisation in the data analysis process undertaken by the human analyst. One of the important principles and benefits of visual analytics is to **not take computational analysis results for granted**.

Computational methods include those that summarise (e.g. summary statistics), find relationships (e.g. correlations), and identify formally specified types of patterns (e.g. groups of objects that have similarities according to a certain computable criterion). Often, the analyst has to make decisions about how these methods are run: the input data (e.g. which subsets of data to use and how the data are processed) and the values of method parameters. Depending on the analytical question and the prior experience of the analyst, it may be more or less obvious as to which of these should be. Visual analytics can help the analyst try different parameter values and see what effects these have on the analysis.

As datasets become larger, computational methods get increasingly important for helping analysts. We cannot just plot everything and look at it, like with a small dataset. Recent advances in machine learning techniques (often referred to as Artificial Intelligence) raise high expectations concerning their power to identify important relationships and useful patterns in extremely large datasets. Many of these newer techniques are black boxes that are often taken for granted. There are plenty of reasons why this is problematic. Visual analytics approaches can be applied to these black boxes as to phenomena whose behaviour needs to be studied and understood. This is similar to studying real-world phenomena by analysing data related to them. In this way, visual analytics can help humans to come to some degree of understanding of these models and judge their appropriateness and reliability.

Interactive visual interfaces enable human analysts to see and interpret the outputs of computational models in the context of the input data and parameters. Crucially, they also offer the ability to tweak parameters or data subsets used by the computational analysis techniques and the ability to show what effects these input data and parameters have on the analysis.

The **visual analytics process** involves human interpretation of visually represented data or model outputs and taking decisions concerning the next analytical steps. The next steps will depend on the interpretation, the current knowledge, and the analytical goals. It might be to vary model inputs or parameters to see how sensitive it is to this variation. It might be more directed, refining the model, for example, by removing one of the less important factors. Visualisation would than help indicate the difference this made. It might prompt an investigation of subsets of the data – e.g., by category, for a particular place, by hour, or according to a natural break in the distribution of one of the variables. If a computational method has identified groups of similar (in some formally defined sense) data items, visual representations characterising each group may help the analyst judge whether these groups are helpful or whether different similarity criteria need to be used.

However, visual analytics is not only the science and conduct of careful and effective use of computational techniques, but it is, first and foremost, the science of human **analytical reasoning**, which does not necessarily require the involvement of sophisticated computing but does require appropriate representation of information to the human, so that it can be used in the reasoning. Visual representation is acknowledged to be the best for this purpose, and it is the task of computers to generate these representations from available data and results of computations. In the following, we present and discuss an example of an analysis process in which visual representations inform human reasoning.

1.2 A motivating example: Investigating an epidemic outbreak

Some readers of this book may have quite vague idea of what visual analytics is and may not understand why and how it can be useful for data scientists. Others may believe that visualisation can be good for communicating ideas and presenting analysis results but may not think of visual displays as tools for doing analysis. To help both categories of readers to grasp the basic idea of visual analytics, we shall start with an example showing visual analytics approaches at work. We shall then discuss this example and outline in a more general way where and how visual analytics fits in the data science workflows. This example comes from the IEEE VAST Challenge 2011 [1]. Although the data are synthetic, they were carefully constructed to resemble real data as much as possible. A good feature of these data is that they contain an interesting and even dramatic story, while similar real data may be either uninteresting or unavailable. Let's dive into the story.

1.2.1 Data and task description

Vastopolis is a major metropolitan area with a population of approximately two million residents. During the last few days, health professionals at local hospitals have noticed a dramatic increase in reported illnesses. Observed symptoms are largely flu-like and include fever, chills, sweats, aches and pains, fatigue, coughing, breathing difficulty, nausea and vomiting, diarrhoea, and enlarged lymph nodes. More recently, there have been several deaths believed to be associated with the current outbreak. City officials fear a possible epidemic and are mobilising emergency management resources to mitigate the impact.

We have two datasets. The first one contains microblog messages collected from various devices with GPS capabilities, including laptop computers, hand-held computers, and cellular phones. The second one contains map information for the entire metropolitan area. The map dataset contains a satellite image with labelled highways, hospitals, important landmarks, and water bodies (Fig. 1.1). There are also supplemental tables for population statistics and observed weather data.

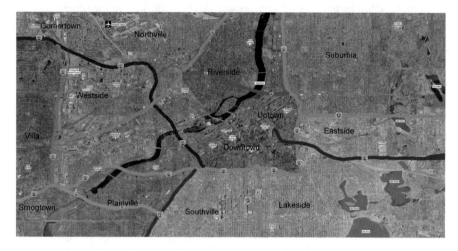

Fig. 1.1: A satellite image-based map of Vastopolis.

We need to find answers to the following questions:

- Identify approximately where the outbreak started (ground zero location). Outline the affected area.

- Present a hypothesis on how the infection is being transmitted. Is the method of transmission person-to-person, airborne, waterborne, or something else?

- Is the outbreak contained? Is it necessary for emergency management personnel to deploy treatment resources outside the affected area?

1.2.2 Data properties

We need to acknowledge that the available data do not directly represent disease occurrences; they just contain texts that may mention disease symptoms. We should not assume that the locations and times specified in the microblog records mentioning disease symptoms are the actual locations and times of disease occurrences. People may write about their health condition not necessarily immediately after getting sick and not necessarily from the location where they first felt some health problems. We should also keep in mind that not everyone who gets sick would send a message about it, whereas some people may send more than one message. People may also write about someone else being sick. Besides, messages mentioning disease symptoms may appear not only during the time of epidemic outbreak but also at any other time. These specifics have the following implications for the investigation we need to do:

- We should keep in mind that the distribution of the microblog posts can give us only a rough approximation of the distribution of the disease cases.

- The epidemics may be manifested as patterns of increased temporal frequency and spatial density of disease-mentioning messages. This is what we shall try to find.

Hence, among all data records contained in the dataset, we need to identify the subset of the data that are related to the epidemic. This subset has two major characteristics: first, the texts of the messages include disease-related terms; second, the temporal frequency of posting such messages is notably higher than usual.

1.2.3 Data preparation

First we need to select the data that are potentially relevant to the analysis goals, that is, the messages mentioning health disorders. The task description lists some symptoms that were observed: fever, chills, sweats, aches and pains, fatigue, coughing, breathing difficulty, nausea and vomiting, diarrhoea, and enlarged lymph nodes. These keywords can be used in a query for extracting potentially relevant data records.

We perform querying in an interactive way. We start with putting the keywords from the task description in the query condition. After the query selects a subset of messages that include any of these keywords, we apply a tool that extracts the most frequent terms from these messages (excluding so called "stop words" like articles, prepositions, pronouns, etc.) and creates a visual display called text cloud, or word cloud (Fig. 1.2) using font size to represent word frequencies. In this display, we find other disease-related terms (e.g., flu, stomach, sick, doctor) that occur in the selected messages together with the terms that have been used in the query condition. We extend the query condition by adding these terms; the query extracts additional messages; in response, the word cloud display is updated to show the frequent words and word combinations from the extended subset of messages. We also find that some frequently used words shown in the word cloud are irrelevant (e.g., come, case, today, day, night, etc.), add them to the list of stop words, and make the word cloud update after exclusion of these words (Fig. 1.3).

Now we notice word combinations that appear irrelevant to the epidemic: "chicken flu" and "fried chicken flu" (Fig. 1.3). We apply another query to the selected sub-

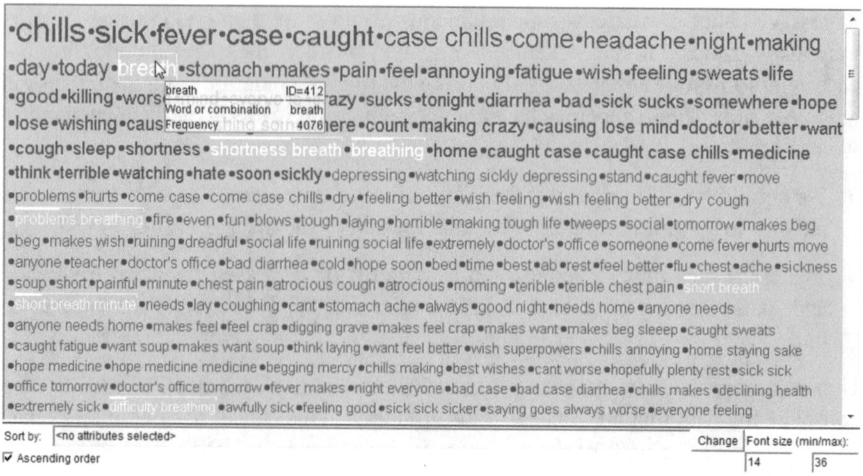

Fig. 1.2: Frequent terms extracted from the messages satisfying filter conditions.

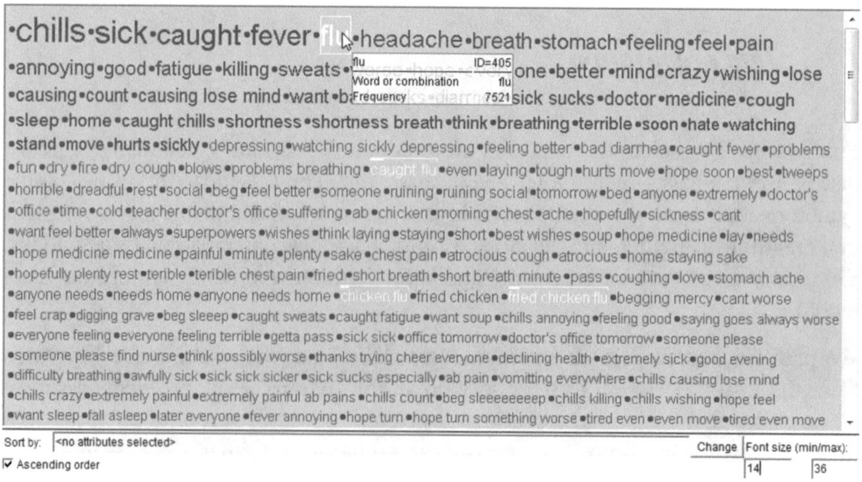

Fig. 1.3: The word cloud display has been updated in response to changing the query condition and extending the list of stop words.

set of messages, which selects only the messages containing the terms 'chicken' and 'flu'. The word cloud changes as can be seen in Fig. 1.4. We also compare the temporal frequency distribution of all messages containing some disease-related terms and the messages containing the terms 'chicken' and 'flu'. For this purpose, we use an interactive filter-aware time histogram, as in Fig. 1.5. The upper image shows the state of the time histogram after selecting the subset of messages containing any of the disease-related terms. Each bar corresponds to one day. The whole bar height is proportional to the total number of messages posted on that day whereas the dark

Fig. 1.4: Frequent words appearing in the messages containing the terms 'chicken' and 'flu'.

segment represents the number of messages satisfying the current filter condition, i.e., containing any of the disease-related terms. We see that the frequency of such messages notably increases in the last three days. This corresponds to the statement in the task description: "During the last few days, health professionals at local hospitals have noticed a dramatic increase in reported illnesses".

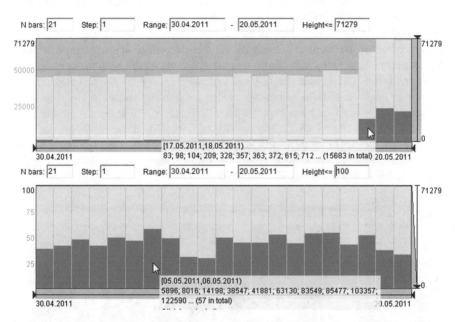

Fig. 1.5: Top: The time histogram shows the temporal frequency distribution of the massages containing disease-related terms. Bottom: The time histogram shows the temporal frequency distribution of the messages containing the terms 'chicken' and 'flu'.

The lower image in Fig. 1.5 shows the state of the time histogram after selecting the messages containing the words 'chicken' and 'flu'. Since the dark segments were small and highly visible (because of low proportions of the selected messages among all messages), we have changed the vertical scale of the histogram using an interactive focusing operation. We see that the messages related to the chicken flu are distributed more evenly throughout the time period covered by the data. The highest frequency of such messages was attained on the seventh day, i.e., long before the increase of the number of disease-related messages. This indicates that the messages mentioning the chicken flu are indeed irrelevant to the analysis task and should be filtered out. So, we exclude these messages from the further consideration. The remaining set consists of 79,579 messages, which is 7.8% out of the original set of 1,023,077 messages.

1.2.4 Analysing the temporal distribution

The temporal histogram (Fig. 1.5, top) shows us that the epidemic happened in the last three days, which are represented by the three rightmost bars of the histogram. More specifically, 59,761 out of the 79,579 disease-related messages (75%) were posted in the last three days. We can conclude that the epidemic happened in the last 3 days; however, we want to identify the time of the epidemic start more precisely. We use a time histogram with hourly temporal resolution, i.e., each bar corresponds to a time interval of 1 hour length (Fig. 1.6), where we see that the temporal frequency of the disease-related messages increased starting from 1 o'clock on May 18, then a very high increase happened at 9 o'clock of the same day, and a high peak occurred at 18 o'clock of that day. In the remaining days, the frequency was stably high except for drops in the night times (between 0 and 2 o'clock), which give us some evidence that the observed frequency increase at 1 o'clock on May 18 is indeed due to the epidemic outbreak start.

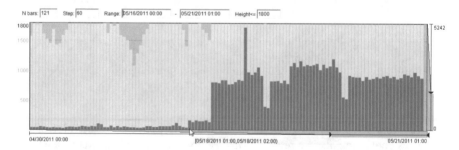

Fig. 1.6: A time histogram shows the temporal frequency distribution of the disease-related messages in the last 5 days by hourly intervals.

1.2.5 Analysing the spatial distribution

To analyse the spatial distribution of the outbreak-related messages (i.e., the messages mentioning disease symptoms that were posted in the last 3 days), we use a *dot map* (Fig. 1.7, top) in which the messages are represented by dots (small circles) in yellow. We observe quite prominent spatial patterns, namely, spatial clusters, which appear as areas with high density of the circle symbols. *Please note that in this and following maps of the message distributions we adjust the transparency of the symbols so that the patterns are best visible.*

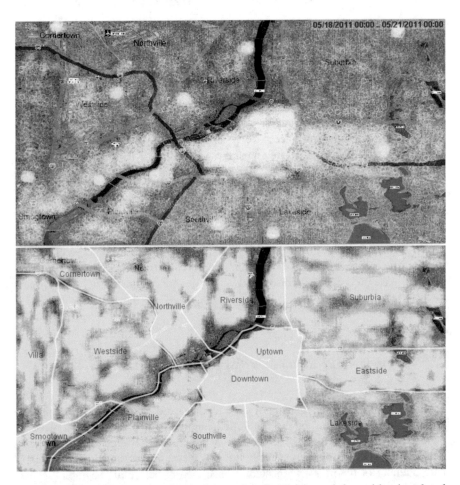

Fig. 1.7: Top: The dot map shows the spatial distribution of the epidemic-related messages. Bottom: The dot map shows the spatial distribution of the disease-unrelated messages.

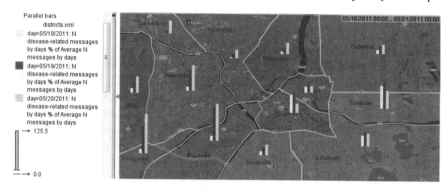

Fig. 1.8: The bar diagrams drawn within district boundaries show the ratios of the numbers of the disease-related messages in the three days of the epidemics to the average daily numbers of the messages posted in the districts before the epidemic outbreak.

Can the distribution of the outbreak-related messages be indicative of the distribution of the disease occurrences? To check this, we need to compare the spatial distribution of the outbreak-related messages to the spatial distribution of the unrelated messages. If these distributions look very similar, there would be no ground for taking the distribution of the messages as a proxy for the distribution of the disease occurrences. The lower dot map in Fig. 1.7 shows the spatial distribution of the disease-unrelated messages, which differs much from the distribution in the upper map. However, we notice a similarity: the density of the messages in the city centre is very high in both maps. Therefore, we cannot be sure that the dense cluster of disease-related messages in the city centre observed in the upper map is a hot spot of the disease outbreak or it corresponds to the generally high density of messages posted in this area.

1.2.6 Transforming data to verify observed patterns

To check whether the high density of the outbreak-related messages in the centre is due to the high spread of the disease in this area or due to the usual high message posting activity, we perform some calculations based on the available data. Using the boundaries of the city districts (visible in Fig. 1.7, bottom), we compute the average daily number of the messages posted in each district before the beginning of the outbreak. We also compute the number of disease-related messages posted in each of the three days of the epidemics. From these numbers, we compute the ratios of the numbers of the epidemic-related messages to the average daily message counts. The computed numbers are represented by bar diagrams in the map in Fig. 1.8.

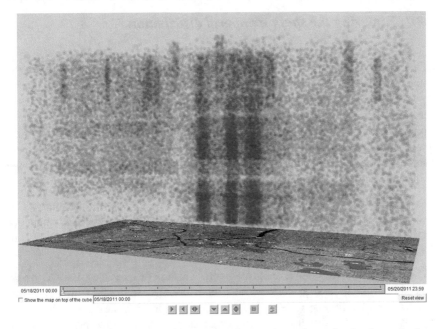

Fig. 1.9: The red dots are put in a space-time cube, where the horizontal plane represents the geographic space and the vertical dimension time, according to the spatial locations and posting times of the outbreak-related messages. The cube thus shows the spatio-temporal distribution of the messages.

The yellow, red, and cyan bars correspond to the first, second, and third day of the outbreak, respectively.

We see that one of the two central districts, called Downtown, has notably higher relative numbers of disease-related messages in the first day of the outbreak than the other districts, except the ones on the east and southeast. Hence, it can be concluded that this district was indeed hit by the outbreak on the first day. The other central district, called Uptown (northeast of Downtown), has only a slightly higher relative number of outbreak-related messages than other districts. However, this district covers a relatively small part of the dense cluster of disease-related messages. Hence, we can conclude that we see the cluster in the city centre because this area was hit by the outbreak and not just because it usually has high message posting activity.

1.2.7 Exploring the spatio-temporal distribution

Already the bar diagram map in Fig. 1.8, indicates that the spatial distribution of the outbreak-related messages was not the same during the three days of the epidemic. To see the evolution of the spatial distribution in more detail, we use a space-time cube (STC) display (Fig. 1.9). It is a perspective view of a 3D scene where the horizontal plane represents the geographic space and the vertical dimension represents time going in the direction from the bottom to the top. The epidemics-related messages are represented by dots (in red, drawn with low degree of opacity) positioned in the cube according to their spatial locations and posting times.

The observed gaps along the vertical dimension of the STC (i.e., time intervals of low density of the dots) correspond to the night drops in the message numbers observed earlier in a time histogram (Fig. 1.6). These gaps separate the three days of the outbreak.

We see that three very dense spatio-temporal clusters of messages, i.e., very high concentrations of messages in space and time, emerged on the first day of the epidemics. We use a dot map to see better the spatial footprints of the clusters (Fig. 1.10, top). It appears that the disease might originate from these three places, or, even more probably, these were areas visited by many people on the first day of the outbreak. Relatively high message density was also on the east of the three central clusters. By the end of day 1 and during day 2, the spatial spread of the messages increased; in particular, the density of the messages increased on the southwest of the city. In the third day, multiple spatially compact clusters emerged. The map in Fig. 1.10, bottom, shows that these clusters are located around hospitals, which indicates that ill people came to hospitals.

1.2.8 Revealing the patterns of the disease spread

Ill people might have posted messages concerning their health condition multiple times. To see how the disease spread, it is reasonable to look only at the distribution of the messages where disease symptoms were mentioned for the first time. To separate such messages from the rest, we apply another transformation to the data. Each record contains an identifier of the person who posted the message. We link the disease-mentioning messages of each person into a chronological sequence. There are 27,446 such sequences, and this is the number of individuals who supposedly got sick (it is 37% of the 73,928 distinct individual's identifiers occurring in the dataset). The lengths of the sequences vary from 1 to 6. Now we take only the first message from each sequence and look at the spatio-temporal distribution of these messages using a space-time cube display, as in Fig.1.11. The distribution differs from that of all outbreak-related messages (Fig.1.9). Most notably, we don't

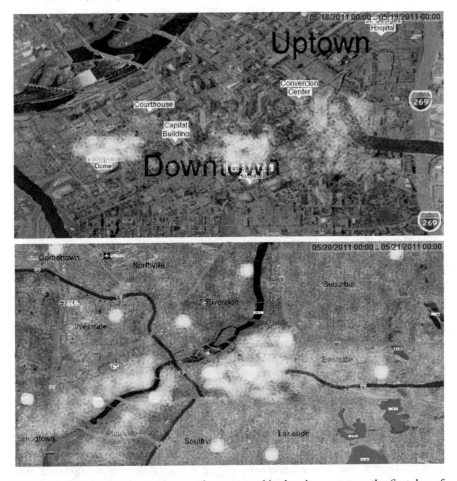

Fig. 1.10: Top: Three dense clusters that emerged in the city centre on the first day of the outbreak. Bottom: The distribution of the disease-related messages on the third day of the outbreak.

see the hospital-centred clusters on the third day. The three very dense clusters that emerged in the city centre on the first day dissolved on the second day. We see a zone of increased message density stretching from the centre to the east on the first day and another zone that formed on the second day on the southwest of the city.

Based on these observations, we conclude that the disease started in the centre and spread to the east on the first day. On the second day, the outbreak hit the southwestern part of the city. In the city centre, the frequency of the disease cases remained quite high during the second and third days. Beyond the observed clusters, the remaining messages were scattered over the whole territory.

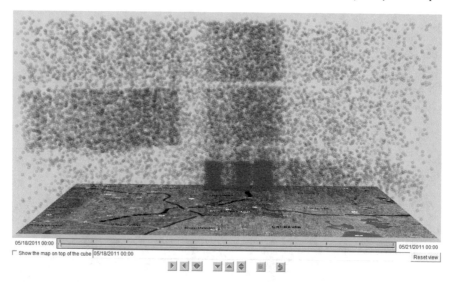

Fig. 1.11: The space-time cube shows the spatio-temporal distribution of the messages where people mention disease symptoms for the first time.

1.2.9 Identifying the mechanisms of the disease transmission

If we consider the first two days of the epidemic, when the disease was spreading (we can disregard the third day when the messages mostly concentrated around the hospitals), we basically see two major areas highly affected by the outbreak: the centre and east of the city and the southwest. The latter area was affected later than the former. We need to understand why it was so. From the weather data provided together with the messages, we learn that on May 18 (i.e., on the first day of the outbreak), there was wind from the west and on the next day from the west and northwest. This could explain the propagation of the disease to the east but not to the southwest. We wonder whether the disease symptoms were the same in the two areas. The illustrations in Fig. 1.12 show that this is not so. We have used an interactive spatial filtering tool for selecting the messages from the central-eastern area and from the southwestern area and looked at the corresponding frequent keywords in the word cloud display. The most frequent keywords in the centre and on the east were chills, fever, caught, headache and other words indicating flu-like symptoms. In the southwestern area, the most frequent symptoms were stomach disorders. Hence, these two areas were hit by different diseases, which, probably, have different transmission mechanisms. It is the most likely that the flu-like disease was transmitted from the centre to the east by the western wind, but this does not apply to the stomach disorders.

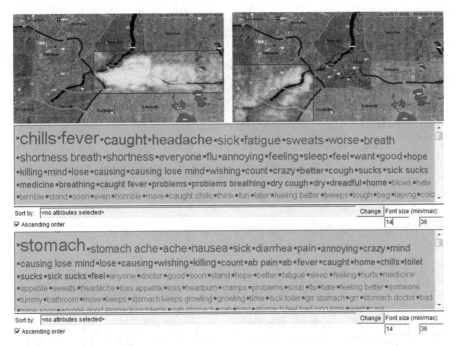

Fig. 1.12: Top: Selecting messages from two outbreak-affected areas by a spatial filter. Middle and bottom: The frequent keywords that occurred in the messages in these two areas.

On the map, we can observe that the dense message clusters on the southeast are aligned along a river; hence, the stomach disease could be transmitted by the water flow in the river.

Could both diseases have a common origin? On the upper map in Fig. 1.7, the central and southwestern clusters seem to emanate from a point where a motorway represented by a thick dark red line crosses the river. It is probable that there was a common reason for both diseases. We come to a hypothesis that some event might have happened on or near the motorway bridge before the 18th of May causing toxic or infectious substances to be discharged in the air and in the river. This probable event might leave traces in the microblog messages. To check this, we apply spatial filtering to select the area around the bridge and temporal filtering to select the day before the outbreak started. The word cloud display (Fig. 1.13) indicates that a truck accident occurred in this place causing a fire and spilling of cargo. Evidently, the fire produced some toxic gas that contaminated the air, and the spilled cargo contained some toxic substance that contaminated the water.

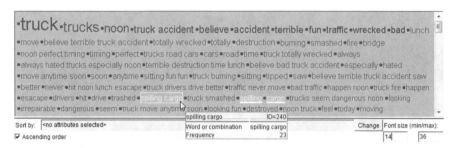

Fig. 1.13: The frequent words and combinations from the messages posted near the motorway bridge on May 17.

1.2.10 Identifying the epidemic development trend

What is the tendency in the outbreak development? Does the disease continue spreading? Are any actions required to stop the spread, or the health professionals mainly need to help people who already got sick? To answer these questions, we first look at the time histogram of the frequencies of the messages that mention disease symptoms for the first time (Fig. 1.14). The histogram bars are divided into segments of two colours, red for the messages mentioning digestive disorders and blue for the remaining messages. We see that the overall frequency of the messages gradually decreases, meaning that the outbreak goes down. This also indicates that the disease, most likely, is not transmitted from person to person; otherwise, we would observe an increasing rather than decreasing trend.

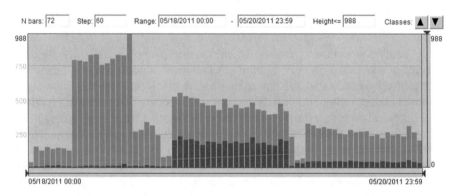

Fig. 1.14: The time histogram of the frequencies of the first mentioning of the disease symptoms. The red bar segments correspond to the messages mentioning digestive disorders and the blue segments to the remaining messages.

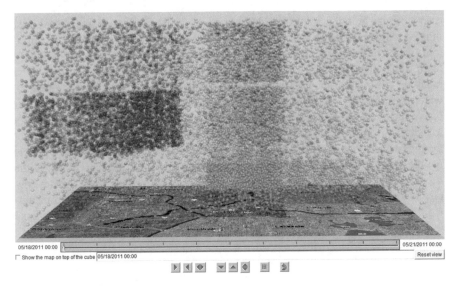

05/18/2011 00:00

☐ Show the map on top of the cube 05/18/2011 00:00

05/21/2011 00:00

Reset view

Fig. 1.15: The spatio-temporal distribution of the messages mentioning health problems for the first time. The messages mentioning digestive disorders are represented in red and the remaining messages in blue.

Looking separately at the red and blue segments, we can notice that the frequencies of the new messages mentioning digestive disorders (red) are much lower on the third day than on the second day, whereas the frequencies of the messages mentioning flu-like symptoms are almost as high as on the second day, which means that the morbidity rate does not decrease as fast as it would be desired. We propagate the red-blue colouring also to the space-time cube (Fig. 1.15). We see that new mentions of digestive disorders appear mainly on the southwest, as in the second day, which means that the water remains contaminated. The new mentions of the flu-like symptoms are scattered everywhere but the highest concentration is in the centre. It may mean that some traces of contamination still remain in this area, and it would be good to clean it somehow. It is also reasonable to warn the population about the risks of contact with the water in the river.

1.2.11 Summary: The story reconstructed

We seem to have plausibly reconstructed the story of what happened in Vastopolis. On May 17, around the noon time, a traffic accident on the motorway bridge in the city centre caused a fire on a truck carrying a toxic substance. The western wind moved the smoke containing toxic particles to the central and eastern areas of the city. The cargo from the damaged car spilled into the river and was moved by the river flow towards the southwest. Many people in the city centre and on the east inhaled toxic particles on the next day after the accident and got ill. The main symptoms were chills, fever, caught, headache, sweats, and shortness of breath, similarly to flu. Unfortunately, since the city centre is the busiest and most crowded area, quite many people were affected. The toxic spill in the river also had sad consequences, which were mainly observed on May 19. It affected people who were on the river banks and, possibly, had direct contacts with the water. Toxic particles somehow got to their stomachs and led to disorders of the digestive system. The morbidity rate of the digestive system disease has notably decreased on May 20. New cases of people feeling flu-like symptoms continued to appear on the second and third day after the accident. The morbidity rate decreases quite slowly, calling for some measures to clean the territory. Fortunately, there is no evidence that the disease can be transmitted through personal contacts; therefore, there is no need to isolate affected people from others and examine everyone who was in contact with them.

1.3 Discussion: How visual analytics has helped us

We have reconstructed the story and answered the questions of the challenge by means of analytic reasoning, which is the principal component of any analysis. To be able to reason, we need to ingest information into our brain. What is the best way to do this? Can we just read all data records? Even if we could read the available 1,023,077 records in a reasonable time, would this help us to understand what was going on? It seems doubtful. Throughout the whole process of analysis, we used visual representations, simply speaking, pictures.

You probably heard the idiom "A picture is worth a thousand words"[1]. In our example, a picture can be worth more than a million records. Instead of reading the records, we could catch useful information in just one look. Pictures were the main sources providing material for our reasoning. This material was put in such a form that could be very efficiently transferred to our brain using the great power of our vision. What our vision does is not just transmitting pixels. Psychological studies show that human vision involves abstraction [26]. Seeing actually means subconsciously constructing patterns and extracting high-level features, and it is these patterns and features that we use as material for our reasoning. Moreover, perceiving patterns and features inevitably triggers our reasoning. Hence, whenever a task cannot be fulfilled by routine computer processing but requires human reasoning, visual representations can effectively convey relevant information to the human mind. Of course, the visual representations need to be carefully and skilfully constructed to be really effective and by no means mislead the viewers by conveying false pat-

---

[1]   https://en.wikipedia.org/wiki/A_picture_is_worth_a_thousand_words

terns. It is one of the goals of this book to explain how to construct such representations.

Although one picture can be worth million records, a single picture may not be sufficient for solving a non-trivial problem. Thus, our analysis consisted of multiple steps:

- data preparation, in which we selected the subset of potentially relevant records;
- analysis of the temporal distribution of the records, in which we identified the start time of the epidemic;
- analysis of the spatial distribution, in which we identified the most affected areas;
- verification of observed patterns, in which we checked whether the high density of the disease-related messages is not just proportional to the usual density of the messages;
- analysis of the spatio-temporal distribution, in which we identified how the outbreak evolved, discovered differences between the temporal patterns in two most affected areas, and came to a hypothesis that they could be affected by different diseases;
- comparison between the texts of the messages posted in the two most affected areas, in which we confirmed our hypothesis of the existence of two different diseases;
- reasoning about the disease transmission mechanisms, in which we related the observed patterns to the context information concerning the weather (the wind) and the geographic features (the river);
- hypothesising about a common source of the two diseases based on the observation of the spatial patterns;
- finding relevant information for explaining the reasons for the epidemic outbreak;
- putting our findings together into a story that gives answers to the questions of the challenge.

In each step of this process, we used visual aids for our thinking. Besides, we used various operations: data selection, spatial and temporal filtering, extraction of frequent terms, derivation of secondary data, such as district-based aggregates, constructing record sequences, and extracting sequence heads. Throughout the process, we continuously interacted with our computer, which performed these operations upon our request and also produced the visual representations that we used for our reasoning. The whole process corresponds to the definition of visual analytics: *"the science of analytical reasoning facilitated by interactive visual interfaces"* [133, p. 4]. An essential idea of visual analytics is to combine the power of human reasoning with the power of computational processing for solving complex problems.

At the same time, our analysis process also corresponds well to what data scientists usually do: select and process data, explore the data to identify patterns, verify the patterns, develop models, and communicate results. These activities cannot be done without human reasoning, and, as we showed and discussed earlier, visual representations can be of great utility. Perhaps, the main difference between visual analytics and data science is their focusing on different aspects of the analytical process, which is performed by a joint effort of the human and the computer. Data science focuses on the computer side, on computational processing and derivation of computer models (here and further throughout the book, we use the term *computer model* to refer to any kind of model that is meant to be executed by computers, typically for the purpose of prediction). Visual analytics focuses on the human side, on reasoning and derivation of knowledge.

Visual analytics considers the **knowledge** generated by the human to be an essential result of the analytical process, irrespective of whether a computer model is built or not. This knowledge can also be seen as a kind of model of the subject that has been analysed. It is a **mental model**[2], that is, a representation of the subject in the mind of the human analyst. In our example, we have constructed such a model in the process of the analysis. It can not only explain what happened but also predict how the situation will develop and tell what actions should be taken. Hence, it may not always be necessary to develop a computer model, yet whenever a computer model is required, visual analytics can greatly help in building it. Moreover, a computer model cannot be appropriately used without human knowledge of what it is supposed to do and when and how to apply it. This knowledge of the computer model is, like the knowledge of the phenomenon that is analysed and modelled, an important outcome of the analytical process. Therefore, the scope of visual analytics includes not only derivation of mental models (i.e., new knowledge represented in the analyst's mind) but also conscious development of computer models that are well considered and well understood.

As you see, visual analytics aptly complements data science, and, moreover, it is instrumental for doing **good data science**, because non-trivial and non-routine analysis tasks require joint efforts of computers and humans. Visual analytics provides a way to perform data science so that the power of human vision and reasoning are effectively utilised.

1.4 General definition of visual analytics

As we already mentioned, visual analytics has been defined as "the science of analytical reasoning facilitated by interactive visual interfaces" [133, p. 4]. This definition emphasises a certain kind of activity (analytical reasoning) and a certain tech-

---

[2] https://en.wikipedia.org/wiki/Mental_model

nology (interactive visual interfaces) supporting this activity. We have discussed in the previous section why visual representations are essential for human analytical reasoning. However, the definition also contains the keyword "interactive". What does it mean, and why do visual interfaces need to be interactive?

When we consider the use of a picture representing data, it is quite likely that this picture does not include all information that we may need for our reasoning, or some information may not be easily perceivable because our attention is attracted to stronger visual stimuli. In such cases, we need to modify the picture or to obtain something in addition to it. Here are some examples of what we may need:

- get a more detailed view of some part of the picture (*zooming*);

- hide information that we currently do not need for our reasoning (*filtering*);

- see exactly what data records stand behind some elements of the picture (*querying*);

- create additional representations showing different facets of the same data (*multiple views*);

- find corresponding pieces of information in two or more displays (*linking multiple views*).

All these are examples of interactive operations. Usually, to gain knowledge from non-trivial data, it is not enough just to look at a single static (non-interactive) picture, even when it is perfectly designed. Therefore, the definition of visual analytics emphasises the importance of interaction for analytical reasoning.

While the human brain is a powerful instrument for analysis and knowledge building, it has its limitations, mostly regarding the memory capacity and the speed of operation. In these respects, computers are immensely more powerful. Then, why not to combine the strengths of the humans and computers? This is what visual analytics aims at! It develops such approaches to data analysis and knowledge building in which the labour is distributed between humans and computers so that they can effectively and synergistically collaborate utilising their unique capabilities, some of which are listed below.

Humans	Computers
flexible and inventive, can deal with new situations and problems	can handle huge amounts of data
can associate diverse information pieces and "see the forest for the trees"	can do fast search
can solve problems that are hard to formalise	can perform fast data processing
can cope with incomplete/inconsistent information	can interlink to extend their capacities
can see and recognise things that are hard to compute or formalise	can render high quality graphics

So, it is sensible to put these great capabilities together and let them work jointly. This requires communication between the human and the computer, and the most convenient way for the human is to do this through an interactive visual interface. The book "Mastering the Information Age : Solving Problems with Visual Analytics" [79] says: "The visual analytics process combines automatic and visual analysis methods with a tight coupling through human interaction in order to gain knowledge from data". As a schematic representation of this statement, Figure 1.16 shows how a human and a computer work together to analyse data and generate knowledge. The computer performs various kinds of automated data processing and derives some artefacts, such as transformed data, results of queries and calculations, statistical summaries, patterns, or models. The computer also produces visualisations enabling the human to perceive original data as well as any further data and information derived by means of computational processing. The human uses the information perceived for reasoning and knowledge construction. The human determines and controls what the computer does by selecting data subsets to work on, choosing suitable methods, and setting parameters for processing. Based on the current knowledge, the human may refine what the computer has produced, for example, discard some artefacts as uninteresting and apply further processing to interesting stuff, or partition the input data into subsets to be processed separately.

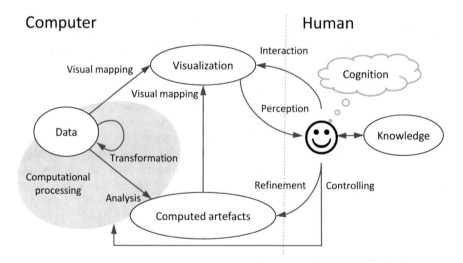

Fig. 1.16: Schematic representation of the visual analytic activity in which human cognition is combined with computational processing. Adapted from [78, 79]

During this activity, the human constantly uses and enriches the knowledge existing in the mind. The human begins the analysis having some prior knowledge and constructs new knowledge as the analysis goes on. The activities of the humans are supported by interactive visual representations created by the computers, and the computers also help the humans by handling data and deriving various kinds of information by means of algorithmic methods. The visual analytics technology thus includes principles and methods of visual representation of information, techniques for human-computer interaction, and algorithmic methods for computational processing. The following chapter provides an overview of these technical means.

Chapter 2
General Concepts

Abstract Analysis is always focused on a certain subject, which is a thing or phenomenon that needs to be understood and, possibly, modelled. The data science process involves analysis of three different subjects: data, real world phenomena portrayed in the data, and computer models derived from the data. A subject can be seen as a system composed of multiple components linked by relationships. Understanding of a subject requires understanding of the relationships between its components. Relationships are reflected in data, which specify correspondences between elements of different components. Relationships are studied by identifying patterns in distributions of elements of some components over elements of other components. A pattern consists of multiple correspondences between elements that are perceived or represented in an integrated way as a single object. In visual displays, patterns appear as shapes or arrangements composed of visual marks. Depending on the kind of components (discrete entities, time, space, numeric measures, or qualities), different types of distributions can be considered: frequency distribution, temporal and spatial distributions, and joint distribution of several components. Each type of distribution may have its specific types of patterns. We give special attention to spatial distribution, noting that the distribution of visual marks positioned in a display space is perceived as a spatial distribution, in which a human observer can intuitively identify spatial patterns. This inclination of human perception to seeing spatial patterns can be exploited by creating artificial spaces where distances represent the strengths of some non-spatial relationships.

2.1 Subjects of analysis

The goal of analysis is to gain understanding of a certain thing or phenomenon. We shall call this thing or phenomenon *the subject* of analysis, or, shorter, analysis subject. This does not imply that analysis is applied directly to the subject. Typically,

© Springer Nature Switzerland AG 2020
N. Andrienko et al., *Visual Analytics for Data Scientists*,
https://doi.org/10.1007/978-3-030-56146-8_2

what we want to understand is a piece of the objectively existing world, but what we analyse is *data*, which are recorded observations and measurements of some aspects of this real world piece. We use the term "subject" to refer to the thing we want to understand. Hence, in this case, the subject of our analysis is the piece of the real world. The data that we use may be called *the object* of the analysis, because we apply analytical operations to the items (records) of the data.

However, we may also want to understand the data as such: the structure of the records, the meanings of the fields and relationships between them, the accuracy, degree of detail, amount, scope, completeness, etc. Moreover, understanding of data is absolutely necessary for performing valid analysis and gaining correct understanding of the thing that is reflected in the data. When we seek understanding of data as such, the data are our analysis subject. In this case, the analysis is applied directly to the subject; in other words, the subject and object of the analysis coincide.

The goal of analysing a real-world phenomenon (here and further on, the word "phenomenon" stands for the longer phrase "thing or phenomenon") may be not only to gain understanding but also to create a computer model of this phenomenon. The purpose of a computer model is to characterise the phenomenon and predict potential observations beyond the available data. Naturally, an analyst strives at creating a good model: correct, complete, prediction-capable, accurate, and so on. To understand how good a computer model is (in terms of various criteria) and in what situations it is valid to use this model, it is necessary to analyse its properties and behaviour. The model becomes the subject of the analysis. For analysing a model, the analyst uses specific data reflecting its performance, such as correspondences between inputs, parameter settings, and outputs. These data will be the object of the analysis.

It is not usual that main goal of analysis is to understand properties of a data set or behaviour of a computer model, so that the data set or the model is the only subject of the analysis. More commonly, the overall goal of analysis is to understand and, possibly, model a real-world phenomenon. The phenomenon is thus the primary subject of the analysis. The expected outcome of the analysis is understanding, or knowledge, of the phenomenon[1] However, to achieve the overall goal, it is necessary to understand not only the phenomenon but also the available data and the computer model if it is built.

Generally, analysis is a multi-step process with the steps focusing on different subjects. At the initial stage of the process, the subject is the data. The goal is to understand properties of the data and assess how well the data represent the phenomenon, which refers both to the data quality and the nature of the phenomenon. At the next stage, the subject is the phenomenon reflected in the data, and the goal is to obtain knowledge and, possibly, derive a computer model of the phenomenon. Once a model is built, it becomes the analysis subject. Visual analytics approaches can be applied to all three subjects: data, phenomena, and computer models.

---

[1] In this book, the terms *knowledge* (of something) and *understanding* (of something) are used interchangeably as synonyms.

In our investigation of the microblog data in Chapter 1, the primary subject of the analysis was the epidemic outbreak, as we aimed at understanding and characterising the outbreak. In respect to this overall goal, the microblog messages were the object of our analysis. However, at the initial stage, it was necessary to select a relevant portion of the data. We made a preliminary selection based on our background knowledge, that is, we extracted texts including particular keywords. To understand whether all these texts are relevant, we investigated the words that frequently occurred in the selected texts and detected frequent occurrences of the phrase "fried chicken flu". We suspected that the texts including this phrase were irrelevant to the outbreak. To be sure, we compared the times of the "fried chicken flu" messages with the temporal distribution of the remaining selected messages. At this stage, the subject of our analysis was the data. In the following steps, the subject was the real-world phenomenon, that is, the epidemic outbreak. Although we did not build a computer model in our example, it was possible to build a model predicting the further spread of the flu-like disease or the numbers of people expected to come to hospitals in the next days. In this case, it would be necessary to study the predictive behaviour of this model, and the model would be the analysis subject.

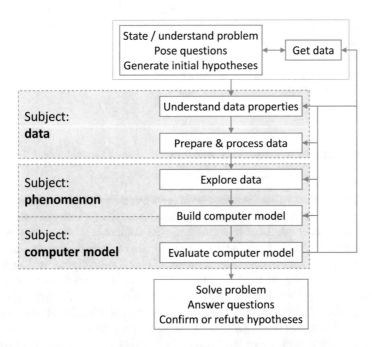

Fig. 2.1: The block diagram represents a general data science workflow. The grey boxes specify the analysis subjects dealt with at different stages of the data science process. Visual analytics approaches can be applied to each of the three subjects.

The view of analysis as a multi-step process with a changing subject corresponds well to what is commonly seen as the data science process, which is schematically represented in Fig. 2.1. It begins with exploring properties of data to understand whether the data are suitable and how they should be prepared to the analysis. Hence, at this stage, the data as such are the subject of analysis. At the next stage, the prepared data are analysed, typically with the aim to derive a computer model that can support decision making. The subject of this analysis is not the data anymore but the phenomenon that is reflected in the data. When a model is derived, it becomes the subject of the following analysis that evaluates model appropriateness. When the model is not good enough, the model building activity needs to be continued. It may involve gaining further understanding of both the phenomenon and the model intended to represent it, which means that these two analysis subjects may be intertwined.

It can be noted that the data science workflow has much in common with the cross-industry standard process for data mining known as CRISP-DM [123], which is a widely used conceptual model describing common approaches used by data mining experts[2]. The CRISP-DM process includes the following phases: business understanding, data understanding, data preparation, modelling, evaluation, and deployment. It can be seen as a business-focused view of the data science process. Also in this view, different steps of the process involve analysis of different subjects: data in the data understanding step, phenomenon in the modelling step, and model in the evaluation step. Visual analytics approaches, which support human reasoning, can be very helpful in each of these steps.

2.2 Structure of an analysis subject

A non-trivial subject of analysis can be considered as a **system** composed of **aspects**, or **components**, linked by **relationships**. Analysis typically aims at understanding and modelling these relationships. The components are usually not just simple, indivisible units. They, in turn, are composed of some **elements**, which may be objects of any nature: concrete or abstract, physical or digital, real or imaginary. Thus, the subject of the VAST Challenge 2011 (Chapter 1) is the epidemic outbreak in Vastopolis whose components are the population of Vastopolis, disease symptoms, posted tweets, space, time, keywords, etc., as schematically represented in Fig. 2.2. The analysis goal is to understand how the disease symptoms are related to the population, space, and time. The population is composed of people, disease symptoms is a set containing multiple possible symptoms, the elements of the space are distinct geographical locations, and the time is composed of moments (instants).

---

[2] https://en.wikipedia.org/wiki/Cross-industry_standard_process_for_data_mining

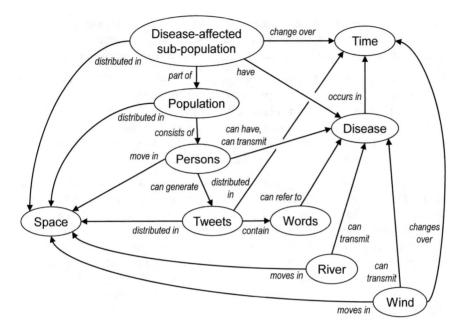

Fig. 2.2: A schematic representation of the structure of the subject in the epidemic outbreak analysis (Chapter 1).

Elements of subject components may also be linked by **intrinsic relationships**, that is, relationships that exist due to the very nature of the components. Based on the existence of certain kinds of intrinsic relationships, different types of components can be distinguished. In our example, there are following types of components, which are common for many actual and potential analysis subjects:

- set of *discrete entities*, such as people, messages, keywords, and disease symptoms;

- *space*, which is a set of locations;

- *time*, which is a set of instants (moments);

- set of values of an *attribute*, which can be a numeric measure, such as amount, frequency, duration, size, and distance, or a non-measurable (qualitative or categorical) characteristic, such as health condition (healthy or sick) and disease type (respiratory or digestive).

Space is a set of locations linked by intrinsic relationships of distance and neighbourhood; the latter can be defined as small distance and absence of barriers. Time is a set of instants linked by intrinsic relationships of distance and order. Besides, time may also be considered as consisting of recurring cycles, such as daily, weekly, and yearly. In such cases, there exist various cycle-based distance relationships between time instances, which are the distances between their positions within the cycles. Similarly to time, values of numeric measures have intrinsic distance and order relationships among them but there are no cycle-based relationships.

Intrinsic relationships, e.g., ordering between time moments or differences between numeric values, are usually not represented in data explicitly. When an explicit representation is needed, it can be obtained by means of routine calculations. For example, distances between values of a numeric attribute or between time instants can be determined by simple subtraction, and there are formulas for determining spatial distances between locations based on their coordinates.

Discrete entities and values of qualitative attributes have no intrinsic relationships among them, that is, relationships that exist by nature and thus cannot be absent. This does not mean, however, that they may not have any relationships. In a specific set of discrete entities, there may be specific relationships, such as social relationships between people and 'reply' relationships between messages. Such specific relationships may be explicitly defined in data or inferred from data. For example, social relationships between people can sometimes be inferred from data portraying people's movements based on repeated simultaneous appearances of two or more individuals at the same places.

2.3 Using data to understand a subject

Understanding a subject requires understanding the relationships between its components. As we said previously, the components are usually composed of elements: people, texts, words, space, time, and so on. Data consist of records that specify correspondences between some (usually not all) *elements* of the components of a subject. The correspondences between elements of the components (i.e., elementary correspondences) are determined by the higher-level relationships that exist between the whole components. Hence, the elementary correspondences specified in the data reflect the higher-level relationships between components.

Analysis is usually concerned not with the elementary correspondences but with the relationships between the components *as the wholes*. Understanding of these higher-level relationships requires *abstraction* from elements to groups of elements and from particular correspondences to relationships between the groups. Such abstraction is metaphorically called "seeing the forest for the trees".

2.3.1 Distribution

Relationships between two components can often be understood by examining the *distribution* of the elements of one of them across the elements of the other. Among multiple definitions of the term "distribution" that can be encountered in dictionaries, the following are appropriate, i.e., correspond to what we mean: "the position, arrangement, or frequency of occurrence (as of the members of a group) over an area or throughout a space or unit of time" (Merriam-Webster's Dictionary[3]) and "the way that something is shared or exists over a particular area or among a particular group of people" (Oxford Advanced Learner's Dictionary[4]).

To understand the concept of distribution, you may think of one component as a *base* (i.e., as a container or a carrier) for elements of another component. In this view, the elements of the base play the role of *positions* that can be filled by elements of the other component. Thus, space and time can be quite easily seen as bases for entities and values of attributes: they provide, respectively, spatial and temporal positions in which entities or attribute values can be put. However, a set of entities can also be seen as a base for attribute values: each entity can be regarded, at a certain level of abstraction, as a kind of "position" for attribute values. The definitions from the dictionaries refer to space (or area), time, and a group of people as possible bases of a distribution. However, these can be extended in an obvious way to groups of any entities and, generally, to a set of elements of any nature. The thing (component of a subject or data) distributed over some base can be metaphorically seen as lying over the base and thus can be called the *overlay* of the distribution.

Based on these considerations, the **distribution** of a component O over (or in) another component B, which is regarded as a base, can be defined as the relationship between the elements of O and the positions in B, that is, which positions in B are occupied by which elements of O. A distribution is composed of the correspondences between the elements of B and O specified in the data.

The types of distributions that are often analysed include:

- distribution of entities over space, such as the spatial distribution of the disease-related microblog posts in our example in Chapter 1 (base: space, overlay: entities);

- distribution of entities over time, such as the temporal distribution of the posted messages (base: time, overlay: entities);

- distribution of values of attributes over a set of entities, such as the distribution of the message lengths over the set of the posted messages (base: entities, overlay: attribute values);

- distribution of values of attributes over space, such as the distribution of the numbers of the disease-related messages over the city districts (base: space, overlay: attribute values);

- distribution of values of attributes over time, such as the distribution of the proportions of the disease-related messages over the days (base: time, overlay: attribute values);

- distribution of one sort of entities over a set of entities of another sort, such as the distribution of the keywords over the messages (base: entities, overlay: entities);

- distribution of specific relationships over a set of entities, such as the distribution of friendship relationships over the set of the microblog users or the distribution of the 'reply' and 're-post' relationships over the set of the messages (base: entities, overlay: relationships).

In a distribution, there are two aspects to consider:

- **correspondence**: associations between elements of the base and the corresponding elements of the overlay;

- **variation**: similarity-difference relationships *between the elements of the overlay* corresponding to different elements of the base.

If we imagine the overlay lying on top of the base, the correspondence consists of the "vertical" relationships between the base and the overlay, and the variation consists of the "horizontal" relationships within the overlay. These concepts are schematically illustrated in Fig. 2.3.

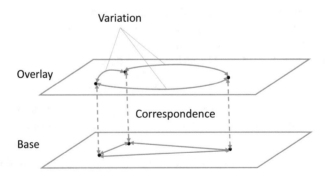

Fig. 2.3: A schematic illustration of the concept of distribution. The dots represent elements of two components and the arrows represent relationships between elements.

Let us give a few examples:

- Distribution: microblog messages over space.
  - Correspondence: which and how many messages were posted in each place.
  - Variation: similarities and differences between the amounts of the messages posted in different places.
- Distribution: contents (keywords) of the microblog messages over space.
  - Correspondence: what keywords occurred in each place.
  - Variation: similarities and differences between the combinations and frequencies of the keywords that occurred in different places.
- Distribution: contents (keywords) of the microblog messages over the set of microblog users.
  - Correspondence: what keywords were used by each individual.
  - Variation: similarities and differences between the combinations and frequencies of the keywords used by different individuals.

Generally, to understand relationships between components, multiple relationships between individual elements need to be considered all together. This implies that the analyst should be able to perceive multiple data items jointly as a meaningful whole. Such a whole can be called a *'pattern'*.

2.3.2 Patterns and outliers

While being widely used in the context of data analysis, the term 'pattern' has no explicit common definition. We propose the following working definition of a pattern:

> A *pattern* is a group of items of any kind that can be perceived or represented as an integrated whole having its specific properties that are not mere compositions of properties of the constituent items.

Particularly, when studying relationships between components of a subject, the analyst tries to identify patterns made by multiple relationships between elements of the components, including correspondences and variation relationships. While there exist many computational methods for automated detection of patterns in data, these methods require a precise specification of the patterns they will seek for. A human analyst can discover patterns by looking at visual representations of data, and no prior specification is required for this purpose. The analyst simply *sees* patterns, which can be manifested, for example, as particular shapes or arrangements of visual marks. Once noticed, patterns need to be interpreted (what do they mean?), verified (do they really exist in data or only emerge in the visualisation?), and, possibly, explained (why do they exist?). Verified patterns can be incorporated in the model that is being built in the analyst's mind and, possibly, in the computer.

There may be items that cannot be put together with any other items because they are very different from everything else. Such items are called *outliers*. Detecting outliers is an important activity in studying relationships. In visual displays, outliers are often manifested as isolated visual marks (such as dots) that are not integrated in any shape formed by other marks. The analyst needs to understand whether outliers are errors in data, or they represent exceptional but real cases. For building a general computer model, outliers may need to be removed, unless they are data errors that can be corrected.

2.3.3 Patterns in different kinds of distributions

As noted in Section 2.3.1 and schematically illustrated in Fig. 2.3, two aspects are considered in exploring distributions, correspondence and variation. We focus on the *correspondence* when we look which or how many elements of one component are associated with particular elements or groups of elements of the other component, for example, how many messages were posted in the city centre and which keywords occurred in these messages. We focus on the *variation* when we look at the differences between the elements of one component associated with different elements or groups of elements of the other component. An example is the differences in the number and contents (keywords) of the messages posted in different areas in the city. A pattern that can be found in a distribution may refer to the correspondence, or variation, or to both.

Frequency distributions. The most basic and common characteristic of a distribution is the frequency of the occurrences of distinct elements of one component over the elements of the other. This characteristic is called *frequency distribution*. In our example analysis in Chapter 1, we looked at the frequencies of the occurrences of distinct keywords in sets of messages. To observe and explore this distribution, we represented it visually in the form of a word cloud.

Frequency distributions of values of numeric attributes are commonly represented by frequency histograms, as in Fig. 2.4, in which the values are grouped into intervals, often called "bins". The general shape of a histogram tells us what values are more usual and thus expectable and what values occur rarely. Thus, the upper two histograms in Fig. 2.4 represent distributions in which medium values are more usual than small and high values, and the lower two histograms exhibit the prevalence of lower values. Still, there are also shape differences between the upper two histograms, as well as between the lower two histograms. On the top left, the range of frequently occurring values is wider, and the frequencies gradually decrease towards the left and right sides. On the top right, there is a narrow interval the values

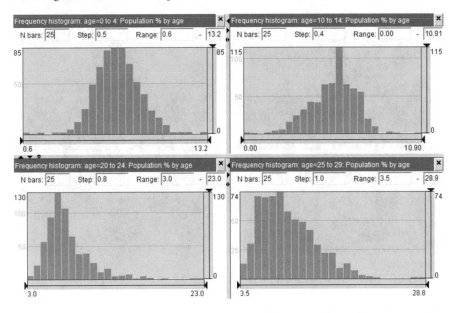

Fig. 2.4: The frequency distributions of four distinct numeric attributes are represented graphically by frequency histograms.

from which are almost twice as frequent as their closest neighbours on the left and on the right. In the lower row, the range of frequent values on the left is much narrower than on the right, and the frequency decrease is steeper. Please note that, when speaking about the differences between the frequencies of different values, we describe *variation patterns*, more specifically, patterns of frequency variation.

A histogram may also reveal the presence of outliers. Thus, in the lower histograms in Fig. 2.4, the very short bars at the right edges represent several outliers.

In exploring a frequency distribution of numeric values with the use of a histogram, it is important to remember that the histogram shape may change depending on the binning, that is, the division of the value range into intervals. Thus, Figure 2.5 demonstrates four histograms representing the same distribution, but each histogram has a different number of bins, which makes the shapes differ. None of these histograms is wrong or better than others, and it is pointless to seek for optimal binning. In order to have better understanding of a distribution, it is reasonable to try several variants of binning and observe whether and how the shape changes.

Temporal distributions. The intrinsic properties of time are ordering and distance relationships between elements, which are time instants (moments). Besides, temporal cycles may be relevant for many application domains and analysis tasks. Some of the types of possible patterns in temporal distributions emerge due to these intrinsic properties. For a component distributed over time, an important concept is *change*, which is a difference between an element related to some time moment and another

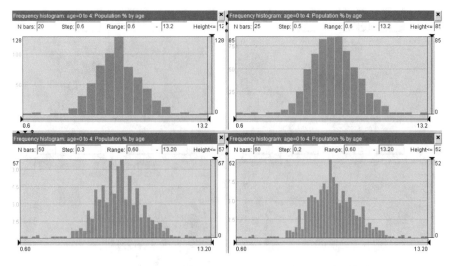

Fig. 2.5: Dependency of the visible pattern on the sizes of the histogram bins.

element related to an earlier time moment. The concept of change is involved in the meanings of many types of temporal distribution patterns. Thus, a constancy pattern means absence of change, and a fluctuation pattern means frequent changes. When a component distributed over time has ordering relationships between elements, a change may be either increase or decrease. A trend pattern means multiple increases or multiple decreases following one another along the sequence of time moments. Since the concept of change refers to the variation in a distribution (Section 2.3.1), all these kinds of patterns can be categorised as *variation patterns*, more specifically, as patterns of temporal variation.

Changes of a set of time-based discrete entities, which can be called *events*, consist of disappearances of existing entities and appearances of new entities. Apart from or instead of the existence of particular entities, analysts are often interested in the total number of entities existing at different times, which can be called the *temporal frequency* of the entities (events); see Figs. 1.5 and 1.6. When we consider the changes of the temporal frequency over time, we are looking for *variation patterns* of the temporal frequency. The temporal frequency can be expressed as a time-based numeric measure whose possible changes are increase and decrease, like for any numeric attribute. It should be noted that the absence of increase or decrease in the number of entities does not necessarily mean the absence of changes in the set of entities, where some or even all entities might have been replaced by other entities. In particular, this is always true for instant events, that is, entities that only exist at the moment of their appearance or entities whose life duration is negligibly short or not relevant to the analysis goals. Thus, the microblog message posts in our example in Chapter 1 are considered as instant events. Examples of non-instant events, i.e., events with duration, are the epidemic outbreak, which lasted for at least three days, according to the data, the bad health condition of each affected person, the truck accident, and the resulting traffic congestion.

A pattern may include other patterns. Thus, a peak pattern consists of an increasing trend pattern followed by a decreasing trend pattern.

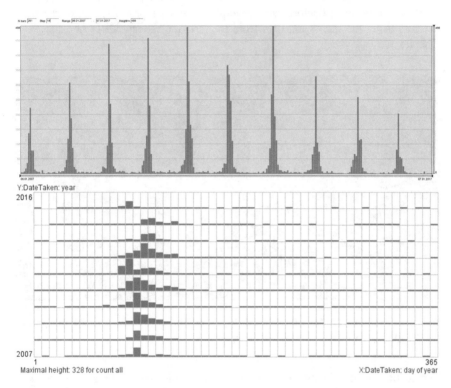

Fig. 2.6: A periodic pattern in a temporal frequency distribution is demonstrated by a linear time histogram (top) and a two-dimensional time histogram (bottom).

Due to the ordering and distance relationships between time moments, a periodicity pattern may take place, which means repeated appearance of particular entities, values, changes, or patterns at (approximately) equal distances in time. A periodicity pattern may be related to time cycles; in this case, repeated appearances of particular entities, values, changes, or patterns have the same or close positions in consecutive temporal cycles, as shown in Fig. 2.6.

Spatial distributions. The intrinsic property of space is the existence of distance relationships among the elements, which are spatial locations. Based on distances, neighbourhood relationships can be defined. Neighbourhood usually means small distance between locations, but, when the space is geographic, two close locations may not be considered as neighbours when there is a barrier between them, such as a river, a mountain range, or a state border.

Space can be a base for discrete entities and for attribute values. For discrete entities, possible types of spatial distribution patterns refer to the distances between the spatial positions of the entities. A distribution of entities in space can be characterised in terms of *spatial density*, which is conceptually similar to the temporal frequency and means the number of entities located in a part of space of a certain fixed size. Figure 2.7 demonstrates examples of some types of spatial distribution patterns using the microblog data from the VAST Challenge 2011 considered in Chapter 1.

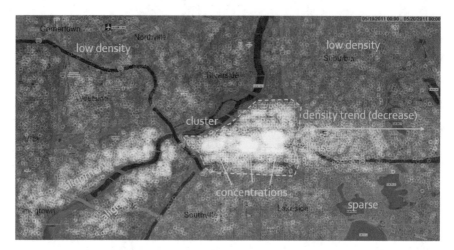

Fig. 2.7: Types of patterns observable in the spatial distribution of the microblog posts in the VAST Challenge 2011 (Chapter 1).

The spatial density of entities can be characterised by a numeric measure, which is a space-based attribute. Types of patterns in spatial distribution of attribute values mostly refer to differences between attribute values associated with neighbouring locations; hence, these patterns can also be called *spatial variation patterns*. Some of such pattern types are illustrated in Fig. 2.8.

Spatio-temporal distributions. Entities or attribute values can be distributed in both space and time, i.e, the base of the distribution is the composition of the space and time. Thus, in Chapter 1, we analysed the spatio-temporal distribution of the microblog posts. Possible patterns in a spatio-temporal distribution can be defined and characterised in terms of changes of the spatial distribution over time. For example, there may be a temporal trend or periodicity in changes of spatial density. Changes of spatial clusters can be described in terms of appearance/disappearance and changes of density, spatial extent, shape, and position. The term *spatio-temporal cluster* refers to a spatial cluster that existed only during some time interval within the time range under analysis. Spatio-temporal clusters of entities can be seen in the space-time cubes presented in Figs.1.9, 1.11, and 1.15. In Figs. 1.11 and 1.15, it is also possible to see temporal trends of decreasing spatial density.

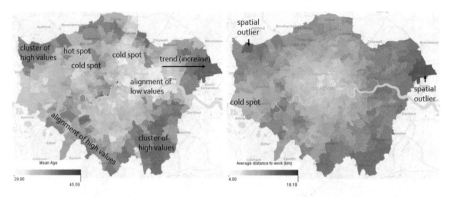

Fig. 2.8: Illustration of some types of patterns that can exist in a spatial distribution of numeric attribute values. Left: A distribution with multiple spatial patterns. Right: The overall pattern is an increasing trend from the centre to the periphery. It is disrupted in some parts by a few local patterns.

2.3.4 Co-distributions

It may be possible to detect and understand a relationship between two components by studying their joint distribution, or co-distribution, over a third component. A tendency of some elements of one component to co-occur with particular elements of the other component indicates the existence of a relationship and tells about the character of the relationship. Such a tendency could be called *correlation*, but this term is very often used in the narrow sense of statistical correlation. Instead, we shall use the term *interrelation*, implying a broader scope.

An easiest way to illustrate the concept of interrelation is to consider joint distribution of values of two numeric attributes over a set of entities, or over space, or over time (the nature of the third set is irrelevant). The existing combinations of their values can be visually represented in a scatterplot, as in Fig. 2.9. A scatterplot reveals interrelations by showing what values of one attribute tend to co-occur with high, medium, and low values of the other attribute.

When two numeric attributes are somehow interrelated, the dots on the scatterplot are arranged in some discernible shape or several shapes. In Fig.2.9, the dots form clouds elongated in oblique directions. On the left, the distribution of the dots over the plot area tells us that lower values of the attribute "mean age", which is represented by the horizontal dimension, co-occur with any values of the attribute "population density" (vertical dimension), whereas higher values of the mean age tend to co-occur with lower values of the population density. On the right, we can see that higher values of the mean age (horizontal dimension) tend to co-occur with higher percentages of the white ethnic group in the population (vertical dimension).

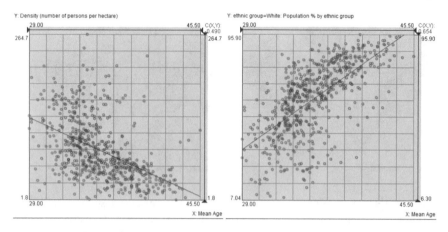

Fig. 2.9: A scatterplot presents the distribution of entities over an abstract space made of all possible combinations of values of two numeric attributes.

When we look at a scatterplot, we, in fact, see a *spatial distribution*, specifically, a distribution of dots in the space of a plot. We are trying to detect *spatial patterns* in this distribution, and these patterns have certain meanings:

- **Spatial clusters** correspond to frequent co-occurrences of values from some intervals.

- Oblique **alignments** and **elongated clusters** indicate specific interrelations when higher values of one attribute tend to co-occur with higher or lower values of the other attribute.

- **Spatial outliers**, which are usually well visible, correspond to exceptional value combinations.

- The variation of **spatial density** tells us about the joint frequency distribution.

Generally, spatial patterns are readily perceived by human eyes; therefore, it is a good idea to use distributions of visual marks, such as dots, over a display space for detecting and investigating interrelations. The question is: How can we apply this idea to more than two numeric attributes? (And the next question is: How to extend this idea beyond numeric attributes?)

We should keep in mind that spatial patterns base primarily on the distances between items located in the space. Thus, multiple items can be perceived together as some shape when the distances among them are small. Spatial patterns are meaningful only when the distances between items have a clear meaning. In a scatterplot, distances between dots represent differences between attribute values; that is why patterns of dot distribution are meaningful and can be interpreted. If we want to use other kinds of artificial spaces, we need to construct them so that the spatial distances represent degrees of similarity or relatedness: the more similar or related, the closer in the space.

In a case of multiple numeric attributes, artificial spaces corresponding to this requirement can be built using dimensionality reduction methods (Section 4.4), which are capable to represent combinations of attribute values by positions in a space with two or three dimensions. This technique is called *embedding* or *projection*, and the space so created is called the embedding space or projection space. Ideally, all distances in a projection space should be proportional to the corresponding combined differences between the values of the multiple attributes; however, this condition is typically impossible to fulfil. The dimensionality reduction methods strive to minimise the distortions of the distances in the projection space, but it should never be forgotten that distortions are inevitable. Hence, distances in a projection space should be treated as approximate. One should look for general patterns and outliers but avoid making judgements based on distances between individual items.

A projection can be visualised in a display that looks similar to a scatterplot and can be called *projection plot*. Unlike in a scatterplot, the dimensions of a projection plot have no specific meanings. Then, how to understand the meanings of spatial patterns that can be exhibited by a projection plot? This requires additional visual and interactive techniques, as, for example, shown in Fig. 2.10. Here, a projection plot (on the left) has been constructed based on the combinations of values of 16 attributes expressing the percentages of different age groups in population of administrative districts. These combinations are represented by lines on a parallel coordinates plot (on the right), where the parallel axes correspond to the attributes. For each dot in the projection plot, there is a corresponding line in the parallel coordinates plot. The projection display is visually linked to the parallel coordinates plot by colour propagation. For this purpose, continuous colouring is applied to the background of the projection display. As a result, each position in the display receives a particular colour, and close positions receive similar colours. The background colour of each dot in the projection display is propagated to the corresponding line in the parallel coordinates plot. This allows us to understand what kinds of value combinations correspond to differently coloured parts of the projection display. The association between the two displays is additionally supported by highlighting of corresponding visual elements in both displays in response to interactive selection of elements in one of them. In Fig. 2.10, bottom, a group of closely located dots has been selected in the projection display. The selected dots and the corresponding lines in the parallel coordinates plot are marked in black.

Another approach to studying interrelations between multiple components is demonstrated in Fig. 2.11. The approach involves multiple displays representing the components and interactive selection of items in one of the displays that causes simultaneous marking (highlighting) of corresponding parts in all displays. The example in Fig. 2.11 involves multiple numeric attributes represented by frequency histograms. The analyst can select high, low, or medium values of one attribute and see what values of the other attributes correspond to these. In principle, this idea can be extended

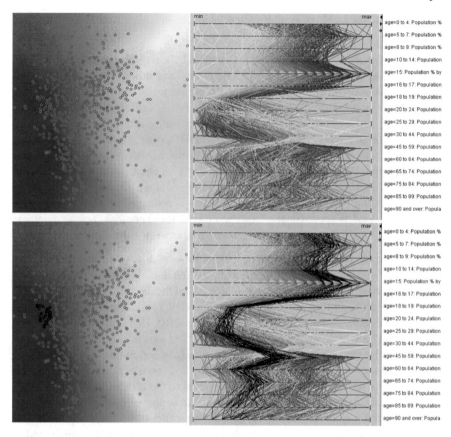

Fig. 2.10: Interpretation of a 2D projection of multi-attribute value combinations
can be supported by an additional display, such as a parallel coordinates plot, representing the original value combinations. The displays can be linked by means of
colour propagation and highlighting of corresponding items.

to other kinds of display representing other kinds of components. A disadvantage of
this approach is the need to perform many interactive operations.

Sometimes, it is possible to visualise joint distributions and thereby reveal interrelations by overlaying two or more information layers within a single display. This
approach is typically applied in map displays, where one or more layers representing data are laid over a background representing features the geographic space. A
layer that is put over another layer should be constructed so that the underlying layer
remains sufficiently well visible. Thus, in the map layers representing the distributions of microblog message posts in Chapter 1 (and also in Fig. 2.7), the messages
are represented by highly transparent visual marks. This allows seeing interrelations
between the messages and the entities existing in the geographic space, such as the
river, the downtown, the major roads, and the bridges.

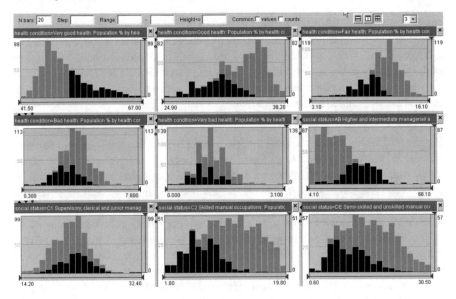

Fig. 2.11: Interrelations among multiple numeric attributes can be studied using interactive histogram displays. Here, the analyst has selected a sequence of bars in the right part of the first histogram (top left). The black segments indicate where the selected data fit in each histogram.

2.3.5 Spatialisation

In the previous section we have seen how an artificial space can be constructed from combinations of values of multiple numeric attributes using a dimensionality reduction algorithm. The idea of creating artificial spaces and spatial distributions that represent non-spatial relationships can be extended to other kinds of data. The process is called *spatialisation*. In general terms, spatialisation can refer to the use of spatial metaphors to make sense of an abstract concept. In visualisation, "spatialisation" refers to arranging visual objects within the display space in such a way that the distances between them reflect the degree of similarity or relatedness between the data items they represent. Spatialisation exploits the innate capability of humans to perceive multiple things located closely as being united in a larger shape or structure, that is, in some pattern. It also exploits the intuitive perception of spatially close things to be more related than distant things.

Dimensionality reduction (also known as data embedding) algorithms is a class of tools that can be used for spatialisation. Some of these algorithms can be applied only to data consisting of combinations of multiple numeric values. If we wish to use them for another kind of data, we need to transform the data to the suitable form. Other methods can be applied to a previously created matrix of pairwise distances between items that need to be spatialised, irrespective of where the distances come from. Hence, for any kind of items, we need to define a suitable way to express the degree of their similarity or relatedness by a numeric value that can be interpreted as a distance. The principle is: the distance value is inversely proportional to the degree of the similarity or the strength of the relationship, and it equals zero when two items are identical. A procedure that measures the degree of similarity or relatedness for pairs of items is called a *distance function*.

Examples of applying spatialisation to different kinds of items can be seen in Fig. 2.12. The examples originate from a paper [28]. In the upper image, the spatialisation has been applied to different versions of the Wikipedia article "Chocolate", which was edited many times. In the middle image, the dots in the artificial space correspond to frames of a surveillance video, and in the lower image, to spatial distributions of precipitation over the territory of the USA. In all cases, the items represent states of some dynamic (that is, temporally evolving) phenomena or processes. The similarities between the states were measured by means of distance functions suitable for the respective data types, i.e., texts in the Wikipedia example and images in the other two examples. The states are represented by dots arranged in the projection spaces according to the distances between them. The dots are connected by curved line segments according to the chronological order of the respective states. The relative times of the states are represented by shading of the dots and connecting line segments from light (earlier times) to dark (later times). In these displays, small distances between consecutively connected points indicate small changes while large distances correspond to large changes.

In the visualisation of the editing process of the Wikipedia article, it is possible to see a kind of "war" between two authors, in which two versions of the article repeatedly replaced each other. In the visualisation of the surveillance video, we see a large cluster of dots representing the states when nothing happened while the remaining dots correspond to movements of pedestrians in front of the camera. In the lower image, long alignments of dots indicate long-term trends in the evolution of the spatial distribution of precipitation. The shape of the overall curve indicates a cyclic pattern in the temporal evolution of the precipitation distribution: by the end of the year, the distribution returns to states that are similar to those at the beginning of the year. In these examples, interpretable patterns are created not only by arrangements of marks in display spaces but also by shapes of curves connecting the marks.

Generally, spatialisation is a powerful tool that can be used in analysing various kinds of relationships when there is a reasonable way to represent the strengths of these relationships numerically.

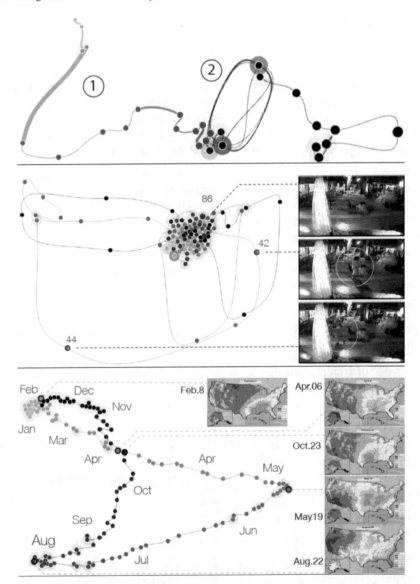

Fig. 2.12: Examples of the use of spatialisation for representing temporal evolution of dynamic phenomena and processes [28]. Top: editing of the Wikipedia article "Chocolate". Middle: a surveillance video of a street. Bottom: spatial distribution of precipitation.

2.4 Concluding remarks

The process of non-trivial data analysis and problem solving essentially involves human reasoning. The goal of analysis is to gain understanding and, possibly, to build a computer model of some subject, such as a phenomenon existing in the real world. A non-trivial subject can be seen as a complex system including multiple components linked by relationships. These relationships are reflected in data, which specify correspondences between elements of different components.

Relationships between components can be understood by studying distributions of elements of some components over elements of other components. To proceed from correspondences between individual elements to understanding of relationships between components as wholes, one needs to find and interpret patterns in the distributions. A pattern consist of multiple correspondences that can be perceived and/or represented together as a single object. The power of human vision gives humans the capability to notice various patterns in visual representations of distributions. Unlike computers, humans do not require a precise definition of patterns to search for. Human perception is especially highly inclined to seeing patterns in spatial distributions. Therefore, it can be helpful to create artificial spaces in which various non-spatial relationships are represented by relationships of spatial distance and neighbourhood.

The types of patterns that may exist in distributions and the meanings of these patterns depend on the types of components whose relationships are analysed. In the following chapters, we shall consider classes of phenomena involving different types of components.

At the beginning of this chapter, we stated that visual analytics is relevant to different steps of the data science workflow. Visual analytics can be seen as a way to fulfil the **principles of good data science**, which can be summarised as follows:

- You should always remember that the human, not the machine, is the leading force of the data science process.

- You need to see and understand your data before trying to do something with it.

- You should not take blindly what machine gives you; look, understand, think, experiment: what will happen if I change something?

- While analysing and modelling, you should also think about presenting your results and how you obtained them to other people. You must be able to explain and justify the analytical process and its outcomes.

Given these principles, the relevance of visual analytics to data science becomes clear. Visual analytics assumes that the human analyst **interacts** with the machine, not just views what it has produced. The analyst tries different parameter settings, different computational methods, different data transformations, feature selections, data divisions, etc. The analyst provides background knowledge and evaluation feedback when the algorithms are designed to accept these. The analyst takes care about explaining and justifying the analytical procedure and its outcomes to others. In the following chapters, we shall demonstrate how visual analytics as an approach, and also as a particular attitude to the analysis process and a discipline of mind, can help you to do good data science.

Chapter 3
Principles of Interactive Visualisation

Abstract We introduce the basic principles and rules of the visual representation of information. Any visualisation involves so-called visual variables, such as position along an axis, size, colour hue and lightness, and shape of a graphical element. The variables differ by their perceptual properties, and it is important to choose appropriate variables depending on the properties of the data they are intended to represent. We discuss the commonly used types of plots, graphs, charts, and other displays telling what kinds of information each of them is suited for and what visual variables it involves. We explain the necessity of supporting the use of visual representations by tools for interacting with the displays and the information they represent and introduce the common interaction techniques. We also discuss the limitations of visualisation and interaction, substantiating the need to combine them with computational operations for data processing and analysis.

3.1 Preliminary notes

The definition of visual analytics emphasises that visual interfaces need to be interactive [133]. It is because visualisation and interaction complement each other in supporting human analytical reasoning. Visualisation is the best way to supply information to the human's mind, and interaction helps to get additional information that is not available immediately and to make use of computational operations.

Visualisation has always been used for studying phenomena taking place in the geographic space. It is therefore not surprising that the fundamentals of visual representation of information were first formulated by cartographers. Statistical graphics, which appeared in the field of statistics, focuses on representing quantitative information. The research field of information visualisation emerged as computer technologies created opportunities for automated generation of visual displays from data.

© Springer Nature Switzerland AG 2020
N. Andrienko et al., *Visual Analytics for Data Scientists*,
https://doi.org/10.1007/978-3-030-56146-8_3

Interaction is the way of getting more information from a visual display than it is possible by mere viewing. It is also the "glue" that enables the tight-coupling of computational and visualisation techniques. It can enable the analyst to interactively request more contextual data on-demand. It can also enable the analyst to make a selection that feeds into the next analytical step, for instance, run the computational method on a portion of the data or change a parameter. It enables analysts to investigate different ways of configuring computational methods and alternative models for more informed analysis. The increasing feasibility of highly interactive analytical interfaces is driven by advances in computation speed and software frameworks for their implementation, particularly on the web.

In the following sections, we shall introduce and discuss the main principles and techniques of data visualisation and interaction.

3.2 Visualisation

3.2.1 A motivating example

A well-designed visualisation can present data in a way that helps analysts to understand the data and the phenomenon that the data reflect and promote analytical reasoning and knowledge building. **Statistical graphics** is a branch of statistics that advocates the use of graphics to help characterise statistical properties of data. This is often illustrated by the well-known Anscombe's Quartet[1]. It contains four different datasets consisting of pairs of numeric values. These datasets are characterised by exactly equal summary statistics that are commonly used for summarising distributions – mean, median, standard deviation, r^2 – suggesting that the distributions are comparable. However, the four scatterplots of these data in Fig. 3.1 show that this is not the case. This carefully-constructed example illustrates that high-level statistical summaries can be misleading, and that visualisation can provide a richer summary in an interpretable form.

In the Anscombe's Quartet, the datasets are small enough to represent every data item by an individual visual mark. In most real-world datasets there are too many data items to plot each of them separately. In these cases, computational methods are still required for summarising data to a manageable set of items that can be visually represented. One of possible ways to present a visual summary of a set of numeric values is a *box plot*, which is built based on descriptive statistics, namely, the quartiles of the distribution. An example of a box plot is shown in Fig. 3.2, top right. The main part of such a plot is a rectangle (i.e., a box) whose two opposite sides represent the lower and upper quartiles, and a line drawn inside the box parallel

---

[1] https://en.wikipedia.org/wiki/Anscombe%27s_quartet

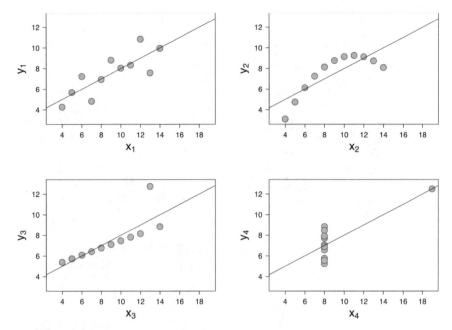

Fig. 3.1: These four datasets have identical summary statistics indicating that they probably have similar distributions but the scatterplots indicate that their distributions are very different.

to these sides represents the median. A box plot may also have lines extending from the box indicating variability outside the upper and lower quartiles. These lines resemble whiskers; therefore, a plot containing such "whiskers" is also called *box-and-whisker plot*. Outliers (determined by means of statistical measures) may be plotted as individual points; in this case, the "whiskers" show the data range excluding the outliers.

The example box plot in Fig. 3.2, top right, summarises the distribution of tips in a restaurant[2]. For comparison, the same data are shown in a *dot plot* on the top left, where the individual values are represented by dots positioned along a vertical axis. Horizontal jittering is used to separate multiple dots representing identical or very similar values and thus having the same vertical position. The same distribution is also represented by a *violin plot* in the lower part of Fig. 3.2. The latter can be seen as a variant of a frequency histogram drawn in a particular way – symmetric and smoothed.

When data are differentiated by categories (e.g., by gender, occupation, or age group), it is reasonable to compare the distributions of values of numeric attributes for these categories. For example, in Fig. 3.3, the box plots summarise the tip dis-

---

[2] https://www.kaggle.com/ranjeetjain3/seaborn-tips-dataset

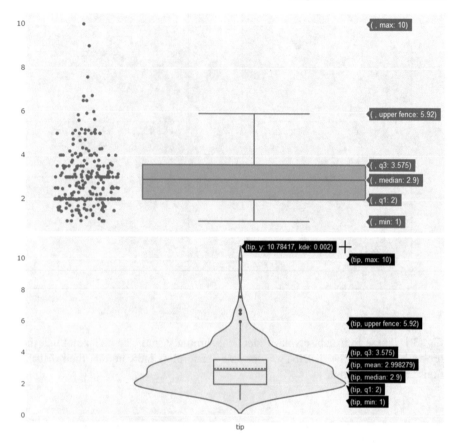

Fig. 3.2: A box plot (top) and a violin plot (bottom) represent the distribution of the tip amounts.

tributions for the lunch and dinner (top) and for the female and male customers (bottom). The same distributions are represented in more detail in dot plots.

All visualisations in the Figs. 3.1 to 3.3 represent numbers, either original values of numeric attributes or summary statistics, by positions along axes; in other words, they employ the visual variable 'position'. The different types of display utilise this variable in different ways and in application to different graphical elements (visual marks): dots in a dot plot, box plot components (box boundaries, median line, and whiskers), and curved outlines in a violin plot. Let us consider the whole set of visual variables and see examples of different ways of using them in the most common types of visual display.

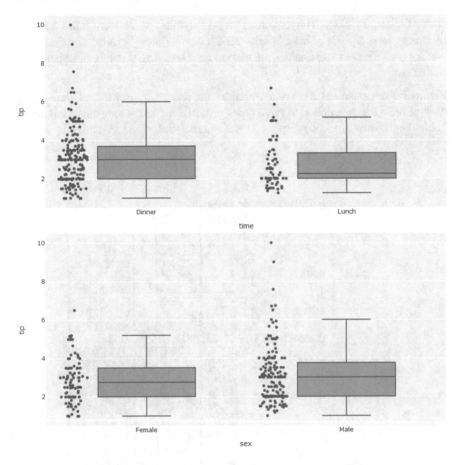

Fig. 3.3: Comparison of the distributions of the tips for the lunch and dinner (top) and for the female and male customers (bottom).

3.2.2 Visualisation theory in a nutshell

According to the seminal theory of Jacques Bertin [31], graphical representation is the encoding of components of data by means of **visual variables**. The most important visual variable is *'position'*. The values of this variable are positions along some axis. Many types of visualisation use horizontal and vertical positions within a two-dimensional display space; hence, they employ two instances of the variable 'position'. These two instances are often called *'planar dimensions'*. Data items are represented by visual **marks**, i.e., graphical elements, such as dots, lines, or symbols. There are three types of marks, called, in accordance with their form, *point*, *line*, and *area* marks. The visual variable 'position' is used for placing the marks within the display space. The remaining visual variables, which were called 'retinal' by Bertin, include *'size'*, *'value'* (lightness), *'colour'* (hue), *'texture'*, *'orientation'*, and *'shape'*. The retinal variables are used to represent data through the appearance of marks.

Figure 3.4 demonstrates the visual variables introduced by Bertin and shows how each of them can be applied to three types of visual marks, points, lines, and areas. Table 3.1 illustrates the concepts of visual marks and visual variables by examples.

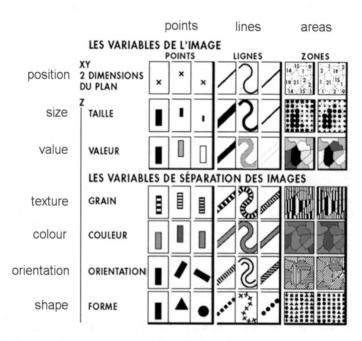

Fig. 3.4: Jacques Bertin's visual variables and how they are applied to 3 types of visual marks, points, lines, and areas. Source: [31].

The visual variables differ in their perceptual properties. Values of some variables can be arranged in a certain logical order. Thus, values of the variable 'position' are ordered in the direction of the axis, values of the variable 'size' can be ordered from smaller to bigger, and values of the variable 'value' (i.e., lightness) can be ordered from lighter to darker. For values of other variables, such as colour hue or shape, there is no logical, intuitively understandable complete ordering, but it may be sometimes possible to construct an ordered set of selected values. For example, it is possible to construct a set of textures differing by their density, which can be ordered from sparser to denser. Values of the variable 'orientation' can be arranged in a cycle from 0° to 360°, which is the same as 0°.

Visualisation	Mark	Visual variables
Scatter plot	Dots represent individual data items.	Values of two numeric attributes are encoded by x- and y-positions; the resulting 2D position represents the association between these two values.
Histogram	Bars represent groups of data items defined based on values of a numeric attribute.	Intervals of attribute values (bins) are represented by positions along an axis. Bar sizes (lengths) encode the sizes of the groups.
Choropleth map	Polygons represent geographic areas characterised by values of a numeric attribute, usually statistical summaries for the areas.	Positions and shapes of the polygons represent the geographic positions and shapes of the areas. Values of the attribute are encoded by the lightness of the painting of the polygon interiors.

Table 3.1: Examples of visual marks and visual variables in visualisations

Some of the variables with ordered values also allow judgement of the amount of difference, or the distance, between any two values. Thus, we can judge the distance between two positions and the difference between the lengths of two bars or between the sizes of two squares. However, it is much harder to estimate how much lighter or darker one shade of grey is from another. Values of the variable 'size' allow one to judge not only differences but also ratios; e.g., we can see that one bar is twice as long as another bar. Judgements of ratios are also possible for positions when the axis origin is specified; in this case, the ratio of the distances from the axis origin can be estimated. Bertin did not distinguish perception of amounts of difference or distances and perception of ratios but applied the same term 'quantitative' to the variables allowing numeric assessments of the differences between the values.

Fig. 3.5: Selectivity test: Can attention be focused on one value of the variable, excluding other variables and values? Association test: Can marks with the same value of the visual variable be perceived simultaneously?

Apart from the ordering, distance, and ratio relationships between values of visual variables, important perceptual properties are selectivity and associativity, that is, how easily distinct values of a variable can be perceptually discerned and how easily multiple marks with the same value of a variable can be perceived all together and differentiated from the remaining marks. The image in Fig. 3.5 can help you to understand what these properties mean. To check the selectivity of a visual variable, look at the image and try to focus your attention on one value of the variable, excluding other variables and values. For example, try to find all letters on the left half of the image ('position'), all small letters ('size'), all pink letters ('colour'), all letters K ('shape'). If such marks are easy to find, the variable is selective.

To check the associativity of a visual variable, try to perceive simultaneously all marks with the same value of the visual variable. For example, try to see simultaneously all letters located on the left half of the image ('position'), all small letters ('size'), all pink letters ('colour'), all letters K ('shape'). If such marks are easy to perceive together as a group, the variable is associative.

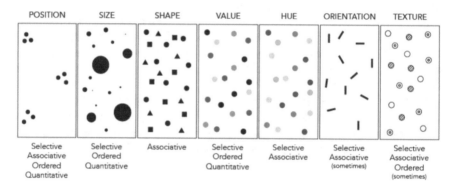

Fig. 3.6: Perceptual properties of Bertin's visual variables. Source: `https://www.axismaps.com/guide/general/visual-variables/`

The perceptual properties of Bertin's visual variables are summarised in Fig. 3.6. Please note that the variable 'value' is claimed to be quantitative, which contradicts to what was stated by Bertin. The reason is that the source of the image in Fig. 3.6 is a web site focusing on map design[3]. The variable 'value' is extensively used in cartography for representing quantitative information, in particular, in choropleth maps. This is because the variable 'position' is used for showing geographic locations, and it is impossible to apply the variable 'size' to geographic areas for showing associated numeric values without distorting geographic information. Therefore, cartography has to employ variables that may not be perfectly suited to representing the type of information that needs to be shown. Besides, as will be discussed later on, some limitations of visual variables can be compensated by interactive operations.

---

[3] `https://www.axismaps.com/guide/general/visual-variables/`

The set of visual variables originally introduced by Bertin was later extended by other researchers. Among others, they introduced dynamic visual variables, such as motion and flicker [60], which can be used in displays shown on a computer screen. The "new" variables are not used as actively as the "old stuff" from Bertin. An exception is, perhaps, display animation, which is quite popular. It can be said that an animated display employs the visual variable 'display time' whose values are the time moments or intervals at which different states of the display (frames) are presented. Obviously, 'display time' is ordered and allows quantitative judgements.

A frequently encountered extension of the original set of visual variables is the use of the third display dimension in addition to the two planar dimensions considered by Bertin. It is used in perspective representations of 3D displays on a computer screen, such as the space-time cube in Section 1.2.7, and also in virtual 3D environments. These displays involve three instances of the visual variable 'position' (certainly, the other visual variables can be used as well, like in two-dimensional displays).

As we have seen, values of some visual variables are linked by relationships of ordering, distance, and/or quantitative difference (i.e., ratio). As we discussed in the previous chapter, the same types of relationships can exist between elements of components of an analysis subject and, consequently, between corresponding elements of data reflecting this subject. The **fundamental rule of visualisation** is to strive at representing components of data by visual variables so that the relationships existing between the values of the visual variables correspond to the relationships existing between the elements of the data components. For example, values of a numeric attribute should be represented using the variable 'position' or 'size', whereas 'value' is less suitable, and 'colour' and 'shape' are completely unsuitable.

Bertin characterised the perceptual properties of visual variables in terms of the **levels of organisation**: associative (the lowest level), selective, ordered, and quantitative (the highest level). The general principle of data presentation, according to Bertin, is: **"The visual variables must have a level of organisation at least equal to that of the components which they represent"**. The level of organisation, also known as *scale*, of a data component may be *nominal*, *ordinal*, or *numeric*. Associative and selective variables correspond to the nominal scale, ordered variables to the ordinal scale, and quantitative variables to the numeric scale. If, for example, the goal is to represent values of a numeric attribute graphically so that a viewer can extract ratios from the visualisation, for example, to see immediately that one value is twice another, one must choose a visual variable with quantitative organisation, that is, either 'position' or 'size'.

Another consideration in choosing visual variables for representing data concerns the "length" of the visual variables. The **length** is the number of categories or steps that can be distinguished visually, for example, discernibly different colours or lightness levels. A visual variable should have a length equal to or greater than the number of distinct values in the component that it represents. If the length of the variable is insufficient, the observer will perceive some of the different data values as being identical.

Tables 3.2 and 3.3 summarise perceptual properties of visual variables, as proposed by Bertin [31] and updated by Green [60] from a psychologist's perspective and by MacEachren [93] as a cartographer. Further details and discussion on this can be found in [17], section 4.3.

Visual variable	Associative	Selective	Ordered	Quantitative
Planar dimensions	yes	yes	yes	yes
Size		yes	yes	yes
Brightness (value)		yes	yes	yes, if scaled
Texture	yes	yes	yes	
Colour (hue)	yes	yes	yes (limited)	
Orientation	yes	yes		
Shape	yes	yes		
Motion: velocity		yes	yes	yes, if scaled
Motion: direction		yes		
Flicker: frequency		yes	yes	yes, if scaled
Flicker: phase		yes		
Disparity		yes	yes	

Table 3.2: Perceptual properties of visual variables, according to Green [60].

Visual variable	Numerical	Ordinal	Nominal	Visual isolation	Visual levels
Location	++	++	++	++	
Size	++	++	++	++	++
Crispness		++		++	++
Resolution		++		++	++
Transparency		++	+	++	++
Colour value	+	++		++	++
Colour saturation	+	++		+	++
Colour hue	+	+	++	++	+
Texture	+	+	++	++	++
Orientation	+	+	++	++	
Arrangement			+	+	
Shape			++		

Table 3.3: Perceptual properties of visual variables, extended by MacEachren [93].

In Section 2.3, we explained the importance of abstraction for understanding relationships between components of a subject. Abstraction is facilitated when multiple elements of a display can be perceived all together. Therefore, associative visual variables play an important role in supporting abstraction. From all visual variables, 'position' and 'colour' (hue) have the strongest associative power. Despite its low organisation level, the variable 'colour' is very intensively exploited in visualisation thanks to its associative property.

While the visual variable 'display time' used in animated displays seems to have high level of organisation, its expressive power is strongly reduced by the fact that the viewer cannot see all information at once. Comparison between display states and extraction of patterns relies on the viewer's memory, which has limited capacity. Due to the short duration of presenting each state, viewer's attention can miss important changes; this phenomenon is known as "change blindness"[4]. However, animation is not always a poor choice [52]. Animation may be good for representing continuously developing phenomena and processes, when the changes from one state to another are gradual rather than abrupt or chaotic (although abrupt and chaotic changes can be noticed, the viewer will hardly be able to link them into any kinds of patterns). Animation can help viewers perceive the general trend of the development and may also be good for revealing coherent changes at multiple locations or coherent movements of multiple objects. Generally, animation may be suitable for supporting an overall view of how something develops over time, but it is much less supportive for tasks requiring attention to details and detection of even small changes.

3.2.3 The use of display space

The position in the display, or the dimensions of the display, are the most important visual variables having the greatest expressive power regarding both the organisation level and the length, which is limited only by the display size and resolution. Respectively, this is a valuable resource for visualisation. To increase the capacity of the display space, different arrangements of marks or complex visual objects comprising multiple marks are frequently used in constructing visual displays, including the following:

- **Space partitioning**: often called juxtaposition, or "small multiples", this technique is used for comparing pieces of information related to different locations, times, objects, categories, attributes, etc., or for showing alternative views of the same data. For example, the display space in Fig. 2.4 is partitioned to accommodate four frequency histograms of different attributes. In Fig. 3.2, top, the display space is divided between two alternative views, and in Fig. 3.3, we see space division providing alternative views and enabling comparison of categories.

- **Space embedding**: used for creating complex representations. A widely known example is diagrams on a map: a diagram has its internal space, which is embedded in the space of the map (see Fig. 1.8).

- **Space sharing**, a.k.a. **superposition**, or **overlay**: representation of two or more data components in a common display space, e.g., multiple information layers on a map, multiple comparable attributes or multiple time series in the same

---

[4] https://en.wikipedia.org/wiki/Change_blindness

coordinate frame, etc. The lower part of Fig. 3.2 contains an example of a violin plot and a box plot sharing the same display space.

- **Fusing data components**: use of two or three display dimensions for representing a larger number of data components, as introduced in Section 2.3.5 and will be discussed in more detail in Section 4.4. An example can be seen in Fig. 2.10, where the 2D display space is used to represent combinations of values of multiple attributes.

As we mentioned in Section 2.3.5, human vision is especially well adapted to perceiving spatial patterns, such as groups of close objects, linear arrangements, and other shapes composed of multiple objects. Regarding visual displays, these capabilities are supported by the strong associative power of the variable 'position'. Hence, this variable should be of primary choice in representing distributions and co-distributions. Many examples of displays representing distributions in Chapters 1 and 2 employ the variable 'position'. Temporal distributions are portrayed by time histograms, where times are represented by positions along the x-axis. Spatial distributions are viewed using maps, where spatial locations of objects are represented by positions on the maps. Frequency distributions are shown in frequency histograms, where the positions on the x-axis correspond to attribute values. For spatio-temporal distributions, space-time cubes are used, where the positions in the display represent spatial locations and times. A scatterplot represents a co-distribution of two numeric attributes; the positions in the display correspond to value pairs. Similarly, a projection plot representing a result of spatialisation supports perception of spatial patterns made by positions of marks corresponding to complex data items.

It is important to understand that perception and interpretation of spatial patterns formed by visual marks in a display makes sense only when the distances between the positions of the marks represent relevant relationships between the corresponding data items. If it is not so, any spatial patterns that can occasionally emerge in such a display may be misleading and need to be ignored. Let us consider, for example, a dataset consisting of texts that include various keywords. In the distribution of the keywords over the texts, both the base (the set of texts) and the overlay (i.e., the keywords occurring in the texts) consist of discrete entities that are not linked by any intrinsic relationships. According to the Bertin's principle "The visual variables must have a level of organisation at least equal to that of the components which they represent", the visual variable 'position' can be used to represent such components. However, the distances between the positions assigned to the elements will not correspond to any intrinsic relationships between the elements, and any possible spatial patterns that are based on distances, such as spatial clustering, density, or sparsity, will have no meaning.

A visualisation technique that is often used for representing frequency distributions of keywords in texts is word cloud. We used such displays in Chapter 1; see, for example, Fig. 1.2. In this visualisation, word frequencies are represented using the visual variable 'size' (namely, font size). In our examples in Chapter 1, the words were arranged in the order of decreasing frequency, which means that the visual variable 'position' was also used for representing the word frequencies. This is an example of a redundant use of two visual variables to represent the same information for improving display readability or convenience of use. In this case, the variable 'position' conveys meaningful information (frequency rank), and the distances thus also have meanings (differences in the frequency ranks). Hence, spatial patterns, such as sequences of closely positioned words, are meaningful as well. However, it is not typical for the general word cloud technique and its widely available implementations that the words are arranged in some meaningful order. Rather, it is more typical to apply so-called *space filling layouts* to pack more words in the limited display space. In such layouts, spatial positions of marks and spatial relationships between them have no meanings; therefore, viewers should ignore any spatial patterns that may emerge.

Hence, we have discussed two principally different approaches to utilising the display space. One approach is to employ the visual variable 'position' for representing elements of data components and/or relationships between them. It is especially appropriate to use this variable for representing data components with intrinsic distance relationships between elements. 'Position' is also suitable for representing sets of data items linked by other kinds of relationships that can be expressed as distances (e.g., social links between people). The other approach is to fill the display space with visual marks so that their positions do not represent any semantically meaningful information. In fact, the visual variable 'position' is not utilised in this case, and any patterns formed by positions of marks in the display need to be ignored.

There is also a third approach: create a spatial layout that represents some kind of relationships between items (e.g., hierarchical) but does not exploit spatial distances between marks for this purpose. In this case, meaningful spatial patterns in the display can be formed by configurations of visual marks. An example is a nested layout, in which smaller marks are put inside bigger ones to represent part-whole relationships.

In the next section, we shall consider the standard, widely used visualisation techniques, which use the display space and the other visual variables in different ways for representing different kinds of information.

3.2.4 Commonly used visualisations

In this section, we list visualisations that are used commonly. It is easy to find them in a variety of implementations, including Python, R, JavaScript, and Java libraries.

3.2.4.1 Representing distributions of numeric values

Bar chart: Each bar represents a discrete entity or category, and the length of the bar represents a numerical value that relates to that entity or category. Where categories are ordinal (i.e., have an intrinsic order), one would usually put the bars in this order, but there are reasons why they might be ordered differently, for example, in the order of decreasing or increasing bar lengths.

Pie chart: can represent the same information as bar charts when categories are parts of some whole. While bar lengths can be compared more precisely than angular sized of pie sectors, the latter are much more supportive for estimating the fractions of the whole covered by the components.

Dot plot: depicts a distribution of numeric values along an axis by representing each value by a dot. When the density of the dots is very high, overlapping of dots can be decreased by jittering in the dimension perpendicular to the axis, as in Figs. 3.2 and 3.3. The use of semi-transparent dots is helpful, but the maximum opacity may be reached very quickly. Interactively altering transparency is useful.

Line graph, or **line plot**, or **line chart**: represents a series of numeric values by a sequence of points positioned along an axis with the distances from the axis proportional to the values; the consecutive points in the sequence are connected by lines. Unlike a bar chart, a line chart emphasises that the data refer to some ordered and, usually, continuous base, such as time (see **time graph**). An example of data with a non-temporal base may be data describing the dependency of the electric conductivity of some material on its temperature.

Heat map, or **heatmap**: Representation of variation of values of a numeric attribute along an axis or over a 2D display space by encoding the values by colours or degrees of lightness. The colouring or shading is applied to uniform elements (pixels) into which the display space is divided. A typical example is cells of a matrix. The heatmap technique can be applied, in particular, on a cartographic map, in which the space is divided into uniform compartments by means of a regular rectangular or hexagonal grid. Examples of heatmaps can be seen in Fig. 4.10.

Histogram (frequency histogram) (Fig. 2.4): Depicts a distribution of numeric attribute values by binning the data along an axis and depicting frequency with bar lengths. Altering the bin size is a useful interaction (Fig. 2.5). Because aggregation is used, it scales better than dot plots. Bars can be divided into segments to show frequencies for subsets of data items, e.g., according to some categories or classes identified by hues of the segments.

Violin plot (Fig. 3.2, bottom): The idea is similar to frequency histogram: one axis represents the range of attribute values, and the other display dimension is used for representing corresponding frequencies by distances from the axis. Instead of discrete bars with their lengths, the frequencies are represented by the distances between two continuous curves symmetric with respect to the attribute axis.

Box plot, a.k.a. **box-and-whiskers plot**: (Figs. 3.2, 3.3): summarises a distribution of numeric values by depicting the quartiles, which are particular values dividing the ordered sequence of the values occurring in the data into four approximately equal parts. The graph is drawn along an axis representing the value range. The lower (first) and upper (third) quartiles ($Q1$ and $Q3$) are represented by the positions of the two sides of the box and the median (i.e., the second quartile, $Q2$) by the position of the line inside the box. The whiskers can connect the box sides to several possible alternative values, such as the minimum and maximum of the data, the 9th percentile and the 91st percentile, the 2nd percentile and the 98th percentile, or the *lower fence* ($= Q1 - 1.5 \cdot IQR$) and the *upper fence* ($= Q3 + 1.5 \cdot IQR$), where $IQR = Q3 - Q1$ is the interquartile range. When a whisker does not reach the minimum or maximum, the graph may also include dots to depict outliers, as in the box plot superposed on the violin plot in Fig. 3.2, bottom.

3.2.4.2 Representing temporal information

Time graph or **time plot**: a line chart representing time-referenced numeric data. The axis along which the points are placed represents in this case time.

A **2D time chart**: represents time-referenced values of an attribute regarding the linear and cyclic components of time or two time cycles. It is a matrix with one dimension corresponding to one time cycle (e.g., hours of a day) and the other to the linear component consisting of repetitions of this cycle (e.g., multiple consecutive days) or to another time cycle, e.g., days of a week. The values of the attribute are represented by marks in the cells of the matrix.

Timeline view: shows occurrences of some events or activities over time [117]. One of the display dimensions (most often horizontal) represents time, and the events are represented by bars or other symbols placed along the time axis according to the times when the events happened. The other display dimension is used for distinguishing between events or activities that occur in parallel or overlap in time. A **Gantt chart**, which is commonly used for portraying activities in project management, is a form of timeline in which the bars representing the activities may be connected by lines or arrows representing relationships, such as dependencies.

Time histogram (Figs. 1.5, 1.6, 2.6 top): depicts a distribution of discrete entities (events) over time. Contains an axis representing times and a sequence of bars corresponding to time intervals; the bar lengths are proportional to the numbers of the entities or events that existed or occurred within the time intervals. The bars can be divided into distinctly coloured segments corresponding to different categories of the entities or events (Fig. 1.14). A **2D time histogram** (Fig. 2.6, bottom) portrays a temporal distribution with respect to a recurring time cycle. Similarly to a 2D time graph, it is a matrix with one dimension corresponding to a time cycle (e.g., hours of a day) and the other to a sequence of repetitions of this cycle (e.g., multiple consecutive days) or to another time cycle, e.g., days of a week. The counts of entities or events are represented by proportional sizes of bars or other kind of marks in the cells of the matrix.

3.2.4.3 Representing spatial information

Dot map: depicts geographic distributions of discrete entities, which are represented by dots positioned on a map according to the spatial locations of the entities (e.g., as in Fig. 1.7) . It can be seen as a 2D version of a dot plot. Because of the nature of geographical data, occlusion is often a significant problem.

Choropleth map (Fig. 2.8, Fig. 3.9, top): portrays the spatial distribution of attribute values referring to discrete geographical areas, such as units of a territory division. It is common to use the visual variable 'colour' (i.e., hue) to represent values of a categorical (qualitative) attribute and the variable 'value' (i.e., lightness) to represent numerical variable values. A common criticism is that it that the size of an area, while having little to do with the data, affects the visual salience and may introduce bias. Diagram maps or area cartograms (see later) are possible alternatives.

Diagram map, or **chart map** (Fig. 1.8): represents a spatial distribution of values of one or several numeric attributes referring to particular geographic locations or areas. The values are represented by proportional sized of symbols, such as circles or rectangles, or components of diagrams, such as bars of bar charts or sectors of pie charts. When the values refer to areas, the symbols or diagrams are drawn inside the areas.

Area cartogram: A map-like display where where 2D positions, shapes, and sizes of geographic areas are distorted in order to represent by the sizes some numeric values. An example is a population cartogram, in which areas of countries or regions are sized proportionally to their population. Such representations may be beneficial for portraying phenomena related to population, such as the poverty rate.

3.2.4.4 Representing co-distributions and interrelations

Scatterplot (Fig. 2.9): has two perpendicular axes representing value ranges of two numeric attributes. Pairs of attribute values are represented by dots in this coordinate system. See Section 2.3.4 for a discussion of possible patterns in a scatterplot and their interpretations.

Scatterplot matrix: A tabular arrangement of multiple scatterplots representing different attribute pairs. Often, the diagonal is used to depict the distributions of the individual attributes, e.g., by frequency histograms.

2D histogram: A scalable extension of the idea of scatterplot, where the attribute axes are binned. For each pair of bins, the number of pairs with the values contained in these bins is represented by a proportional size of some symbol, such as bar, square, or circle. The appearance is similar to that of the 2D time histogram, as in Fig. 2.6, bottom.

Parallel coordinates (Fig. 2.10, right): A graphic depicting combinations of values of multiple attributes. The attributes are represented by parallel axes and attribute values by positions on the axes. Each combination of values is represented by a polygonal line connecting the value positions on neighbouring axes. Occlusion is often a big problem, which can be addressed by optimal ordering and flipping of axes and the use of semi-transparent lines. **Spider plot**, also called **radar chart**, is a radial version, in which the axes of the attributes emanate from a common origin and go in different directions. It may fit better in some composite layouts.

3.2.4.5 Representing relationships between entities

Node-link diagram: A graph representation with discrete marks, typically circles or dots, representing entities (graph nodes), which are connected by lines representing pairwise relationships between the entities (graph edges). Additional node and edge attributes can be encoded using visual variables like the size and colour of the node marks and the stroke thickness and colour of the link marks. Node-link diagrams get cluttered relatively fast even for moderately-sized graphs, and the visibility of interesting substructures or patterns is highly dependent on a suitable layout, that is, node positioning and edge routing. Finding a good layout is a non-trivial task with a large number of methods researched for different application contexts.

Connectivity matrix: a representation where graph nodes (entities) are encoded as the rows and columns of a square matrix, and a link between nodes n_i and n_j is encoded in the matrix cell of the corresponding row i, column j of the matrix. Link attributes can be represented in the matrix cells using visual variables like symbol size, colour, or value (lightness). Matrix representations are free from clutter and occlusions and thus can handle larger graphs compared to node-link diagrams; however, the number of the nodes is limited by the display size and resolution. As with node-link diagrams, finding interesting substructures or patterns typically requires effective sorting of matrix rows and columns to produce prominent groupings of filled cells.

Treemap[5]: displays hierarchical data using nested figures, usually rectangles. Each branch of the hierarchy is given a rectangle, which is then tiled with smaller rectangles representing sub-branches. The lowest level (terminal) rectangles have areas proportional to values of some numeric attribute. Often these rectangles are also coloured to represent values of another attribute. For example, the data describing

---

[5] https://en.wikipedia.org/wiki/Treemapping

the world population by continents and countries could be portrayed by a treemap where the terminal rectangles representing the population counts of the countries are arranged into upper level rectangles representing the counts for the continents. The rectangles may be coloured according to the continent, or shaded according to the average income or GDP per capita. A treemap uses a space-filling layout to optimise the utilisation of the display space.

Many of the visualisations that are listed in this section can be parts of **composite graphics** depicting complementary portions of information or multiple simultaneous perspectives of the data. Composite graphics can be created using different strategies of the display space utilisation (partitioning, embedding, or sharing) as described in Section 3.2.3. As well as potentially providing a richer visual display, composite graphics usually facilitate **comparisons**. For example, placing line graphs or bar diagrams on a map where each graph or diagram depicts a temporal trend for a region can help compare temporal trends between different regions but can also help establish whether there might be a geographical pattern of these temporal trends (see Fig. 1.8).

3.2.5 General principles of visualisation

Although there are so many standard visualisations, not always you find the one that is suitable for your data and analysis goals. You may need to design your own displays, and you will want them to be effective and not misleading. Based on the literature and our own experience, we can suggest the following principles of creating good data visualisations:

1. **Utilise space at first**: Taking into account the analysis tasks, represent the most relevant data components by display dimensions (variable 'position') or suitable spatial layouts depending on the intrinsic properties of the components or task-relevant relationships between data items.

2. **Respect properties**: Properties of the visual variables must be consistent with the properties of the value domains of the data components.

3. **Respect semantics**: Data and corresponding phenomena may have characteristics and relationships that are not explicitly specified but supposed to be known to everybody or to domain specialists. Domain knowledge and common sense should be used in creation of visual displays.

4. **Enable seeing the whole**: Design the visualisation so that it facilitates gaining an overview of all data items that are shown. Care about completeness (all relevant items should be present in the display while overlapping and visual clutter should be avoided) and possibilities for associating multiple items in groups, which is prerequisite for extracting patterns.

5. **Include titles, labels, and legends**: Interpretation of patterns strongly depends on appropriate labelling of data components and explanation of how they are encoded by visual means.

6. **Avoid excessive display ink**: Avoid visual components that do not convey useful information (e.g. decorations); they may distract attention and clutter the display. However, relevant contextual information needs to be present.

7. **Consider employing redundancy**: It may be appropriate to represent the same component of the data using two different visual variables if this can improve display readability, understanding, and/or perception of patterns. This may seem to contradict to the previous principle, but it is not so. It is quite appropriate to use redundant ways of conveying important information, but excessive display ink should be minimised by hiding irrelevant information and by avoiding non-informative elements and components of visualisation.

8. **Enable looking at data from multiple perspectives**: When the number of data components is higher that the number of effective visual variables, create multiple representations for looking at the data from different perspectives. Data items on multiple visualisations may be linked by interactive operations as described in Section 3.3.

9. **Rely on interactivity**: Interactivity changes the ways in which visualisations are used. For example, instead of trying to decode exact data values from their visual representation, we can simply access these values through interaction with the display. Hence, it is possible to use visual variables that are not very effective for decoding exact values but have other useful perceptual properties. Interactive change of symbolisation may facilitate comparison and complement visual representation in many other ways, as demonstrated in Section 3.3.

Let us demonstrate a few examples in which some of the principles are violated. In Fig. 3.7, both variants of a pie chart contradict to the principle "respect semantics". The sectors represent three components 'male', 'female', 'and 'total'. Semantically, the latter component includes the first two, but this relationship has not been accounted for in constructing the diagram. Besides, the chart on the right violates the principle "avoid excessive display ink", because the third dimension does not represent any meaningful information.

In accord with the principle "respect properties", radial layouts and rainbow colour scales should be applied only to data components having cyclic organisation of elements. For example, the radar chart on the right of Fig. 3.8 (right) is incorrect as the temporal component has linear and not cyclic organisation in this dataset. A radar chart would be more appropriate when, for example, the values were statistical summaries for months over multiple years (e.g. average monthly temperature). Another example is a rainbow colour scale, which is quite popular for its colourfulness and is often misused in visualisations. Applying it to a non-cyclic attribute may destroy patterns and complicate interpretation, as happens with patterns of the gender structure of populations on Figure 3.9. Applying the rainbow colour scale to cyclic data

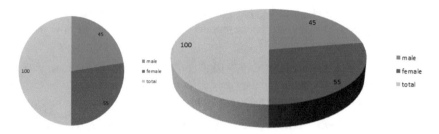

Fig. 3.7: Bad design: Pie charts, either plain (left) or fancy 3D (right), are not applicable to data with part-of relationships.

(e.g., prevailing direction of the wind or traffic movement) would be preferable over the diverging colour scale, however.

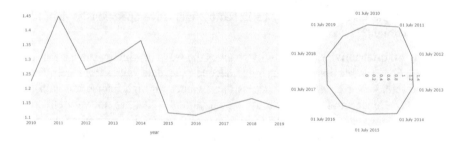

Fig. 3.8: Good design: Line chart (left) shows 10-years dynamics of € to $ exchange rates. Bad design: Radar chart (right) representing the same data delivers a false cyclic pattern.

3.2.6 Benefits of visualisation

It is common to believe that "A picture is worth a thousand words"[6]. Indeed, properly designed visualisations provide numerous benefits that can be grouped into the following categories:

Overview: An appropriate visual representation gives a compact and easy to understand summary of the data. Based on this representation, it is possible to understand

---
6   https://en.wikipedia.org/wiki/A_picture_is_worth_a_thousand_words

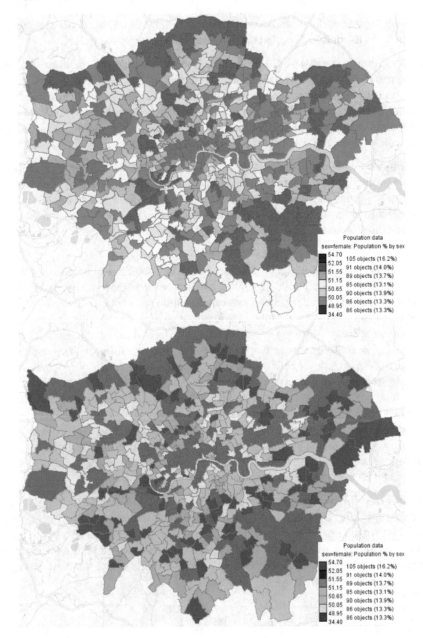

Fig. 3.9: Two maps show the same attribute (proportion of female population in London wards) with a diverging colour scale (top), which is appropriate for these data, and a rainbow colour scale (bottom), which is not appropriate.

the overall pattern, if any, and exceptions, or abnormalities. A good visual represen-tation helps to confirm expected and discover unexpected patterns.

Understanding the distribution: Visualisation is often much more informative and understandable than numerical summaries of data. For numeric data, visual repre-sentations of distributions can effectively indicate the type of the distribution, e.g. normal, skewed, bimodal, logarithmic, as well as reveal outliers. Spatial and tem-poral distributions cannot be adequately expressed through any numeric summaries and require visual representations for understanding.

Seeing relationships: Visual representation of multiple data items that respects in-trinsic properties of the data (Section 3.2.5) can enable the viewer to perceive im-mediately a multitude of relationships between all these items, such as their relative positions in space and/or time. Computational extraction and expression of these relationships would require more time, and the results would be much harder to interpret.

Comparison: A well-designed graphical encoding facilitates comparisons much more effectively than a table of numbers. Gleicher et al. [55] identify three visual methods of comparison: juxtaposition, superposition and explicit encoding. Aligned **juxtaposition** places visual representations next to each other applying the strategy of display space partitioning (Section 3.2.3). Juxtaposition can be good for com-parison of overall patterns in two or more distributions. **Superposition** facilitates comparison by overlaying things to be compared in the same display space (space sharing strategy), such as overlaying multiple annual trends for different years.. This technique can be good for detailed exploration of commonalities and differences. **Explicit encoding of distinctions** is where a comparison metric is calculated (e.g., difference) and visually encoded. For example, original numeric values can be com-pared to some baseline such as the value for a particular year or a global or local average. This technique, however, should be used as a complement rather than re-placement to showing the original data.

Discovering the unexpected: Faithful visual encoding of data as they are (letting the data speak for themselves [58]) can help identify patterns and properties of the data that we did not necessarily expect, whereas numeric summaries and measures often base on certain assumptions and may be meaningless or misleading when these assumptions do not hold. For example, the statistical mean and standard deviation are informative only for a normal distribution.

3.2.7 Limitations of visualisation

In using visualisation methods, their limitations need to be taken into account. Some of the limitations are related to the increasing volumes of data. It is generally recog-nised that visualisation alone is usually not sufficient for data analyses in real-world applications. That is why visual analytics science is developing approaches that combine statistical and computational analysis techniques these with interactive visualisation.

Implicitly defined and unchecked assumptions: Since many statistical summaries implicitly assume certain properties of the data, visual representations involving these summaries inherit these assumptions, which are not necessarily checked before creating visual displays. An example is the box plot representing the quartiles of the data. The quartiles can provide an adequate statistical summary only when the data distribution is unimodal; hence, the box plot should not be used for distributions other than unimodal. Unfortunately, a box plot does not show any indication that it is not applicable to the data it represents. This limitation, however, is not intrinsic to visualisation but is related to a wrong choice of the visualisation technique. Fortunately, visualisation provides alternative ways of representing the same data; for example, a histogram can show whether a distribution is unimodal.

Lack of visual resolution: Attempts to accommodate all data items in the display space often results in the majority of the marks being visually clustered, so that they cannot be visually distinguished. In addition, this can draw viewer's attention to outlying values, which may not be very important. The problem can be alleviated, to some extent, by enabling zoom and pan interactions in an implementation, and, computationally, by performing pixel-based aggregations or using data-reduction techniques to select a small number of representative points.

Visual clutter: Display clutter is usually associated with the abundance of information, especially when some information is irrelevant to the task. A disorderly, crowded image can also be a result of an improper arrangement of marks in the display space. Display clutter can bury important information, complicates visual perception and cognition, and decreases analytical performance of the user. Lack of visual resolution can exacerbate visual clutter, as s higher ratio of data items to pixels increases the chance of overplotting (occlusion). Zoom and filter interactions may help, giving users the ability to remove items or details irrelevant to the task at hand. However, this can lead to losing context. Therefore, it is appropriate to combine a zoomed-in view of a part of the data with a more aggregated view of the whole data indicating what part of it is currently explored. A computational solution is to apply data reduction/sampling methods; in particular, context-aware (semantic) filtering.

Obvious patterns obliterate more interesting patterns: Even for uncluttered visualisations, observable patterns are often dominated by very obvious and expected patterns, obliterating more interesting and important but also more subtle patterns. Adding zoom and pan interactions and filtering is an obvious approach to overcome this problem; however, it may be hard to guess where interesting patterns can be found. A more sophisticated approach is to use a model for removing the expected patterns and revealing deviations from this. For example, normalising geographical data by population or area may result in that the data are not dominated by uninteresting structures. Another example is time series; removing expected trends and seasonal components highlights deviations and is able to uncover surprising patterns.

False patterns appear when a user tries to attach meaning (thus adding a semantic interpretation) to artefacts created by a visualisation technique. This may happen, in particular, when some techniques are used for improving the aesthetic of display appearance. As an example, consider connecting a series of points on a 2D plane by a smooth line. Usually, Bezier curves[7] and other kinds of curves are used for this purpose. Interestingly, most of the methods don't guarantee that the control points remain on the resulting smooth line. Smoothing algorithms apply various heuristics, optimising some kinds of a cost function which is not transparent and not understood by users of the produced visualisations. In the result, unexpected and unwanted patterns like self-intersections, loops, parallel segments may appear. The viewer may try to interpret them, although they are not based on the represented data. To eliminate such effects, visualisations should explicitly highlight distortions. A useful technique is an interactive control of the processing degree, as, for example, proposed in [146] for edge bundling. False patterns may also emerge due to truncation or non-linear scaling of axes or due to improper use of visual variables, as demonstrated in Section 3.2.5.

Slow rendering for large datasets: Both computational pattern detection and visual exploration are more efficient when their outputs appear rapidly. Generally, increasing the data volume slows down data processing. Typically used approach to alleviate this problem is data reduction (discussed further on in Section 5.3.7), including aggregation, sampling, dimensionality reduction, and feature selection methods. It is always necessary to ensure that a right data reduction procedure is taken, and its parameters are set properly. Thus, if a sample is taken, it is necessary to ensure that it reflects all important properties of the complete data set. Another approach is design of so-called anytime methods. An anytime algorithm is an algorithm that can return a valid solution to a problem even if it is interrupted before it ends. The algorithm is expected to find better and better solutions the longer it keeps running[8]. In visual analytics research, there exist approaches that follow a progressive visual analytics paradigm [129, 25]. In this paradigm, visual displays adapt to continuously refined patterns, allowing users to make decisions as early as possible, without long waiting for final precise results of calculations.

In some applications, visualisations need to be applied to **streaming data**. Usually the amount of streaming data does not allow storing everything, therefore a visualisation needs to be able to represent the recent data in the context of patterns extracted from older data that are not available anymore. This requires a special design of visualisation and interaction. From the human perspective, it is necessary to take into account the **change blindness**. Respectively, a visual component of a system dealing with streaming data must ensure that important changes are detected

---

[7] Bezier curve: https://en.wikipedia.org/wiki/B%C3%A9zier_curve
[8] Anytime algorithm: https://en.wikipedia.org/wiki/Anytime_algorithm

automatically and highlighted in visual representations to attract the observer's attention.

It could be noticed that the words "interaction" and "interactive" appeared several times in this section. Let us consider **interaction** and **interactive visualisations** in more detail.

3.3 Interaction

Although a picture may be worth a thousand words, a single static picture is in most cases insufficient for a valid analysis and for understanding of a complex subject. It is usual that an analyst needs to see different aspects or parts of data and look at the data from different perspectives. This means that the analyst needs to interact with the data and with the system that generates visual displays of the data: select data components and subsets for viewing, select and tune visualisation techniques, transform the views, transform the data, and so on. We can distinguish the following types of interaction with graphical representations:

1. Changing data representation.

2. Focusing and getting details.

3. Data transformation.

4. Data selection and filtering.

5. Finding corresponding information pieces in multiple views.

3.3.1 Interaction for changing data representation

Changing visual representation is useful for looking at the same data from different task-specific perspectives. Let's illustrate this on an example of a time series of 21 years monthly temperature data collected at the Tegel airport in Berlin, Germany. Weather measurements are usually done at high temporal resolution (e.g., every hour), but for analysing long-term patterns there data are aggregated, e.g., into monthly averages of the daily minimum, maximum and average temperatures. In this illustrative example, we consider a single attribute, TNM, which represents monthly averages of the daily average temperatures. Figure 3.10 shows this time series on a time plot. The time plot clearly shows annual repetition of similar seasonal patterns. Respectively, it is useful to change the representation and look at the same data with regard to two components of time, the months of a year and the sequence of the years. This view is shown in Fig. 3.11 by 2D time charts where the vertical dimension represents the months of a year and the horizontal dimension represents the sequence of the years. The charts employ three variants of rendering: horizontal and vertical bars (left and right) for enabling comparisons across the years and the months, respectively, and squares (centre) for enabling comparisons across months and years.

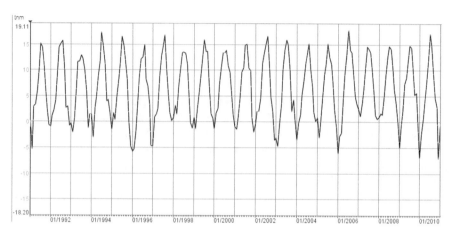

Fig. 3.10: Time plot of monthly averages of average daily temperatures at the Tegel airport (Berlin, Germany).

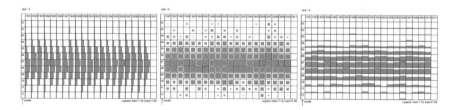

Fig. 3.11: The same data as in Fig. 3.10 are represented, with regard to time cycles, in a 2D time chart with columns for the years 1991-2011 and rows for the months 01-12 (January to December). Three variants of rendering emphasise annual (left), overall (centre) and monthly (right) patterns.

In these two figures (3.10 and 3.11), we applied the most expressive visual variables 'position' and 'size' for representing the TNM dynamics over time. These visual variables work well when the display space has sufficient size, i.e., provides enough pixels for distinguishing small values from large. They may be less suitable when the display space is small, e.g. as in charts on a chart map. In this case, it may be better to represent the attribute values by means of a *colour scale* combining the colour value and hue, as shown in Fig. 3.12. Theoretically, it could be sufficient to use only the variable 'value', encoding the attribute values linearly by shades from light to dark. However, such encoding is sensitive to outliers: the maximum darkness is assigned to one or a few very high values, whereas the shades assigned to the bulk of the values are hardly distinguishable. Besides, when some value has a special meaning, such as the temperature 0°, it is advisable to encode the values using a *diverging colour scale* with distinct colour hues representing values below and above the special value. In the first chart in Fig. 3.12, shades of blue are used for the values below zero and shades of red for the values above zero, while the values around zero are shown in light yellow.

Please note that the attribute values are not directly translated into colours or shades, but the value range of the attribute is divided into intervals, and the colours are assigned to the intervals. Such intervals are called *classes* or *class intervals* in cartography. This approach may work better than explicit encoding, because human eyes are not able to distinguish small variations of shade. Another advantage is the possibility to reduce the impact of outliers. The division into class intervals can be done in many different ways using task-relevant criteria. Figure 3.12 demonstrates different variants of division applied to the same data. The first two are human-defined and the remaining two were created automatically. In particular, a division that creates approximately equal classes (in terms of the number of data items per class) is called *equal-frequency intervals*.

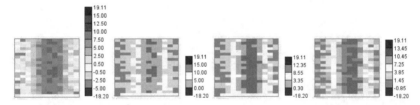

Fig. 3.12: 2D time charts, also called "mosaics plots", represent the same data as in Figure 3.11. Different divisions of the attribute values into class intervals are applied, two variants of human-defined breaks and automatic divisions into 5 and 7 equal-size classes, i.e., intervals containing approximately equal numbers of values.

So far, we have looked at the temporal distribution of the attribute values. Some tasks may require consideration of the frequency distribution of the values. A frequency histogram is suitable for supporting such tasks (Figure 3.13). Similarly to what was discussed in Section 2.3.3, we observe that some patterns either appear or disappear depending on the selected bin size. For example, the bi-modal character of the distribution with two peaks is clearly visible in the variants with small bins, but this pattern disappears in the variant with large bins in the bottom-right image. However, the latter shows a pattern of decreasing frequencies as the attribute values increase, which is not so clearly visible in the other histograms.

A frequency distribution of numeric values can also be visualised in a way that is less dependent on the bin size, as shown in Fig. 3.14. This representation is called

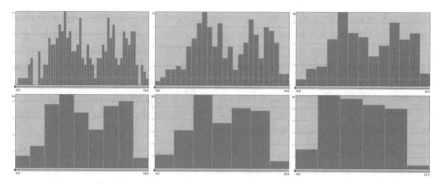

Fig. 3.13: Frequency histograms of the temperature data with the bin sizes equal to 0.5, 1, 2, 3, 4, and 5 degrees Celsius, correspondingly (left to right, top to bottom).

cumulative frequency curve. The x-axis of the display represents the value range [*min*, *max*] of the attribute, as in a histogram. Each position on the x-axis represents a certain value v of the attribute, $min \leq v \leq max$. The corresponding vertical position is proportional to the cumulative frequency of all values from the interval [*min*, *v*]. It can be seen that the shape of the curve does not change when the division of the x-axis into bins change.

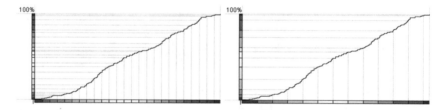

Fig. 3.14: Cumulative frequency curves for the same data as in Fig. 3.13. Grey vertical lines are spaced by steps of 1° (left) and 2° Celsius (right).

The interactions changing the visual representation of data demonstrated in this section are suitable for looking at data under analysis from different perspectives. In the illustrative example, we explored the data from the perspectives of linear time, cyclic time, and the frequency of the values. To adjust visual representations to properties of the data, properties of the display medium, and analysis tasks, display modification operations may be applied, such as changes of display size and aspect ratio, scales. colours, class intervals, and other parameters of a visualisation.

3.3.2 Interaction for focusing and getting details

It is now a standard that graphical displays provide **access to exact data values** upon pointing at visual marks. Mark-related information is often shown in a popup window (Figure 3.15). It may include not only the data values but also additional information, such as references, details of calculation, etc. Popup windows can also contain graphical representations, e.g., showing more details or additional related data.

Fig. 3.15: A popup window activated by mouse hovering shows attribute values and details of their processing.

Other operations that are used for seeing more details are **zooming** and **panning**, which allow a contiguous part of the display to be enlarged, so that the data it contains can be presented in an increased visual resolution. An example is shown in Fig. 3.16, where a fragment of a long time series is shown in higher detail than in Fig. 3.10. There are implementations employing *focus+context* techniques, in which a portion of data is shown in maximal detail, while the remaining data are also presented in the display but shown in a less detailed manner. Examples are time intervals before/after a selected interval, spatial information around a selected area, etc.

In representing values of a numeric attribute by means of a colour scale (i.e., by shades of one or two colour hues), **colour re-scaling** can be used to increase visual discrimination of values in a particular range. One possible approach is to use the full colour scale for the selected range and hide the values beyond this range (Figure 3.17, top right). Another approach may be called "visual comparison": it converts a sequential colour scale to a diverging one by introducing a user-selected reference value (Figure 3.17, bottom), so that the colours show the differences of all values in the display to the reference value. The reference value can be specified explicitly as a number, or implicitly as a value associated with a selected reference object. This operation can be performed, for example, by clicking on the reference object on a map or another graphical display. These techniques were introduced for choropleth maps in interactive cartography and geovisualisation [7]. In a sim-

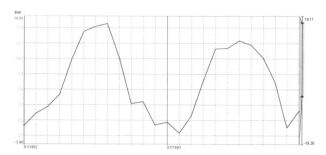

Fig. 3.16: Time plot shown in Fig. 3.10 has been zoomed in to show data for selected two years.

ilar way, re-scaling can be applied to other visual variables, for example, 'size' in bar charts. Another possibility for interactive colour re-scaling is converting from a continuous to a discretised representation of the data by introducing class intervals; see Figure 3.12 and the corresponding discussion.

Ordering is an interactive operation that is often applied to matrices and tables. An example is shown in Fig. 3.18, where ordering of the rows by the values of one of the presented attributes reveals the relationships of this attribute with other attributes. It may be useful to apply reordering to axes in parallel coordinates plots, components of a scatter-plot matrix, table columns, slices of pie charts, and other kinds of display components.

3.3.3 Interaction for data transformation

In the process of exploring data with the use of visual displays, different motives for transforming the data may arise: to get a clearer view, to simplify the display or the data, to reduce the amount of data, to disregard excessive details and facilitate abstraction, and others. Some data transformation operations may be intertwined with operations for display modification. Thus, the division of a range of real numbers into class intervals applied in the 2D time charts in Fig. 3.12 and in the choropleth maps in Fig 3.9 transforms the data from a continuous range of real values to a finite sequence of classes. This kind of transformation is called *discretisation*.

Logarithmic transformation can be applied to values of a numeric attribute when most of the values are small, but there are a few high values. After such transformation, the differences between the small values can be represented visually with higher expressiveness; however, the representation will be harder to interpret than a display of the original values.

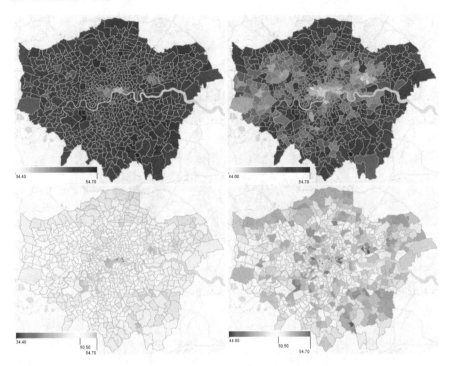

Fig. 3.17: Variants of a choropleth map representing the proportions of females in the population of the London wards (the data are the same as in Fig. 3.9). On the right, the colour scale focuses on a reduced value range, excluding small values. At the bottom, the colour scale has been transformed from sequential to diverging for showing deviations from a selected reference value.

Aggregation is a means to reduce the amount of data to consider and, simultaneously, to facilitate abstraction and gaining an overall view of a subject. Aggregation is done by grouping data records according to different criteria (e.g., by categories, classes of attribute values, time intervals, or areas in space) and deriving attribute values for the groups as statistical summaries (minimum, maximum, mean, median, etc.) of the original values.

Data may include many numeric attributes describing something in excessive detail. For example, sales of different shops may be described at the level of individual articles, which may not be needed for a particular analysis task. Such data can be simplified by merging groups of attributes into singular attributes. In our example with shops, the articles can be grouped into categories, and the corresponding groups of attributes can be integrated by summing.

When data comprise multiple comparable attributes, it may be useful to have a representation showing which attribute is dominating, i.e., has the highest value. For example, data describing geographic regions may consist of the proportions of

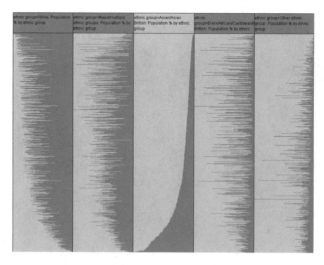

Fig. 3.18: Ethnic structure of the population of the London wards. The rows (each having one pixel height) are ordered by the proportions of the Asian population.

the working population employed in different branches, e.g. industry, agriculture, and services. The original attributes can be transformed to a single attribute specifying the dominant occupation for each region. Such dominant attribute aggregation can be interactively controlled for handling cases when the dominance is very small [8].

For numeric time series, and, more generally, numeric attribute values distributed over a continuous base, value *smoothing* can decrease small fluctuations and make patterns more clearly visible. Data smoothing integrates in some way each attribute value with values from its neighbourhood within a chosen maximal distance and replaces the original value by the integrated one. The simplest variant of integration is taking the mean of the values. A more sophisticated approach accounts for the distances of the neighbouring values to the value in focus, i.e., the one being modified: as the distance increases, the impact on the integrated value decreases. For time series, there are specific smoothing methods, such as exponential smoothing[9], giving higher weights to more recent values when integration is done. Double exponential smoothing is a method suited for time series with overall trend patterns, and triple exponential smoothing is designed for periodic time series, like the one we had in Fig. 3.10.

Another data changing operation for time series is calculation of *changes* with respect to the previous time step or a selected time step. The changes can be expressed in the form of differences or ratios. Transformed data enable efficient *comparison*, as increase is separated from decrease by standard values 0 for differences and 1 for ratios. Examples are shown in Fig. 3.19. The time plot on the left shows the time

---

[9] https://en.wikipedia.org/wiki/Exponential_smoothing

series of the differences to the previous month values. To take into account the periodicity of the variation, it is useful to perform comparison to the values for the same month a year earlier, as on the right image of Fig. 3.19.

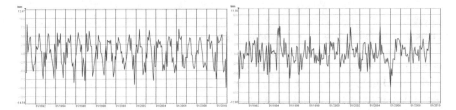

Fig. 3.19: Time plots obtained by transforming the data presented in Fig. 3.10 into the differences to the previous month and previous year values.

3.3.4 Interaction for data selection and filtering

Data **selection** (also called **querying**) and/or data **filtering** are necessary operations for detailed exploration of particular portions of data. Selection means taking a portion of data in order to visualise this portion only. Filtering means temporarily removing the data that are currently out of interest from already existing visual displays. For both selection and filtering, the analyst needs to specify the properties of the data of interest. Selection or filtering can be based on the following criteria:

- attribute values (*attribute-based filter*);
- position in time, either along the time line or in a temporal cycle (*temporal filter*);
- position in space (*spatial filter*);
- references to particular entities (*entity-based filter*);
- relationships to other data (*filter of related data*).

Examples of different filtering operations suitable for spatio-temporal data can be found in the monograph [10], section 4.2 entitled "Interactive filtering". Interactive filtering is usually enabled by UI elements, such as sliders, or can be done directly through a data display, e.g., by encircling a part of the display containing the data of interest. For informed selection or filtering, it is appropriate to use visual displays of the data distribution, such as histograms of value frequencies and time histograms. Spatial filtering is convenient to do through a map display showing the spatial distribution of the data.

The result of data selection or filtering is that only data satisfying query conditions are shown. It is often desirable to see how a specific part of the data relates to other data. This task is supported by **highlighting** the data subset of interest while keeping the remaining data visible; hence, the highlighted items can be viewed in the context of all currently active data (which may have been previously selected or filtered). A common operation for making a group of data items highlighted in a display is **brushing**, which is typically performed by dragging the mouse cursor over an area in the display space. Items can also be highlighted by clicking on the visual marks representing them. When two or more data displays exist simultaneously, highlighting usually happens in all of them in a coordinated manner affecting the appearance of the visual marks representing the same subset of data items in all these displays. Such simultaneous highlighting of corresponding items is essential for the use of multiple complementary displays, as it facilitates linking distinct pieces of information visible in these displays. The use of different colours or styles for representing highlighted items makes it possible to have several subsets highlighted simultaneously. These subsets can thus be compared to each other and to the remaining data.

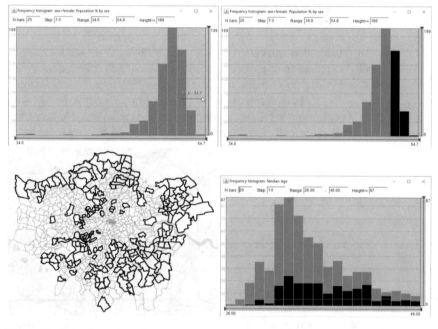

Fig. 3.20: Selection of a subset of London wards by brushing. Top left: selection of the wards belonging to the three rightmost bars of the frequency histogram for the proportion of the female population. Top right: the bars representing the selected items are highlighted in the histogram. Bottom left: the selected wards are highlighted in the map. Bottom right: bar segments corresponding to the selected wards are highlighted in a histogram for the median age.

The use of concurrent highlighting in multiple displays has been demonstrated in Fig. 2.11. We have presented it as a tool for studying interrelations between multiple attributes. Figures 2.10 and 3.20 demonstrate that highlighting may have different appearance depending on the display type. In the Figs. 2.11 and 3.20, highlighting is a result of brushing in a histogram display executed by dragging the mouse cursor over a part of the attribute axis. In Fig. 2.10, brushing has been done by dragging the mouse over a region in the projection space. This has highlighted the dots located in the region and the corresponding lines in the parallel coordinates plot.

Apart from explicit specification of a part of the display containing data to be high-lighted, it may be convenient for analysis to be able to highlight data items that are close or similar to a particular item. There exists an example of using the so-called Mahalonobis distance[10] for selecting a given number of data items around a selected point in multiple dimensions [111]. It is possible, in principle, to use also other distance functions: see Section 4.2.

3.3.5 Relating multiple graphical views

Coordinated multiple views (CMV) is a common approach to visual analysis. It is applied when all data components under consideration cannot be effectively repre-sented in a single display. It can also be applied for looking at the same data from different perspectives, e.g., from the perspectives of linear and cyclic time models, as in Figs. 3.10 and 3.11. The concept of CMV implies that concurrently existing data displays are *coordinated*, which means that certain techniques support finding corresponding pieces of information in the different displays. One of such coordina-tion techniques is brushing discussed in the previous section. Filtering, which was also discussed earlier, can also be a mechanism for coordination, if it consistently affects all data displays. Other frequently used techniques for display coordination are *common symbolisation* and *conditioning*.

Common symbolisation as a coordination mechanisms means propagation of some visual property of marks from one display to other displays. An example of propa-gating colours has been shown in Fig. 2.10: the display background colours below the dots in the projection plot have been propagated to the corresponding lines in the parallel coordinates plot. Another example is demonstrated in Fig. 3.21: the colours assigned to the London wards in the choropleth map shown in Fig. 3.9, top, are propagated to a scatterplot and a frequency histogram. In the scatterplot, the colours ape assigned to the dot marks. In the histogram, the bars are divided into segments painted in the propagated colours. The lengths of the segments are proportional to the corresponding numbers of the wards having these colours in the choropleth map.

---

[10] https://en.wikipedia.org/wiki/Mahalanobis_distance

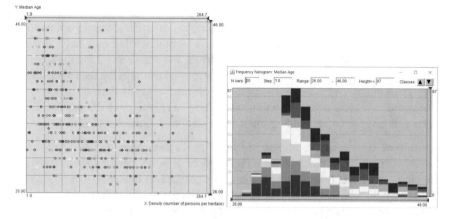

Fig. 3.21: An example of coordinating multiple displays through colour propagation. The colours of the London wards from Fig. 3.9, top, are propagated to corresponding data items shown in a scatterplot of the median age against the population density (left) and in a frequency histogram of the median age (right).

Conditioning means creating several instances of a display showing different data portions. Conditioning as a coordination mechanism means creation of multiple display instances for representing classes or subsets of items defined in another display. For example, Figure 3.22 demonstrates a possible response of a scatterplot to the propagation of the classes of the London wards defined in the choropleth map from Fig. 3.9, top, according to the proportions of the female population. The original appearance of the scatterplot is shown on the left of Fig. 3.21. In response to the propagation of 7 classes, the single scatterplot has been replaced by 7 instances each showing the wards from one of the classes.

After the idea of CMV appeared in early implementations, researchers, developers, and data analysts realised very soon that this is very useful tool for data analysis; moreover, this is often the only viable approach to analysing complex data. However, simultaneous changes in multiple displays in response to every user interaction may not always be desirable. Such changes can distract analyst's attention, and they also can "eat" computer resources and hamper the computer performance. Therefore, there are systems for visual data exploration in which coordination between displays can be controlled by the user. Thus, users of the CDV system [49] can set in each display whether it is allowed to (1) broadcast interaction events to other displays and (2) react to interaction events broadcast from other displays. The system Improvise [149] contains more sophisticated tools for controlling coordination, so that users can explicitly choose which displays will be linked and in what way.

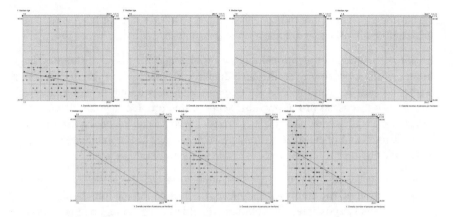

Fig. 3.22: An example of coordination in CMV through conditioning. The division of the London wards into 7 equal-size classes defined in Fig. 3.9 has been propagated to a scatterplot of the median age against the population density. In response, the single display has been replaced by 7 instances showing the different classes of the wards. The correlation coefficients and the trend lines have been computed individually for the classes.

3.3.6 Limitations and disadvantages of interaction

To a large extent, interaction techniques are designed to compensate for the limitations of visualisation, particularly, the impossibility to show many data components in a single display (compensated by CMV), impossibility to show large amounts of data in sufficient detail (compensated by techniques for focusing and obtaining details), and difficulty of finding particular information (compensated by tools for data selection and filtering). However, the existence of these tools and the necessity to apply them complicates the use of visualisations and the process of data analysis. The analyst needs to keep in mind what tools are available, for what purposes they are used, and how they are operated. Substantial parts of the analyst's attention and time have to be given to technical operations at the cost of cognitive activities.

A big problem of visual exploration with a strong emphasis on the use of interactions is the **lack of systematic approach** to the analysis. It is hard to ensure that the entire subject is comprehensively studied in sufficient detail when the analysis is done through interactive selection of particular data portions and aspects for viewing. It is hard to check whether everything what is essential has received sufficient attention and thought of the analyst. Without a systematic approach, interactive exploration mostly relies on serendipity. Although serendipity can have a role, a strong reliance on it is deeply unsatisfactory within any scientific endeavour. For a more systematic interactive exploration, it is useful to keep track of the inspected data components and performed operations and document the visual data analysis process, which helps to identify data components and portions that have not yet been considered. Besides, it facilitates the externalisation of the analysts' reasoning processes.

The use of interaction operations may be problematic for displays showing large amounts of data, because updating such displays in response to every user's action may be too slow for comfortable and efficient analysis without unwanted interruptions. As we discussed in the previous section, this problem can be alleviated by controlling when, where, and what interaction events and reactions are allowed. However, tools and operations for making such settings also complicate the use of visual tools and take a certain amount of the analyst's attention and time.

It is always necessary to remember that interaction consumes the most valuable resource – **time of the analyst**. Therefore, the use of interaction should be well justified. Whenever possible, computations should be used, either on request or proactively, preparing answers to potentially emerging questions.

3.4 Concluding remarks

In this chapter, we have considered the major principles of creating analytical visualisations and the purposes and methods to interact with them. To design an effective and efficient visualisation that facilitates data analysis, it is necessary to understand these principles and apply them carefully. However, we have provided only a short and quite superficial overview of this exciting discipline. We recommend interested audience to read the two books: "Visualization Analysis and Design" by T.Munzner [103], which focuses on the principles of visualisation, and "Interactive Visual Data Analysis" by C.Tominski and H.Schumann [135], focusing on designing and using visualisations for data analysis. Our book has a different focus. It does not discuss the design of new visualisations but explains how to *utilise* the existing visualisation methods that have proved to be useful in analytical workflows together with other kinds of techniques for data processing and analysis.

It is necessary to understand both the benefits and unavoidable costs of interactive visualisations. Whenever possible (i.e., for routine tasks, where human judgement and reasoning are not necessary), computations need to be performed instead of visual analysis, saving time and cognitive efforts of human analysts.

Chapter 4
Computational Techniques in Visual Analytics

Abstract Visual analytics approaches combine interactive visualisations with the use of computational techniques for data processing and analysis. Combining visualisation and computation has two sides. One side is computational support to visual analysis: outcomes of computations are intended to provide input to human cognition; for this purpose, they are represented visually. The other side is visual support to application of computational methods, which includes visual exploration of data properties for preparing data to computations, evaluation of computation outcomes, and comparison of results of different runs of computational techniques. This chapter focuses in more detail on the computational support to visual analysis. A major common purpose for using computational methods in visualisation is enabling an overview of voluminous and complex data. The general approaches are spatialisation, which is achieved by means of data embedding techniques, and grouping, which is achieved using clustering algorithms. Distance functions are auxiliary computational techniques used both for data embedding and clustering. They provide numeric assessments of the dissimilarity between data items. Another class of auxiliary techniques is feature selection. The chapter describes the use of data embedding and clustering. By example of clustering, the general principles of visual analytics support to application of computational methods are demonstrated. Topic modelling is a special group of data embedding techniques originally designed for textual data but applicable also to other kinds of data. The main ideas, properties, and uses of the topic modelling methods are discussed in a separate section.

4.1 Preliminary notes

Visual analytics approaches combine interactive visualisations with various operations performed by computers. Computers are capable to do numerous things that can be useful for analysis: store data in a database and select data satisfying query

© Springer Nature Switzerland AG 2020
N. Andrienko et al., *Visual Analytics for Data Scientists*,
https://doi.org/10.1007/978-3-030-56146-8_4

conditions, compute descriptive statistics, arrange and group data items, transform data, derive new data components from existing ones, detect data pieces with specific properties, generate predictive models, and many others. Computational methods of data processing and analysis are increasingly important due to our desire to extract meaning from ever-larger datasets.

It is hardly feasible to describe in one book chapter how each of the existing computational techniques can be used in analysis. What we are going to do is to look at the whole mass of these techniques from the visual analytics perspective. This reveals two aspects relevant to combining computing with visualisation and interaction:

• Visualisation and interaction can support appropriate use of computational methods.

• Computational methods can enable effective presentation of voluminous and/or complex information to human analysts, mitigating the limitations of visualisation and interaction techniques.

4.1.1 Visualisation for supporting computations

Correct use of computational methods requires first of all selection of right methods depending on the nature and properties of the data under analysis and the current analysis task. Then, the data need to be appropriately prepared to the application of the chosen method. Data preparation also requires thorough investigation and good understanding of the data properties, which can hardly be achieved without the help of visualisation. While the next chapter of this book is fully dedicated to investigating and processing data, here we would like to draw your attention to two features of data that can be uncovered by means of visualisation: existence of outliers and dissimilar subsets.

Outliers can greatly affect the work of computational methods and skew their results making them useless or misleading. It is usually necessary to remove outliers, but it may also be sensible to investigate and understand their impact on the computation results, and visualisation is a suitable tool for this task.

When a dataset consists of highly dissimilar subsets, it may be a good idea to treat them differently by involving different data components in the analysis and/or applying different parameter settings or even different analysis methods. This approach can be especially recommended in using methods for computational modelling. For example, in studying and modelling the variation of the prices for housing or accommodation, it may be reasonable to take separately residential, touristic, and business areas, as the respective prices can be affected by different factors.

After running a computational method, the analyst needs to evaluate its results. For many methods, there are numeric measures of the quality of their results. However, these measures are not sufficient for understanding whether the results are meaningful, i.e., can be interpreted and do not contradict to analyst's knowledge, and whether they are useful for reasoning and drawing conclusions or for performing further steps in the analysis. Whenever human understanding is required, there is a need in visualisation. Besides, it is quite a usual case that the quality of method results is not the same for all parts of the input data. This variation remains unnoticed when only overall quality measures are taken into account, whereas visualisation can reveal the variation and help the analyst to find a way to improve the result quality. Examples of this use of visualisation will be demonstrated in the chapter about visual analytics approaches to building computer models (Chapter 13).

Most of the computational methods have parameters that need to be set for running them. The outputs of the methods can vary depending on the chosen parameter values. It is often not quite clear in advance what settings can be suitable for the data at hand and the analysis goal. Even when the analyst's knowledge and experience suggest sensible choices, it is reasonable to perform sensitivity analysis by testing the robustness of the method results to slight variations of the settings for either gaining more trust and certainty or understanding the degree of uncertainty.

Besides parameters, method outcomes can be affected by variations of input data due to taking different samples or selecting different combinations of attributes. In any case, a method should be run several times with modifying parameter settings and/or input data, and the outcomes of different runs need to be compared.

Another source of possible differences in analysis results is the existence of alternative methods suitable for the same kind of data and the same tasks. When there are no serious reasons for preferring one method over the others, a diligent analyst will try different methods and compare their results.

Hence, good practices of data analysis involve comparisons of results of different runs of the same or different methods. Consistency between the results indicates their validity, whereas inconsistencies require investigation, explanation, and making a justified choice or drawing a justified overall conclusion. For comparing results, it is usual to use various statistical measures. However, these measures are typically not sufficient for supporting understanding and substantiated judgements, as they can tell *how much* difference exists between two results but cannot tell *what* is different. This is where visualisation can greatly help.

The role of visualisation is to present outcomes of different executions of computational methods so that analysts can see where they diverge and in what manner and look for possible reasons. Interaction techniques provide access to details or additional relevant information. In this role, interactive visualisation can be used together with any kind of computational method. While the types of visual displays that are suitable for this purpose depend on the type and properties of method outputs, there is a *general principle*: the visualisation should link the outputs to the input data, so that the analyst can see how the same input data have been dealt with by different runs of computational methods. Further in this chapter, we shall demonstrate the application of this principle by example of clustering.

Some computational methods are designed to work in an iterative manner allowing, in principle, inspection of the course of method execution and the intermediate results of the processing. Visualisation can be of great help in performing such inspections. Analysts can better understand the data processing and method training, which adds transparency to the approach and increases the trust in its results.

Based on our discussion, the **tasks of visualisation** in supporting appropriate use of computational methods can be summarised as follows:

- Before applying computational methods: enable investigation of data properties in order to

  - choose suitable computational methods and make appropriate parameter settings;

  - detect outliers, which may need to be removed;

  - detect disparities between parts of the data, which may require different approaches to analysis.

- During applying computational methods:

  - inspect how the method uses and processes the input data;

  - investigate the intermediate structures constructed by the method;

  - examine the current state of the model being built by the method and understand whether is develops in the right direction.

- After applying computational methods:

  - enable evaluation of method results for understanding their meaningfulness and usefulness and seeing the variation of the quality across the input data set;

  - enable comparisons of different results obtained by varying the input data or values of method parameters, or by choosing alternative methods.

4.1.2 Computations for supporting visual analysis

What we are going to discuss further in this chapter is the computational techniques that are commonly used in visual analytics workflows for enabling effective visual presentation of voluminous and/or complex information to human analysts. These techniques are employed when there is no meaningful way to create a visualisation showing each and every data item in detail: either it is impossible technically, or such a visualisation would be incomprehensible.

One possible approach to dealing with huge data amounts is to visualise only small pieces of data. Such pieces can be extracted from the bulk either in response to queries with explicitly specified conditions or by applying data mining techniques that search for combinations of data items having particular properties or frequently occurring in the dataset. This approach is taken when only such data pieces are of interest and there is no need in getting a "big picture" of the whole dataset. Techniques that are used for extracting data pieces support data analysis but do not play the special role of supporting visualisation.

Visualisation can be supported by computational techniques that organise and/or summarise information so that it becomes better suited for visual representation and human perception and comprehension. The main goal is to enable an overall view of the whole bulk of information. Such an overall view is achieved owing to a high degree of abstraction and omission of many details. The details, when needed, can be obtained by means of interaction.

There are two major approaches to creating an overall view of a dataset:

- **Spatialisation**: position the data items in an artificial space with two or three dimensions according to a certain principle, usually according to their similarity or relatedness. The resulting spatial arrangement of the data items can be visually represented in a two-dimensional plot or in a three-dimensional plot that can be interactively turned for viewing from different perspectives.

- **Grouping**: organise the data items in groups according to their similarity. The resulting groups can be treated as units, i.e., each group is a single object. The groups are characterised by statistical summaries of characteristics of their elements. The statistical summaries can be represented visually.

Spatialisation can be achieved by means of the class of computational techniques called *data embedding* or, more frequently, *dimensionality reduction* or *dimension reduction*. These methods, generally, represent data items by points (i.e., combinations of coordinates) in an abstract "space" with a chosen number of dimensions, which can be 2 or 3, in particular. One category of data embedding methods apply some transformations to the original components of the data, called "dimensions", in order to derive a smaller number of components by which the data items can be described so that substantial differences between them are preserved. This is where the term "dimension reduction" comes from.

Another category of the data embedding methods use numeric measures of the similarity or relatedness between any two data items. Such a measure is expected to be zero when two items are identical or indistinguishable according to chosen criteria of similarity, and, as the similarity decreases, the value of the measure must increase. A measure fulfilling this principle is called *distance*. A mathematical formula or an algorithm for determining the distance between given two items is called a *distance function*. Data embedding methods using distances aim at reproducing the distances between the data items as accurately as possible through the distances between the points representing these data items. Some of these methods ignore the original components, or dimensions, of the data and use only the distances irrespective of how they have been determined. Although the term "dimensionality reduction" appears less apposite in this case, it is customary to apply this term to all data embedding methods.

Grouping can be achieved by means of techniques for *clustering*. As the name suggests, these techniques create or find *clusters*, which are groups of similar or related data items. Clustering techniques also use distances between items, which can be determined by means of distance functions. Hence, distance functions is a class of computational techniques that can be used for both spatialisation and grouping.

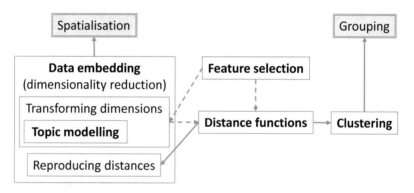

Fig. 4.1: Relationships between classes of computational methods supporting visualisation. The arrows indicate the use of methods; solid arrows mean required use and dashed arrows mean possible use.

In the remainder of this chapter, we shall discuss the classes of techniques that are commonly employed in visual analytics approaches for supporting creation of an overall view of a dataset. The main classes are data embedding and clustering, the methods from which are used for spatialisation and grouping, respectively. As both classes of methods use distance functions, we shall discuss distance functions as well. For defining appropriate distance functions, it is often necessary to perform *feature selection*, which means finding a small non-redundant subset of data components that is sufficient for representing substantial differences between data items (data components are called "features" or "variables" in statistics and data mining). Therefore, the chapter includes a section about feature selection.

We also decided to write a special section about topic modelling. This class of methods was originally created specifically for analysing text data. However, several visual analytics researchers have demonstrated in recent years that topic modelling can have much wider area of applicability than just texts. In essence, topic modelling is a kind of dimension reduction. We discuss it in a separate section because it needs to be described using specific terms, such as documents, words, and topics, which are not relevant to the other dimension reduction methods. The uses of topic modelling for non-textual data are still new; therefore, we shall first describe the traditional use and then show how it translates to other data.

The scheme in Fig. 4.1 shows the classes of computational methods discussed in the remainder of this chapter. The arrows indicate the use of methods. A solid arrow means that methods of one class require the use of methods of another class. Thus, methods for distance-reproducing data embedding and for clustering require the use of distance functions. A dashed arrow means that methods of one class may use methods of another class. Thus, distance functions can be defined using results of feature selection or dimension-transforming data embedding.

Please keep in mind that each of these classes of methods is used not only for supporting visualisation but also for other purposes; moreover, most of these methods are primarily created for other purposes but can be used for visualisation as well. Particularly, feature selection, dimensionality reduction, and clustering can be used for building computer models, and distance functions can be used for searching by similarity.

4.2 Distance functions

There exist many distance functions differing in the types of data they can be applied to and in the approaches to expressing dissimilarities. In the subsections of this section, we shall group the distance functions according to the data types.

4.2.1 Multiple numeric attributes

When two data items differ in values of a single numeric attribute, the arithmetic difference between the two values (the sign being ignored) can serve as a measure of the dissimilarity between the data items. This does not apply, however, to attributes with *cyclic value domain*, such as spatial orientation or time of the day. Take, for example, the spatial orientation, which is an angle with respect to a chosen reference direction, such as the geographical north. It is usually measured in degrees, and $0°$ is the same as $360°$. For the time of the day, the time 00:00, or 0 AM, is the same as 24:00, or 12 PM. When we need to determine the difference between, for example, $15°$ and $350°$, straightforward subtraction of one of them from the other will give the result 335°, which is wrong; the correct result is just 25°. Similarly, the difference between 23:45 and 00:15 is not $23\frac{1}{2}$ hours but only half an hour.

Hence, determining differences between values of an attribute with a cyclic value domain requires the following special approach: if the result of subtracting one value from another exceeds the half of the cycle length, it needs to be subtracted from the cycle length. Thus, the cycle length of the spatial orientation is 360° and the half of it is 180°. The subtraction result for 15° and 350° is 335°. It is greater than 180° and therefore needs to be subtracted from 360° to yield the final result 25°. Similarly, the cycle length of the time of the day is 24 hours and the half is 12 hours; hence, $23\frac{1}{2}$ hours needs to be subtracted from 24 hours to yield the final result $\frac{1}{2}$ hour.

When data items differ in more than one numeric attributes, the dissimilarity can be expressed by combining in some way the arithmetic differences between the values of each attribute. The most popular method for combining multiple arithmetic differences is the **Euclidean distance**, which can be interpreted as the straight line distance between two points in space. For n numeric attributes and two data items (points) denoted x and y, the formula of the Euclidean distance is

$$\sqrt{\sum_{i=1}^{n} |x_i - y_i|^2}, \tag{4.1}$$

where x_i and y_i are the values of the i-th attribute for the items x and y, respectively. Another popular distance function is the **Manhattan distance**, which is merely the sum of the per-attribute differences.

The **Minkowski distance**[1] generalises the possible approaches to combining multiple differences in the formula

$$\sqrt[p]{\sum_{i=1}^{n} |x_i - y_i|^p} == \left(\sum_{i=1}^{n} |x_i - y_i|^p \right)^{1/p}, \tag{4.2}$$

where p is a parameter. When $p = 2$, this general formula transforms to the formula of the Euclidean distance, and the formula of the Manhattan distance is obtained when $p = 1$. Figure 4.2 demonstrates the impact of the parameter p on the resulting distances by example of points in a two-dimensional space. The blue outlines encircle the areas in which the distance of any point to the intersection of the axes is smaller than some constant.

Regarding the Euclidean and Minkowski distances, as well as other possible distance functions that aggregate arithmetic distances between attribute values, there are two potential problems you need to care about. First, it is inappropriate to use such functions when the attributes are not comparable, that is, when the value ranges of the attributes significantly differ. In such cases, it is necessary to apply *normalisation*, as described in Section 4.2.9. Second, these distance functions don't pay

---

[1] https://en.wikipedia.org/wiki/Minkowski_distance

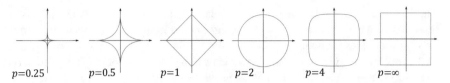

$$p=0.25 \qquad p=0.5 \qquad p=1 \qquad p=2 \qquad p=4 \qquad p=\infty$$

Fig. 4.2: The impact of the parameter p of the Minkowski distance on defining the neighbourhood of a point. The blue outlines encircle the areas where the distances to the intersection point of the axes are below the same threshold value.

attention to relationships between the attributes. If some attributes are correlated, their impact on the results will differ in comparison to unrelated attributes. A common approach to solving this problem is *feature selection* (Section 4.3). A so-called **Mahalonobis distance**[2] addresses both problems by integrating normalisation and reducing the impact of relationships between the attributes involved. This distance, however, is more difficult to compute and, from the human perspective, to interpret.

The use of arithmetic differences between attribute values is not the only possible approach to expressing dissimilarities. Another approach, which is also taken quite frequently, is to treat data items as vectors (rather then points) in a multidimensional space and measure the angles between the vectors irrespective of their magnitudes. This idea underlies the **cosine similarity** function[3]. Using the same notation as for the Euclidean and Minkowski distances, the function is defined by the formula

$$\frac{\sum_{i=1}^{n} x_i y_i}{\sqrt{\sum_{i=1}^{n} x_i^2}\sqrt{\sum_{i=1}^{n} y_i^2}} \tag{4.3}$$

The application of this formula gives the cosine of the angles between the vectors. The cosine equals 1 when two vectors have the same orientation and -1 when two vectors have opposite orientations. Hence, the cosine similarity function *is not a distance function*, because a distance function must return 0 for identical items and positive values otherwise. In a case of comparing vector orientations, a distance function must return 0 when the orientation is the same and the maximal value when the orientations are opposite. To convert the cosine similarity into a distance function, the most obvious approach is to take the arccosine of the result of the cosine similarity calculation and thereby obtain the angle between the vectors. In this way, you define the **angular distance** function. The implementations of the arccosine function available in mathematical computation packages typically return the result in radians in the range from 0 to π. If you wish to have the angle measured in degrees, you need to divide this result by π.

---

[2] https://en.wikipedia.org/wiki/Mahalanobis_distance
[3] https://en.wikipedia.org/wiki/Cosine_similarity

The functions of cosine similarity and angular distance are often used in analysis of texts. Thus, when texts are compared regarding the frequencies of occurring terms or keywords, these functions can assess the similarity or dissimilarity irrespective of the text lengths. Generally, this approach can be useful in application to numeric attributes whose values depend on the sizes of the entities they refer to. For example, in census data, values of many attributes depend on the population of the census districts.

Even more generally, distance functions that disregard absolute values of attributes are useful when data items need to be compared in terms of their structure irrespective of quantitative differences. Such distance functions can be called *scale-independent*. Apart from the angular distance, a scale-independent distance function can be defined based on the **correlation** between two combinations of values of multiple attributes. Like with the cosine similarity, the correlation values, which range from -1 (opposite) through 0 (unrelated) to 1 (absolutely correlated), need to be transformed to distances in the range $[0, max]$. Scale-independent functions are applied when the attributes have similar semantics and can be meaningfully compared. For example, in analysing census data, a correlation-based distance function can be applied to a group of attributes reporting the numbers of the inhabitants by age intervals, or by types of occupation, or by social status, but it would not be reasonable to apply such a function to a mixture of attributes from these different groups, or to the mean age, mean income, and mean distance to the work taken together.

Like the cosine similarity, a correlation coefficient needs to be treated as a measure of similarity and *not as a distance*. It equals 1 when two combinations of values are fully correlated, which is treated as the highest similarity. The coefficient value of -1 means that two combinations are opposite. The **correlation-based distance** is defined by subtracting the correlation coefficient from 1; hence, the value of 0 means the maximal similarity.

There are many formulas for expressing the correlation. The most popular is the Pearson correlation, which measures the degree of a linear relationship between two value combinations. Using the same notation as before, the Pearson correlation is calculated according to the formula

$$\frac{\sum_{i=1}^{n}(x_i - \bar{x})(y_i - \bar{y})}{\sqrt{\sum_{i=1}^{n}(x_i - \bar{x})^2 \sum_{i=1}^{n}(y_i - \bar{y})^2}} \tag{4.4}$$

Here, $\bar{x}$ and $\bar{y}$ are the arithmetic means of the groups of the attribute values belonging to the data items x and y, respectively.

In dealing with multidimensional data, you need to be aware of the problem known as the **curse of dimensionality**[4]. When the number of data dimensions (attributes) is very high, the volume of the space comprising these dimensions is huge. The available data are very sparse in this space, and there is very little difference in

---

[4] https://en.wikipedia.org/wiki/Curse_of_dimensionality

the distances between different data items. In such a case, it is necessary to apply feature selection (Section 4.3) and/or dimensionality reduction (Section 4.4). These is no clear definition of what number of dimensions should be considered high. It is acknowledged to be dependent on the number of available data items. When the number of dimensions exceeds the number of data items, the data should be definitely treated as high-dimensional. However, for analyses involving assessment of similarities or distances, it is admitted that the curse of dimensionality begins much earlier than the number of dimensions starts exceeding the number of data items.

4.2.2 Distributions

Here we refer to different kinds of distributions, such as

- distributions of probabilities or frequencies of values of an attribute in subsets of data; for example, distributions of salary amounts of employees of different departments;
- spatial distributions; for example, distributions of colours over different images, or distributions of vehicle traffic over a city at different times;
- temporal distributions; for example, weekly distributions of the visits of different web sites.

A larger list of types of distributions is given in Section 2.3.1, where we also explained the concept of distribution. Please recall that a distribution involves two sets, one of which is treated as a base for the other. Thus, in a probability or frequency distribution, the set or range of all possible attribute values is the base for the probabilities or frequencies of the values, in a spatial distribution, the base is the space, and in a temporal distribution the time.

To compare distributions, their bases are divided into equal or corresponding bins, so that each bin in one base has a unique corresponding bin in the other base. For example, numeric value ranges in frequency distributions or time ranges in temporal distributions can be divided into the same number of equal-length intervals, images can be divided into the same number of compartments using a regular grid, and the territory of a city can be divided using a regular grid or boundaries of administrative districts, traffic zones, or other predefined areas. A distribution is represented by per-bin statistical summaries, such as counts of value occurrences, counts of entities, or mean attribute values. This representation can be called a *histogram* in a general, mathematical sense: "a function that counts the number of observations that fall into each of the disjoint categories (known as bins)"[5].

---

[5] https://en.wikipedia.org/wiki/Histogram

The sets of per-bin statistics representing two distributions can, in principle, be compared in the same ways as combinations of values of multiple numeric attributes, as discussed in the previous section. Distributions are typically compared using scale-independent distance functions allowing to disregard quantities and compare the distribution profiles, such as the angular distance or correlation distance. Besides, distributions can also be compared using chi-square statistics, Bhattacharyya distance, or Kullback-Leibler divergence. The respective formulas can be easily found in literature (e.g., [29]) or on the web.

In representing distributions by histograms, be mindful of the *curse of dimensionality* mentioned in the previous section. Since you can usually choose how many bins to create, make sure that the number of bins is smaller than the number of distributions you need to analyse, or apply feature selection (Section 4.3) and/or dimensionality reduction (Section 4.4) to the histograms.

4.2.3 Numeric time series

Numeric time series consist of values of a numeric attribute referring to a sequence of consecutive steps, which may be time moments or intervals, or just ordinal numbers. By composition, a numeric time series does not differ from a combination of values of multiple comparable numeric attributes. Therefore, all distance functions that are applicable to multiple numeric attributes are also applicable to time series. Thus, the Euclidean distance is very commonly used for numeric time series when it is necessary to account for the absolute values, and the correlation-based distance is used for scale-independent comparison of the variation patterns.

It should be noted, however, that all these functions require that the time series have the same length, i.e., the same number of steps. If it is not so, one time series can be re-sampled to the length of the other [81] (see a discussion on resampling in Section 5.3.2).

The distance function called **Dynamic Time Warping**[6], abbreviated as **DTW**, was proposed specifically for time series. It is an algorithm that matches each element of one sequence with one or more elements of the other sequence without changing the order of the elements. The algorithm finds such matching that the sum of the distances between the matched elements is minimal. This sum is taken as the distance between the time series. A schematic illustration of how DTW matches points of two rime series is shown in Fig. 4.3.

Since DTW can match an element of one sequence to two or more elements of another sequence, it can be applied to time series whose lengths differ. It has been first applied in speech recognition for comparing pronounced phrases irrespective of

---

[6] https://en.wikipedia.org/wiki/Dynamic_time_warping

the speed of the speech. In research on human-computer interaction, DTW has been applied for comparison of EEG signals of different individuals who performed the same task on a computer at their own speed [154]. In general, DTW can be useful for comparing processes unfolding with different speeds. DTW has also been used for comparing trajectories of moving objects following similar routes.

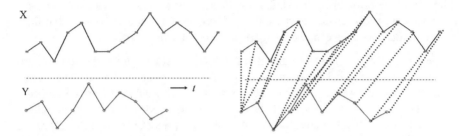

Fig. 4.3: A schematic illustration of the point matching by the Dynamic Time Warping algorithm. Source: [154].

While DTW has become quite popular, it has been argued that the results of using DTW for dissimilarity assessment do not substantially differ from using the Euclidean or Manhattan distance after re-sampling of sequences to the same length [113].

Comparison of time series may be greatly affected by noise (i.e., random fluctuations) in the data. Application of a smoothing method (see Section 5.3.2) can reduce the noisiness and thus help to compare the general patterns of the temporal variation rather than minor details. When time series are long and may involve several variation patterns, such as an overall trend and seasonal or cyclic fluctuation, it makes sense to decompose the time series into components[7] and apply a chosen distance function to the components that are relevant to the analysis task. If two or more components need to be involved, you can compute separately the distance for each component and then take the average or a weighted average of the distances.

For periodic time series, it is also possible to use the Fourier transform[8], which represents a time series as a sum of sinusoidal components, i.e., sine and cosine multiplied by some coefficients, which are called Fourier coefficients. Then, the distance between two time series can be expressed as the distance between the combinations of their Fourier coefficients using one of the functions suitable for multiple numeric attributes (Section 4.2.1).

---

[7] https://en.wikipedia.org/wiki/Decomposition_of_time_series
[8] https://https://en.wikipedia.org/wiki/Fourier_series

4.2.4 Categorical attributes and mixed data

The distance between data items described by multiple categorical attributes (i.e., attributes with categorical values) is obtained from the per-attribute distances, which are expressed by some numbers and combined in a single number by summing or applying the formulas of the Minkowski distance or angular distance that are used for numeric attributes (see Section 4.2.1). The question is how to measure the similarity or distance between two categorical values of a single attribute.

The simplest approach is to assume that the distance between any two distinct values is the same, usually 1. Then, if the per-attribute distances are combined by summing, the total distance will equal the number of attributes in which two data items differ, i.e., the number of mismatches between two combinations of categorical values. This approach to expressing the per-attribute distances is equivalent to replacing each of the categorical attributes by k "dummy" numeric attributes, where k is the number of values of the categorical attribute. Each "dummy" attribute corresponds to one categorical value and has two values, 0 and 1. When a data item has this categorical value of the original attribute, the value of the "dummy" attribute is 1, otherwise it is 0. After such a transformation of the data, it is possible to compute the Minkowski or angular distances in the same way as for usual numeric data.

A more sophisticated approach takes into account the frequencies of different values, assuming that rarely occurring values should be considered as more distant from all others than values occurring frequently. The idea is to compute the probability p of each value, i.e., its frequency divided by the total number of data items, and use $1 - p$ or $1/p$ as the distance of this value to any other, or, equivalently, as the value of the "dummy" numeric attribute corresponding to this value instead of the value 1.

There are also much more sophisticated measures of similarity or distance. A comparative testing of a variety of similarity measures [37] showed that none of them was always superior or inferior to others, but some measures, such as the Goodall's and Lin's similarity (both have been implemented in R), performed consistently better on a large number of test data.

When you have mixed data, i.e., both numeric and categorical attributes, you can take one of the following approaches:

- Convert the categorical attributes into numeric by introducing "dummy" attributes as described above. Than deal with the transformed data as with usual numeric attributes.

- Compute separately the distances for the numeric and for the categorical attributes and combine them into a single measure, for example, by computing a weighted average $(w_1 \cdot d_1 + w_2 \cdot d_2)/(w_1 + w_2)$, where d_1 and d_2 are two partial distances and w_1 and w_2 are the weights given to the two groups of attributes.

The weights can be proportional to the numbers of the attributes, or you can have a different idea concerning the relative importance of each group.

- Use a distance function that was specially created for mixed data, such as the Gower's distance [59]. When deciding to use such a function, it is good to have understanding of how it works.

4.2.5 Sets

The Jaccard index, or Jaccard similarity coefficient[9], is used for comparing sets, for example, sets of students attending different classes. It is defined as the size of the set intersection divided by the size of the set union. Hence, when the sets are identical, the result is 1, and when they have no common elements, the result is 0. To convert this measure of similarity into a distance function, you can subtract the result from 1.

4.2.6 Sequences

Sequences of symbols or any elements represented by symbolic labels (for example, sequences of tourist attractions visited by tourists) are compared by means of one of the **edit distance** functions. It is a family of functions that compute the minimal number of edit operations required for transforming one sequence into another. The possible operations are the deletion, insertion, substitution, and transposition of adjacent symbols. The existing functions differ in which of these operations are allowed. Particularly, the Levenshtein distance allows deletion, insertion and substitution, the Damerau–Levenshtein distance extends the Levenshtein distance by allowing transposition, the Longest common subsequence (LCS) distance allows only insertion and deletion, the Hamming distance allows only substitution, and the Jaro distance allows only transposition. The Hamming and Jaro distances, which do not allow deletions and insertions, can be applied only to sequences of the same length.

---

[9] https://en.wikipedia.org/wiki/Jaccard_index

4.2.7 Graphs

A graph is a structure representing a system of pairwise relationships between entities. A graph consists of a set of *vertices*, which are also called *nodes*, and a set
of *edges*, which are also called *links*. The vertices represent entities and the edges
represent pairwise relationships between the entities. Two vertices connected by an
edge are called *adjacent*. Two graphs are said to be *isomorphic* if it is possible to
find such matching between the vertices and between the edges of the graphs that
the adjacency is preserved: for any two adjacent vertices of one graph, the corresponding vertices of the other graph are also adjacent. The distance between two
isomorphic graphs is considered to be zero.

Similarly to sequences, dissimilarity of two graphs can be assessed by means of an
edit distance function: having specified certain costs of edit operations (e.g., addi-
tion/deletion of nodes and edges), determine the minimum cost transformation from
one graph to another. Other commonly used approaches to assessing graph dissimilarity are based on finding the **maximum common sub-graph** or the **minimum
common super-graph**. The former identifies the largest isomorphic sub-graphs of
two graphs The more nodes and edges are left beyond these sub-graphs, the larger
the distance. The latter identifies the smallest graph that contains both graphs. The
distance is defined by the size of the super-graph, i.e., the number of nodes and
edges in it. There exist a number of algorithms implementing the definitions of
these distance functions, and computer implementations of these algorithms exist in
many packages and libraries for data analysis as well as open-source codes shared
by their developers.

Apart from the distance functions assessing the (dis)similarity of graph structures,
there are statistical methods in which graph (dis)similarity is assessed in terms of
various aggregate features, such as the numbers of the nodes and edges, the diameter (the longest path length), and/or the distribution of centrality indicators[10] of
the nodes or edges. These features are, in essence, numeric attributes; hence, the
distance functions suitable for numeric attributes (described in Section 4.2.1) can
be applied for measuring the distances between graphs in terms of their statistical
features.

Graph-oriented distance functions can be applied to whole graphs or to parts (sub-
graphs) of a single graph, which allows finding structurally similar parts of a
graph.

---

[10] https://en.wikipedia.org/wiki/Centrality

4.2.8 Distances in space and time

To find clusters of objects in space, a clustering tool needs to use the spatial distances between the objects. The spatial distance is easy to compute when the objects can be treated as points in the space. When the spatial positions of the points are specified by coordinates in an orthogonal coordinate system, the spatial distance is computed as the straight-line distance, i.e., the Euclidean distance. This applies, for example, to positions of players and a ball on a sports game field where an internal coordinate system is defined. When the spatial positions are specified by geographic coordinates, i.e., longitudes and latitudes, the distance is calculated as the **great-circle distance**[11] (the formulas are quite complicated and therefore not shown here). Please note that coordinates in geographically referenced data can also be specified in an orthogonal coordinate system obtained after applying one of the methods for map projection[12]. Hence, you should be aware of the kind of coordinates you have in your data in order to choose the right method of computing the spatial distances.

For some analyses, not the straight-line or great-circle distances between spatial positions are important but the lengths of the paths that are taken for getting from one position to another, or the duration of the travelling. There are web services that can compute path lengths and durations of travelling through the street network.

For determining the distances between spatial objects that consist of more than one points, different approaches can be taken depending on the kind of objects and specifics of the analysis that needs to be done. The possible approaches include taking the distance between the centres of the objects, or between the closest points of their boundaries or outlines, or the **Hausdorff distance**[13], which is the longest of all distances from a point of one object to the closest point of the other object.

Finding distances between linear objects, such as paths of movement, may involve matching points from the two lines, similar to matching points from two time series by the Dynamic Time Warping distance function discussed in Section 4.2.3. In fact, DTW can be applied for calculating distances between lines in space. It is also possible to do point matching using DTW or another suitable algorithm [121] and then compute the mean of the distances between the corresponding points. Please note that almost all of the existing point matching algorithms, including DTW, match the first and the last points of one line to, respectively, the first and the last points of the other line. If you want to find clusters of largely overlapping paths that may differ in their lengths, you may need another algorithm, such as the "route similarity" distance function [10, p.147, Algorithm 5.4] or the point matching algorithm proposed for analysis of route variability [23].

---

[11] https://en.wikipedia.org/wiki/Great-circle_distance
[12] https://en.wikipedia.org/wiki/Map_projection
[13] https://en.wikipedia.org/wiki/Hausdorff_distance

The distance in time between two time moments is the time difference between them, which is determined by subtracting the earlier moment from the later one. The distance between two time intervals can be determined, following the approach of the Hausdorff distance, as the maximum of the distances between the start times of the intervals and between their end times. For finding clusters of events in time, it may be reasonable to define the distance as 0 if the intervals overlap and as the difference between the start time of the later interval and the end time of the earlier interval otherwise.

What if you need to find clusters of objects in space and time? In this case, you need a distance function that combines the distance in space and the distance in time. Formally, you could combine the two distances by applying the formula of the Euclidean or Minkowski distance. This idea may not be especially good, however, because it will be very hard to figure out what the result means. A more meaningful approach would be to set an explicit rule of trading time for space, or vice versa. For example, knowing the properties of your data, you can decide that the temporal distance of 5 minutes should be treated equivalently to the temporal distance of 200 metres. On this basis, you can transform the temporal distances into spatial, which can then be meaningfully combined with the "normal" spatial distances.

It may be more intuitive to take a slightly different perspective: think when two objects or events should be treated as neighbours in space and time. For example, your judgement may be that two neighbouring objects should be not more than in k metres away from each other in space and separated by not more than m minutes in time. In essence, it is the same as deciding that m minutes is worth k metres, which gives you an opportunity to transform the temporal distances into spatial.

This approach can be applied not only to time but also to other attributes you may wish to involve in determining distances. For example, for finding traffic jams, i.e., spatio-temporal concentrations of vehicles moving very slowly in a common direction, it is appropriate to have a distance function combining the distance in space, the distance in time, and the difference in the spatial orientation [14]. This can be done by choosing appropriate distance thresholds for defining neighbourhoods in space, time, and orientation.

4.2.9 Data normalisation and standardisation

When you wish to use Euclidean or Minkowski distance for computing distances between combinations of values of multiple attributes, you should take care that the attributes are comparable. The reason is that the function you use combines per-attribute differences between values, and the magnitudes of these differences can differ greatly. For example, attributes characterising countries include population, gross domestic product per capita, and life expectancy. The per-attribute differences may be millions for the first attribute, thousands for the seconds, and quite small numbers for the third. Obviously, the contributions of these differences into the combined distance will be extremely unequal. To give all differences equal treatment, the attributes need to be transformed to a common scale. Such a transformation is commonly called *normalisation*.

One of possible ways of doing this is to re-scale the values of all attributes to the same range, usually from 0 to 1. This may be done by subtracting the minimal attribute value from each value and dividing the result by the difference between the maximal and minimal values of this attribute. This may not be a good idea when the distribution of values of an attribute is very much skewed towards small or large values, or when there are outliers. In such a case, most of the transformed values will be very close to either 0 or 1, and the differences between them will be tiny. In fact, the "effective scale" of the attribute (i.e., the interval containing the bulk of the values) will not be comparable to the scales of other attributes whose values are distributed differently. Hence, this approach may not always fit to the purpose.

Another common approach is to transform the values according to the formula: $(x_i - centre(x))/scale(x)$, where x_i is one of the values of an attribute x, $centre(x)$ is either the mean or the median of the values of x, and $scale(x)$ can be the standard deviation, the interquartile range, or the median absolute deviation, which is the median of the absolute differences of the existing attribute values from the attribute's median value. When the transformation is done based on the mean and the standard deviation, the resulting transformed values are called *z-scores*. The mean of the transformed values is 0 and the standard deviation is 1. A transformation producing data with the mean of zero and the standard deviation of one is called *standardisation*. Since the mean and standard deviation are sensitive to outliers, the transformation based on these statistics may be affected by outliers too. The transformation variants based on the median are more robust.

Non-linear transformations, such as logarithmic, are often applied for dealing with skewed distributions. The problem is that these transformations distort the data. Differences between transformed attribute values are unintelligible. Even worse, the same difference between transformed values may correspond to different amounts of difference between original values. The utility of involving such differences in computing distances is questionable.

There are implementations of distance functions that internally perform some kind of data normalisation, while others may not do this automatically. If you are not sure whether a given implementation includes automatic normalisation, a possible way to check this is to compute pairwise distances between data items using the original attributes and their normalised variants, i.e., obtain two distances for each pair of data items. Then you need to create a scatter plot of these distances. If you see that all dots lie precisely on the diagonal of the scatter plot, you can conclude that normalisation is done automatically.

Existing implementations of clustering and data embedding algorithms may apply some built-in transformation methods to the input data. When you use any tool implemented by someone else, you need to be informed what kind of transformation, if any, it applies to your data. If it is possible to choose, take care to choose the appropriate one with respect to the properties of your data. If the default transformation cannot be changed and is not appropriate, as, for example, scaling to the $[0,1]$ range for skewed data, it is better to find a different implementation.

4.3 Feature selection

The terms "feature" used in data mining and machine learning and "variable" used in statistics have the same meaning as the term "attribute" used in visualisation literature and in our book. Feature selection (or variable selection, or attribute selection) means selecting from a large number of attributes a smaller subset that is sufficient for a certain purpose. The term "feature selection" appears in the literature and materials available on the web predominantly in relation to constructing predictive models. It is typically explained that the purpose of feature selection is being able to construct a better model using less computational resources. Accordingly, the existing methods for feature selection are designed to be used in the context of modelling. Some of them may exclude features from a model one by one and check how this affects the model's performance. Others may assess the importance of input features for predicting the model output using certain numeric measures, such as correlation or mutual information. These methods have no relation to visualisation or analytical reasoning, and we shall not discuss them here.

Apart from modelling, feature selection may also be needed for supporting visual analysis by means of computational methods, such as clustering and data embedding. Specifically, a typical task is to select attributes to be used in a distance function (Section 4.2). One of the things to care about is possible correlations between attributes. All distance functions involving multiple attributes implicitly assume that all attributes are independent from each other, but it is rarely so in real data. When several attributes are highly related, their joint effect on the result of a distance function may dominate the contributions of the remaining attributes.

For example, we have demographic data with 18 attributes expressing the proportion of different age ranges in the population: 0-4 years old, 5-7 years old, 8-9, 10-14, and so on, up to 90 years and more. There are strong positive correlations between the proportions of the lowest six age ranges (i.e., from 0-4 to 16-17) and between the proportions of the highest six age ranges (i.e., from 45-59 to 90 and more). This can be seen in the visualisation of the pairwise linear correlations between the attributes (more precisely, the values of the Pearson's correlation coefficient) shown in Fig. 4.4. Each column and each row in this triangular layout corresponds to one attribute, which is denoted by a label. The blue and red bars in the cells represent,

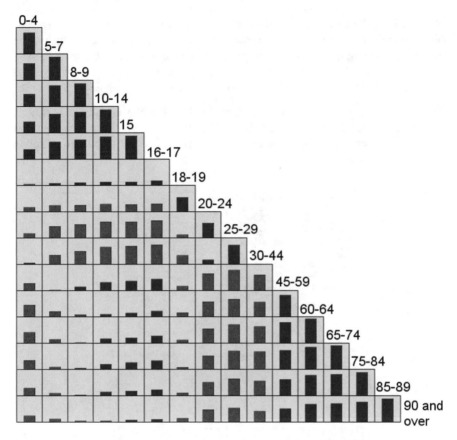

Fig. 4.4: Pairwise correlations between demographic attributes.

respectively, positive and negative correlations between the attributes corresponding to the columns and rows in which the cells are located. The bar heights are proportional to the absolute values of the correlation coefficient. We see that the upper left and the lower right corners of the triangle are filled with high blue bars indicating high positive correlations. The correlations are especially high between adjacent age ranges.

When there is a group of correlated attributes, it often happens that the values of all these attributes are simultaneously high. In such cases, these values will have much higher impact on the result of a distance function than values of any other attributes. In our example, two districts with high proportions of either the first or the last six age ranges may be treated as very similar even when they substantially differ in the proportions of the other age ranges. To give more equal treatment to different age ranges, it is reasonable to select a subset of the attributes instead of taking them all. Thus, we would take the proportions of the ages 0-4 and 16-17 from the first group of correlated attributes. It makes sense to take these two rather than a single attribute from the group because the correlation between these two attributes is not high while their relationships to the other attributes are quite different. This can be seen by comparing the columns corresponding to the age ranges 0-4 and 16-17. The range 0-4 is negatively correlated with the ages starting from 45-59 whereas the range 16-18 has slight positive correlations with the upper age ranges.

The attributes in the second group (i.e., from 45-50 to 90 and over) have quite similar relationships with the remaining attributes; hence, it may be sufficient to select one of them. The best candidate is the range 75-84, which has the highest correlations with the remaining five attributes in the group.

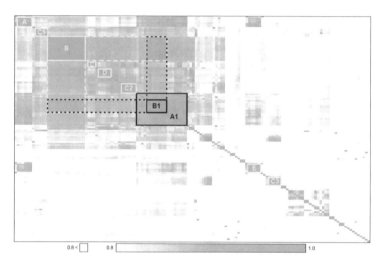

Fig. 4.5: A matrix of correlations between attributes. The ordering of the rows and columns reveals clusters of correlated attributes. Source: [106].

Figure 4.5 shows an example of visualising pairwise correlations for a much larger set of attributes [106]. This is a matrix the rows and columns of which correspond to attributes. Since the attributes are numerous, the cells of the matrix are tiny. The green shading represents positive correlation values exceeding a chosen threshold; in this example, it is 0.8. The rows and columns of the matrix have been arranged in such an order (by means of a hierarchical clustering algorithm) that the rows and columns of correlated attributes are put close to each other. With this ordering, groups of correlated attributes are manifested by dark rectangular areas in the matrix. Some groups are marked and labelled in the figure. Please note that the groups labelled A1 and B1 have high correlations with the groups labelled A and B, respectively. This means that A and A1 belong together, as well as B and B1, but the matrix reordering algorithm failed to put them together.

Apart from impacts of correlated attributes, another thing to care about is the number of attributes that are used for distance computation. One reason why the number of selected attributes should not be large is the curse of dimensionality (see Section 4.2.1): when the number of attributes is very large, the distances between data items tend to be nearly the same. Another reason, which is often ignored in the literature focused on computational methods for analysis and modelling, is the necessity to *understand* outcomes of computational techniques: what do neighbours in a data embedding or members of the same cluster have in common? How do they differ from others? It may not be easy to find answers to such questions when the attributes are many.

Hence, when the aim of using a computational method is to empower your analytical reasoning, it may be a bad idea to throw all available attributes into the method and see what happens. Thus, if you use all existing demographic attributes for clustering of districts, the result, most probably, will not be insightful. What makes more sense is to decompose your analysis into steps in which you deal with subsets of semantically related attributes. In the example with demographic data, you may focus separately on the age structure, education, occupation, health-related attributes, and so on. In each step, you obtain a piece of knowledge concerning some aspect of the phenomenon under study, and at the end you synthesise these pieces into a comprehensive whole.

4.4 Data embedding

4.4.1 Embedding space

Data embedding, also called *data projection*, usually means representing data items by points in an abstract *metric space*, that is, a set of possible locations with some non-negative numeric measure of the *distance between any two locations*. Each point receives a certain location in this space. The distances between the points are supposed to represent certain relationships between the corresponding data items, most often, relationships of similarity, so that stronger relationships (such as higher similarity) are represented by smaller distances. The space that is used for data embedding may be either continuous or discrete. In a continuous space, there is an infinite number of locations within any chosen non-zero distance from any location. In practice, analysts usually strive to embed data items in a continuous *Euclidean space*, which means that Euclidean (straight-line) distances, as defined by the formula 4.1 in Section 4.2.1, exist between the locations. While an abstract Euclidean space may, in principle, have any finite number of dimensions, Euclidean spaces with two or at most three dimensions are of particular interest due to the possibility of visualisation. The most popular methods for embedding data in continuous Euclidean spaces include Principal Component Analysis (PCA)[14], Multidi-

---

[14] https://en.wikipedia.org/wiki/Principal_component_analysis

mensional scaling (MDS)[15], Sammon mapping, also called Sammon projection[16], and T-distributed Stochastic Neighbor Embedding (t-SNE)[17]. The most usual way of visualising results of data embedding in a continuous 2D space is the *projection plot*, in which data items are represented by dots in the plot area.

In a discrete space, each location has a finite number of neighbouring locations (i.e., being within the smallest possible distance from this location), and no intermediate locations exist between any two neighbouring locations. A discrete space can be imagined as a container with compartments inside it. Each data item needs to be put in one of the compartments, and one compartment may accommodate multiple distinct data items. It is supposed that data items that are put together in the same compartment are stronger related (more similar) to each other than to data items put in other compartments. Furthermore, the distances between the compartments need to represent the relationships between the groups of items that have been put in them. For enabling visualisation, the compartments of the embedding space need to be arranged on a plane according to their neighbourhood relationships. A popular method for discrete data embedding is Self-organising Map (SOM), also known as Kohonen map[18][84]. It is a kind of an artificial neural network, and the locations in the embedding space are therefore called "neurons" or "nodes". Results of 2D SOM embedding may be given the form of a regular rectangular or hexagonal grid, which is drawn in a visual display.

4.4.2 Representing strengths of relationships by distances

As we stated earlier (Section 4.1.2), the value of data embedding for visual analytics is that it can be used for creating spatialisations of data. The purpose of spatialisation is to enable a human observer to associate multiple objects into patterns. The fundamental idea is that important patterns (combinations of relationships) existing in the data should translate into spatial patterns that emerge between visual representations of data items (Section 2.3.5). In Section 2.3.3, it is explained that spatial patterns, such as density, sparsity, and spatial concentration (cluster), emerge from distance relationships between objects located in space.

When human eyes see a set of points or other objects located on a plane or in the three-dimensional space, the human brain instinctively judges the straight-line distances (i.e., Euclidean distances) between them. Objects are deemed close or distant based on these straight-line distances, and it is quite hard to change these intuitive

---

[15] https://en.wikipedia.org/wiki/Multidimensional_scaling

[16] https://en.wikipedia.org/wiki/Sammon_mapping

[17]   https://en.wikipedia.org/wiki/T-distributed_stochastic_neighbor_embedding

[18] https://en.wikipedia.org/wiki/Self-organising_map

judgements by introducing another concept of distance. Therefore, when data embedding is supposed to represent some relationships, it is very important that the strengths of the relationships are faithfully represented by the Euclidean distances between the locations given to the data items. This means that, whenever a data item d_i is stronger related to a data item d_j than to the data item d_k, the distances between the locations of the data items d_i and d_j in the embedding space is smaller than the distance between the locations of the d_i and d_k, and two distances are equal only when the corresponding relationships have the same strength. Under this condition, spatial patterns that may be perceived by a human observer in a data spatialisation display will correspond to truly existing relationships in the data; otherwise, such patterns may be spurious. Faithful representation is important not only for two- or three-dimensional embeddings used for visualisation but also for embeddings with any number of dimensions used in computational methods of analysis and for creation of computer models.

However, it is seldom possible to create a fully faithful data embedding. Typically, distortions are unavoidable. While the computational techniques for data embedding are designed so as to minimise distortions, they can be more or less successful in doing this. There is no single technique that always works better or worse than others. When you have several techniques at your disposal, you may try each of them and choose the best result. An overall measure of the distortion in a data embedding is the *stress*, which summarises the deviations of the inter-point distances in the embedding from the distances between the corresponding data items. Both sets of distances need to be re-scaled to a common range. One of possible ways of summarising the deviations is known as Kruskal's stress [87]. It is the square root of the sum of the squared deviations divided by the sum of the squared original distances. Its author suggested the following interpretations of the stress values: 0 means perfect embedding (no distortions), values around 0.025 are excellent, around 0.05 good, around 0.1 fair, and values close to 0.2 or higher indicate poor embedding.

The distortions are usually not equal throughout the embedding space, which makes the overall measure of the distortion insufficiently informative. To provide more information, the visualisation of an embedding, particularly, a projection plot, can be enhanced by showing the cumulative distortions of the original distances for individual data items, for example, as shown in Fig. 4.6 [127].

In interactive settings, it may be possible to select areas in a projection plot and obtain in response the amounts of the distance distortions (i.e., the stress) within these areas. Another possible interactive operation is viewing individual distance distortions of all data items with respect to a selected data item, i.e., a single dot in the projection plot. An example is shown in Fig. 4.7: after an analyst selected a dot in the projection plot, the other dots moved either closer to this dot or farther away from it, depending on the actual distances between the respective data items. The white and grey lines connect the new positions of the dots to the original positions. White corresponds to moving towards the selected dot and grey to moving away.

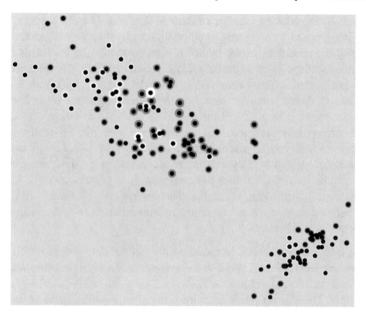

Fig. 4.6: Halos around points in a projection plot represent the cumulative distortions of the original distances of the corresponding data items to the other items. The radii are proportional to the absolute values of the distortions. White colour means that the other points in the projection are mostly farther away than they should be, and grey means the opposite direction of the distance distortion. Source: [127].

For a discrete embedding space, such as a result of the SOM method, a common way to show the distortion is to visualise the distances between data items assigned to neighbouring locations (or neurons, or nodes), which are usually represented as cells of a regular grid. A common approach to visualising the distances is to separate the representations of the locations in the display by spaces and paint these spaces according to the average distances between the data items assigned to the adjacent locations. In a case of using SOM, the average distances between data items corresponding to adjacent SOM nodes are usually represented in the form of a matrix, which is called U-matrix (unified distance matrix) [84]. When discrete locations (nodes) are represented by regular hexagons, the spaces between them can also have the hexagonal shape, as shown in Fig. 4.8. In the example in Fig.4.8, right, we see light and dark zones in the SOM space. In the light zones, there is high similarity between the groups of the data items put in the neighbouring nodes. In the dark zones, the groups of data items in the neighbouring nodes are quite dissimilar.

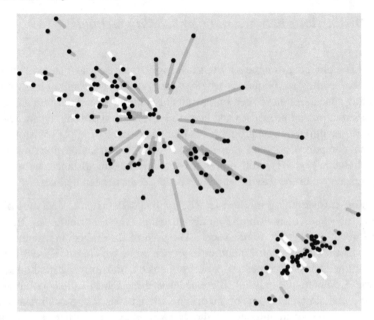

Fig. 4.7: After interactive selection of a dot in the projection plot, the distortions of the distances from the respective data item to all other data items are represented by shifting the dots closer to or farther away from the selected dots. The white and grey lines show the directions and distances of the shifts. Source: [127].

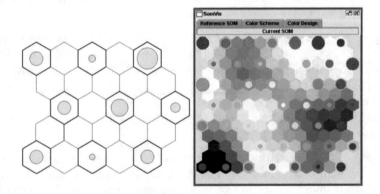

Fig. 4.8: A result of the SOM method is represented as a grid with hexagonal cells. The cells with the coloured circles in them represent the neurons, or nodes, of the neural network. The sizes of the circles depict the numbers of the data items in the nodes. The remaining hexagons are, essentially, spaces between the node cells. The shading of these spaces represents the distances between the data items in the nodes separated by the spaces. Source: [64].

4.4.3 Distinctions between data embedding methods

As should be obvious from the previous discussion, an essential distinction between the data embedding techniques is whether they construct a continuous or discrete embedding space. You have seen that results of continuous and discrete embedding differ in structure and properties and need to be visualised differently. Another distinction we mentioned in Section 4.1.2 is that some methods apply transformations to original components of data (these methods are called "transforming dimensions" in the scheme in Fig. 4.1) while others strive at reproducing distances between data items irrespective of the way in which the distances have been defined.

Dimension-transforming methods can be applied to data items consisting of combinations of values of multiple numeric attributes. These attributes are treated as dimensions of an abstract "data space". The goal of the embedding methods is to create a space with a smaller number of dimensions, which are derived from the original dimensions. The most common dimension transforming method is PCA (Principal Component Analysis). It creates new dimensions, called "components", as linear combinations of some of the original dimensions. The new dimensions are ordered according to the amount of the variation between the data items each of them captures. Hence, the first two components taken together are supposed to capture a large amount of the data variation. It is typical to use these two components for visualisation in the form of a scatter plot with the axes corresponding to these dimensions. This means that the remaining dimensions generated by PCA are ignored, which may not be a very good idea, as the distortions of the distances between the data items in the 2D plot may be very high.

It may be more reasonable to use the complete set of dimensions created by the PCA for computing distances between the data items by means of some distance function suitable for numeric attributes (see Section 4.2.1) and then apply a distance-reproducing data embedding method using the distances thus obtained. The benefit gained by applying the PCA compared to computing the distances based on the original attributes is avoidance of the "curse of dimensionality", which is explained in Section 4.2.1.

The class of methods called "topic modelling" also belongs to the category of dimensions-transforming methods. The new dimensions generated by these methods are called "topics". Each "topic" is defined as a composition of weights of the original attributes. Usually, the number of "topics" to be generated is specified as a parameter of the method. Each data item is described by a combination of weights of the "topics". To achieve spatialisation of the data, these combinations can be used for computing distances between the data items using a suitable distance function (Section 4.2.1). The distances, it turn, can be used as an input to a distance-reproducing embedding method.

Distance-reproducing methods, such as MDS, Sammon mapping, and t-SNE, can be applied to a previously created matrix of pairwise distances between data items, which may be computed by means of any suitable distance function depending on the nature and characteristics of the data (see Section 4.2). Usually, these methods can also be applied to data items consisting of values of multiple numeric attributes. In this case, the methods will use some in-built distance function (such as the Euclidean distance) for computing the distances between the data items. Self-Organising Map (SOM) can only be applied to multiple numeric attributes, which are used for defining the neurons in the neural network.

There are methods, such as Isomap[19] and LLE (Locally Linear Embedding), that use the provided or internally computed distances between data items to determine the neighbours of each data item by taking either all items within a certain distance ε (called ε-neighbours) or the nearest k items (called k nearest neighbours, shortly, k-NN). On this basis, the methods construct a neighbourhood graph, in which each data item is connected to its neighbours, and compute another set of inter-item distances as the lengths of the shortest paths between them in the graph. These secondary distances are then used for creating an embedding [156]. This approach is designed for preserving local neighbourhoods, i.e., small distances between data items, while larger distances may be greatly distorted. The method t-SNE (an abbreviation of T-distributed Stochastic Neighbour Embedding) also gives more importance to small distances by constructing a special probability distribution over pairs of items such that similar items have high probability of being taken together [142]. The class of methods that strive to preserve local neighbourhood is called *neighbourhood-embedding (NE)* methods.

The choice of the best data embedding method for your data is usually not obvious, and it may be reasonable to try a couple of them. Generally, dimensions-transforming methods, particularly, methods applying linear transformations, are good in capturing global patterns in the data whereas distance-reproducing methods can be better in representing local patterns, such as clusters (groups of similar or strongly related items). Neighbourhood-embedding approaches do not preserve long distances between data items; therefore, the arrangement of non-neighbouring groups of dots in a projection plot made from such an embedding is not informative and needs to be ignored.

4.4.4 Interpreting data embeddings

A visual display of an embedding space with data arranged in it allows perceiving spatial patterns that emerge in this arrangement. Spatial outliers and spatial clusters often correspond to outliers and clusters in the data. Areas of high density in a projection plot and nodes representing large groups of data items in a SOM display signify that certain combinations of properties occur more frequently in the data

---

[19] https://en.wikipedia.org/wiki/Isomap

than others. Sparsely filled regions correspond to highly varied data items. A nearly uniform density of the dots throughout the plot area signify large variance among the data items and absence of clusters of similar items.

When data items can be meaningfully ordered, in particular, by the times the data correspond to, data spatialisation can be enhanced by connecting dots (or other visual marks) by lines according to their order. This idea is employed in the Time Curve visualisation technique demonstrated in Fig. 2.12. The connecting lines can show small and large changes along the order, as well as re-occurrences of properties that occurred earlier in the order. However, you should keep in mind that the shapes of the connecting lines between the dots do not convey any meaning. They may be straight or curved, with higher or lower curvature, but do not try to interpret these shapes.

Hence, the properties of the overall spatial distribution of visual marks in a data spatialisation can give us useful information concerning relationships within a set of data items. However, the interpretation of the spatial patterns requires caution due to the unavoidable distortions of the distances. All observations and judgements should be treated as approximate and potentially partly erroneous. For example, a spatial cluster of dots may mostly correspond to a group of highly similar or strongly related data items, but it may also include a few dots corresponding to less similar or related items. Even when neighbouring dots correspond indeed to similar or related items, you should not assume that smaller distances necessarily represent higher similarity or stronger relationships. Beyond small neighbourhoods, the distances in the embedding space should not be used for any estimations of the relationship strengths or amounts of dissimilarity.

The spatial patterns observable in an embedding display give us highly abstract information about the distribution of relationships over the set of data items. A major problem of such a display is that we do not know what data item(s) each visual mark in the display stands for. It is insufficient to just label the marks; it is important to know not the names but the characteristics of the data items. Some implementations of visual tools for data spatialisation allow interactive selection of dots or groups of dots in the data embedding display and show in response information about selected data items. For example, in Fig. 2.12, images corresponding to selected dots are shown beside the projection plot. In Fig. 4.9, information about attribute values is shown for two selected groups of dots and one singular dot. The density plots (smoothed histograms) shown in a popup window enable comparisons of the frequency distributions of the attribute values in the two groups and in the whole dataset. The density plots are also shown in larger sizes in the lower right section of the display. Moreover, the grey-scale thumbnails appearing in the same section left to the attribute names show the distributions of the values of these attributes over the area of the projection plot. Larger images appear upon clicking on the thumbnails.

Visualisation of the distributions of attribute values over an embedding space can also be done for discrete embedding, such as SOM. Particularly for SOM, the term

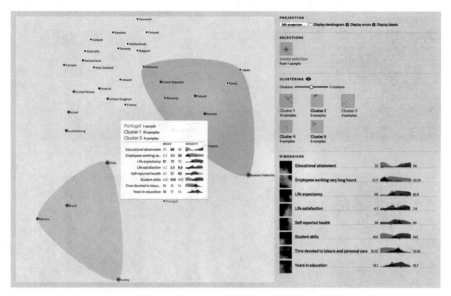

Fig. 4.9: Two groups of dots and one singular dot are selected in a projection plot for viewing and comparison. The corresponding mean values of attributes and density plots of the value distributions are shown in a popup window. Larger density plots are also shown in the lower right section of the display. This section also includes small squares showing the distributions of the values of the different attributes over the plot area as grey-scale heatmaps. Source: [127].

"component plane" refers to visual representation of the distribution of values of a single attribute over the nodes of a SOM, which is often done using the heatmap technique. An example of component planes of a SOM is shown in Fig. 4.10.

Figure 2.10 demonstrates another way of providing information about the data represented in a projection plot. The idea is to paint the plot background so that the colour continuously and smoothly varies from position to position. Then, each data item is associated with the colour corresponding to the point representing this item in the projection plot. The colours assigned to the data items can then be used for painting visual elements representing the items in other display, such as the parallel coordinates plot in Fig. 2.10. This approach does not depend on the use of tools for interaction but does not exclude the possibility of selecting data items interactively though the display, as demonstrated in the lower part of Fig. 2.10.

A similar approach can also be applied to an embedding in a discrete space, such as SOM. In Fig. 4.8, colours have been assigned to the SOM nodes, and Figure 4.11 demonstrates how these colours are used in a parallel coordinates plot for showing information about the groups of data items corresponding to the SOM nodes. Figure 4.12 shows a possible approach to creation of a two-dimensional colour scale, also called "colour space". Two pairs of contrasting colours are assigned to the two

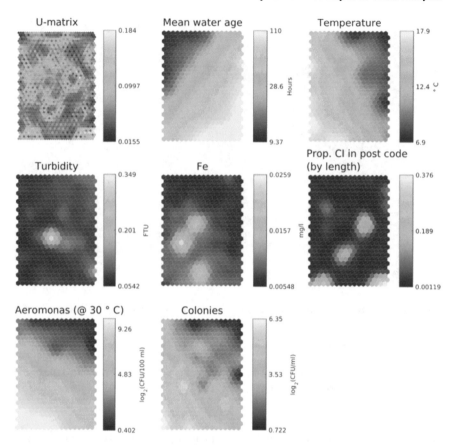

Fig. 4.10: SOM component planes, in which the heatmap technique is employed to show the distribution of values of individual attributes over the discrete space of a SOM. Source: [34].

pairs of opposite corners of a square. The colours for all other positions within the square are obtained by interpolation between the four colour. While simple linear interpolation has been used in Fig. 2.10, the interpolation in Fig. 4.12 is done based on an ellipsoid model. In this model, two dimensions correspond to the colour hues, which are interpolated between the corners of the square, and the third dimension corresponds to the colour lightness, which reaches maximum in the centre of the square. For all other positions, the lightness values are obtained as vertical coordinates of the corresponding points on the surface of the ellipsoid, which are calculated using trigonometric functions. Compared to linear interpolation applied in Fig. 2.10, ellipsoid-based interpolation generates lighter colours in the inner part of the colour space. Hence, when you see a coloured visual mark in another display, the colour lightness suggests in what part of the embedding space the corresponding data item is positioned.

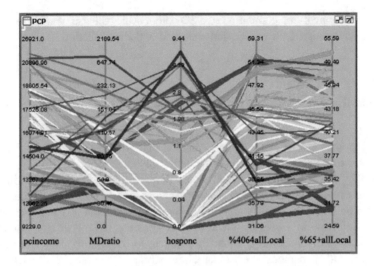

Fig. 4.11: Colours assigned to nodes of a SOM, as shown in Fig. 4.8, are used in a parallel coordinates plot for colouring the lines representing the average attribute values for the data items included in the SOM nodes. Source: [64].

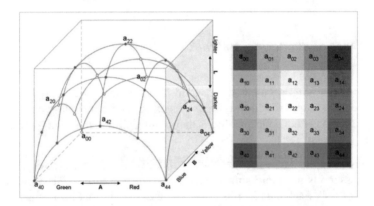

Fig. 4.12: Generation of a two-dimensional colour scale using an ellipsoid model. The horizontal dimensions of the ellipsoid are A (green-red) and B (blue-yellow). The vertical dimension is the lightness. Source: [64].

The approach involving generation of a 2D colour scale can also be used for assigning colours to clusters, as will be demonstrated in the next section.

4.5 Clustering

Clustering is a class of computational method that groups data items by their similarity. An example is customer segmentation in retail, where customers may be grouped by the similarity of the products they buy. The groups of similar data items are called "clusters". As in data embedding, similarities between data items are represented as distances, which are defined using suitable distance functions as discussed in Section 4.2. Some clustering methods use in-built distance functions, most often the Euclidean or Manhattan distance, for computing distances between data items. Other methods allow external definition of the distances, which can be supplied to the methods in the form of a distance matrix. There are also implementations of clustering methods that can call an external distance function for computing distances when required.

4.5.1 Types of clustering methods

Three types of clustering methods are used in data science frequently: **partition-based**, **density-based** and **hierarchical** clustering. These method types differ not just algorithmically but conceptually, as they assume different meanings of the term "cluster".

Partition-based clustering divides the set of items into subsets such that the items within the subsets are more similar than between the subsets. These subsets are called "clusters". Each data item is put in some cluster, so that its average distance to the other members of this cluster (or, alternatively, its distance to the *centre* of this cluster) is smaller than its average distance to the members (or the distance to the centre) of any other cluster. However, an item can be closer (more similar) to *some* members of other clusters than to the nearest members of its cluster. Many such cases can be seen in the example on the left of Fig. 4.13. The most popular method for partition-based clustering is k-means[20].

Density-based clustering involves the concept of *neighbourhood*. Two data items are treated as neighbours if the distance between them is below some chosen threshold. When an item has many neighbours, it means that the *density* of the data around it is high. Density-based clustering methods aim at finding dense groups of data

---

[20] https://en.wikipedia.org/wiki/K-means_clustering

Fig. 4.13: Conceptual differences between the partition-based (left) and density-based (right) types of clustering are demonstrated by clustering the same set of points in the geographical space according to the spatial distances between them. The number of the resulting clusters is the same in both images. The cluster membership is signified by the dot colours (the colours are used independently in each image). On the left, the clusters are enclosed in convex hulls for better visibility.

items, i.e., such groups in which each item has sufficiently many neighbours. These dense groups are called "clusters". Items that don't have enough neighbours are not included in any cluster. All these items are jointly called "*noise*". Unlike in partition-based clustering, items that are put in clusters by density-based clustering cannot be closer to members of other clusters than to neighbouring members of their own clusters; however, the distances between non-neighbouring members of the same cluster may be quite high. The differences between clusters created by partition-based and density-based clustering are demonstrated in Fig. 4.13 by example of clustering a set of points distributed in the geographic space according to the spatial distances between the points. In both images, dots are coloured according to their cluster membership. The grey dots in the right image are the "noise", according to the density-based clustering. On the left, the clusters resulting from partition-based clustering are enclosed in convex hulls to be better visible on the map. The representative method for density-based clustering is DBSCAN[21].

---

[21] https://en.wikipedia.org/wiki/DBSCAN

Hierarchical clustering[22] builds a hierarchy of data subsets, either in a "top-down" divisive approach by recursively partitioning the set of items or in a "bottom-up" agglomerative approach by recursively grouping similar items. Each node in the hierarchy represents a subset of the data, which is called "cluster" irrespective of its position in the hierarchy. At the bottom of the hierarchy, each "cluster" includes a single data item. Clusters at higher levels include lower-level clusters as their parts (subsets). Hence, a data item can be a member of multiple nested clusters. Outcomes of hierarchical clustering are usually represented visually by a *dendrogram*, i.e., a tree-like structure showing the hierarchical relationships between the clusters, as illustrated in Fig. 4.14 by example of clustering of six points on a plane according to the spatial distances between them.

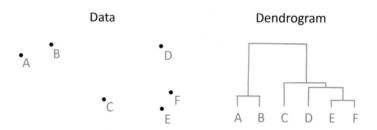

Fig. 4.14: Illustration of the concept of hierarchical clustering.

At each level of the hierarchy, the clusters do not overlap and jointly include all data items, which is analogous to results of partition-based clustering. Hence, by cutting the hierarchy (or the dendrogram representing it) at some chosen level and discarding the levels below, we obtain clusters similar to what can be obtained with partition-based clustering. Thus, if we cut the dendrogram in Fig. 4.14 at level 1 (i.e., the highest), we obtain two clusters $\{A,B\}$ and $\{C,D,E,F\}$, and cutting at the next level gives us $\{A,B\}$, $\{C\}$, and $\{D,E,F\}$.

If the cut is done farther down the hierarchy, a possible result can be that many clusters include singular data items or very few data items. If the members of such very small clusters are treated as "noise", the result will be similar to that of density-based clustering. Hence, hierarchical clustering can be seen as a kind of "hybrid" of partition-based and density-based clustering. For example, cutting the hierarchy in Fig. 4.14 at level 3 gives us two "dense" clusters $\{A,B\}$ and $\{E,F\}$, while the items C and D can be treated as "noise".

When you use a dendrogram representing results of hierarchical clustering, you should keep in mind that only the distances along the hierarchy are informative: they are proportional to the distances between the clusters. The distances in the perpendicular dimension are not meaningful and should be ignored when you interpret the clustering results. Particularly, when a dendrogram has a vertical layout

---

[22] https://en.wikipedia.org/wiki/Hierarchical_clustering

of the levels, as in Fig. 4.14, meaningful are the heights of the branches but not the horizontal distances between the branches or between the nodes at the bottom of the hierarchy. For example, the nodes C and D are very close in the dendrogram in Fig. 4.14 but quite distant from each other in the data space. A dendrogram can also be drawn using a horizontal layout, e.g., with the highest level on the left and the lowest on the right. In this case, the horizontal distances are meaningful and the vertical are not.

This discussion of the principal differences between the types of clustering methods should help you to understand what type of clustering you need to apply in each particular situation. When you want to divide your dataset into subsets consisting of homogeneous (more or less) items, you apply partition-based clustering. When you want to identify groups of highly similar items and assess the degree of variability in the data, you apply density-based clustering. Apparently, hierarchical clustering combines benefits of partition-based and density-based clustering: a dendrogram can show the variability and groups of similar items and also allow division into internally homogeneous subsets. However, practical use of these benefits is possible only for relatively small datasets. One limitation of the hierarchical clustering is the long computation time. Another limitation, which is even more relevant in the context of visual analysis, is the impossibility to represent a very large dendrogram in a visual display and to examine and interpret it.

4.5.2 Interpreting clusters

After you have run some clustering tool and obtained clusters, you need to interpret and compare these clusters in terms of the characteristics of the data items they consist of. To see these characteristics, you need to use suitable visual displays, in which the clusters need to be somehow distinguished. A typical approach is to assign a distinctive colour to each cluster and use these colours to represent cluster affiliations of data items or subsets of data items. If the data items are not very numerous, you can use displays showing them individually. In case of larger data, you will have to visualise aggregated characteristics. Aggregated data visualisations may be clearer and more informative also for smaller datasets. Let us illustrate these approaches by example of clustering of the London wards according to the values of several attributes representing proportions of population with different levels of qualification.

For example, Figures 4.15, 4.16, and 4.17 demonstrate different degrees of aggregation in visualisation of results of partition-based clustering, which was done according to values of multiple numeric attributes. The result consists of four clusters. Each cluster has been assigned a unique colour. Figure 4.15 demonstrates a parallel coordinates plot where the parallel axes correspond to the attributes that were involved in the clustering. Each individual data item is represented by a line in this

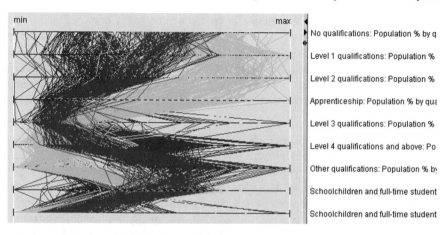

Fig. 4.15: Representation of results of partition-based clustering by colouring of lines in a parallel coordinates plot.

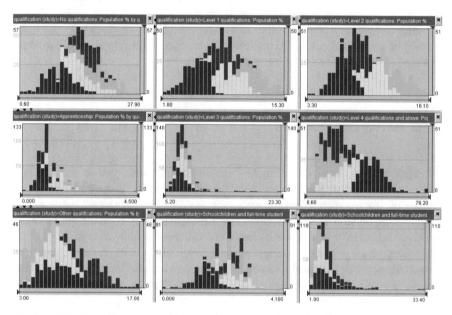

Fig. 4.16: Representation of results of partition-based clustering by colouring of bar segments in histograms

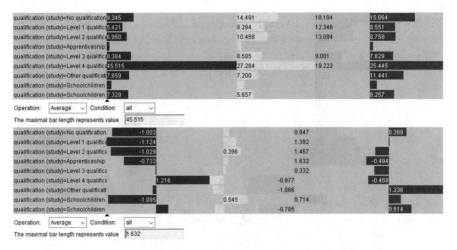

Fig. 4.17: Representation of results of partition-based clustering by coloured bar charts. Top: mean attribute values for the clusters. Bottom: mean z-scores of the attribute values for the clusters.

plot, and the cluster membership of this data item is represented by the line colour. In Fig 4.16, the same data are represented by a set of histograms, each corresponding to one attribute. The bars of the histograms are divided into coloured segments with the heights proportional to the numbers of the data items belonging to the four clusters. Unlike the parallel coordinates plot, the histograms are free from over-plotting and visual clutter, and it is easier to see and compare the ranges of the attribute values and the most frequent values in the clusters. The bar charts in Fig. 4.17 show the characteristics of the clusters in the most aggregated form. In the upper display, the mean values of the attributes for the clusters are represented by the bar length. To facilitate comparison of the cluster profiles, the bar charts in the lower display show the means of the attribute values transformed to z-scores, i.e., differences from the attribute's means divided by the standard deviations.

Of course, such displays as in Figs. 4.15, 4.16, and 4.17 are suitable for representing clustering results when the clustering is done according to values of multiple numeric attributes. If the type of data used for clustering is different, you will need to choose other visualisation techniques, such as time series plot and time histograms, or maps of spatial distributions, or other methods suitable to the data type.

Having understood what the clusters mean, you may wish to see how they are related to other components of the data. In our example, the London wards have been grouped according to the similarity of the qualification profiles of the population. It is interesting to see how these different groups of qualification profiles are distributed over the territory of London. For this purpose, we can use a map with the wards painted in the colours of their clusters, as shown in Fig. 4.18, top. Looking at this map and the bar charts in Fig. 4.17 or histograms in Fig. 4.16 in parallel, we can observe that the wards with high proportions of highly qualified people (painted in purple) are mostly located in the centre, whereas wards having high proportions of people with low to medium levels of qualification as well as schoolchildren and students 16-17 years old (green) are on the periphery, particularly on the east and southeast.

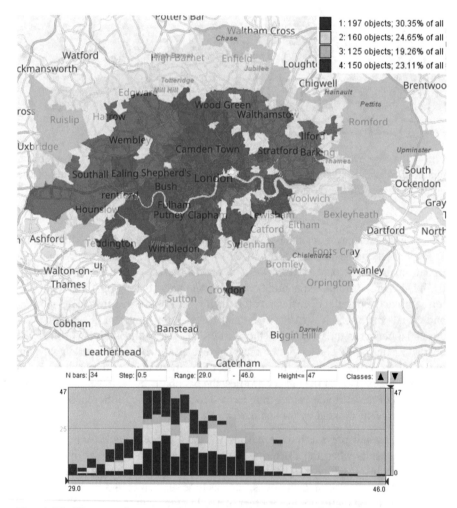

Fig. 4.18: Cluster colours can be propagated to various displays representing the objects that have been clustered, such as a map (top) or a frequency histogram of values of a numeric attribute (bottom).

To relate the qualification profiles to the mean age of the ward inhabitants, we can create a frequency histogram for the attribute 'mean age' and represent the clusters in the same way as it was done in Fig. 4.16 for the attributes involved in the clustering: the bars of the histogram are divided into coloured segments proportionally to the counts of the members of the different clusters in the corresponding intervals of the values of 'mean age', as shown in Fig. 4.18, bottom. In this histogram we see, for example, that the wards from the red cluster have mostly low to medium values of the mean age, while the mean ages in the wards from the green cluster are mostly medium to high.

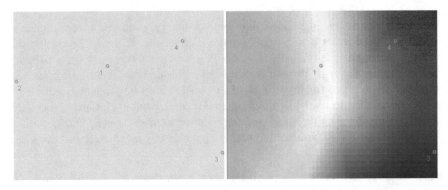

Fig. 4.19: The use of data embedding for assigning colour to clusters. Left: 2D embedding of summarised cluster profiles. Right: Cluster colours are picked from a continuous colour scale spread over the embedding space.

Hence, the use of cluster colours in various visual displays can reveal the relationships between the data components that were involved in defining the clusters and the other components of the data. It is very beneficial, especially when the clusters are more numerous than in our example, to assign colours to the clusters in a meaningful way, so that similarity of colours signifies similarity of the clusters. This can be achieved by applying data embedding techniques to summary characteristics of the clusters (such as the mean values of the attributes involved in the clustering) and spreading a continuous colour scale over the embedding space as shown in Fig. 4.19 (which is similar to Fig. 2.10, where this technique was applied to original data rather than clusters). Each cluster gets a particular position in the embedding space and can be assigned the colour corresponding to this position.

When applying partition-based or hierarchical clustering, you usually have the power to decide how many clusters you wish to have. For practical needs, you will typically not generate too many clusters. With density-based clustering, it is hard to predict how many dense groups of data items the algorithm will find, and it is not unusual to obtain quite many clusters. It may not be possible to give each cluster a sufficiently distinct colour. In this case, it is even more important to assign similar colours to similar clusters. Although you will not differentiate individual clusters in the displays where the colours are used, you will see the characteristics and distributions of groups of similar clusters. To see individual differences within a group, you may investigate this group separately by applying secondary embedding and colouring only to the clusters of this group.

4.5.3 Clustering process

Clustering algorithms, like almost any computational algorithms, have parameters, which you need to set before running them. The results, naturally, depend on the settings made, but it is typically not clear in advance what values for the parameters to choose. Thus, the popular algorithms for partition-based clustering, such as k-means, require you to choose how many clusters must be produced, and this is what you usually do not know in advance. For density-based clustering, you need to choose the distance threshold and the minimal number of neighbours, and for hierarchical clustering, you need to decide upon the way in which items' distances to clusters are defined.

An obvious consequence is that a single run of a clustering algorithm is *never sufficient*, and any analysis based on a single run of clustering is therefore *never valid*. You need to run an algorithm several times with different parameter settings, *compare* the results, and synthesise your observations of the different results into a general characterisation of the data.

The question is: How to compare results of different runs of a clustering algorithm? There exist several numeric measures of cluster quality, such as the average silhouette width (ASW) [118], which is often used for choosing the "right" number of clusters in partition-based clustering. Generally, these measures are intended to show how well the clusters produces by an algorithm correspond to "true clusters" existing in data. "True clusters" are subsets of similar or close data items that are well separated (i.e., quite dissimilar or distant) from other subsets. However, it is not very usual that data consist of such internally homogeneous and clearly distinct subsets. For example, the parallel coordinates plot in Fig. 4.15 shows that there are no well-separated groups of wards with particular profiles in terms of population qualification levels. Hence, any division of this dataset into clusters will be artificial.

This means that the concept of cluster quality may be questionable when there are no "true clusters". Although the numeric measures of cluster quality may still be useful, you should not put absolute trust in them. For analysis and reasoning, clusters need to be first of all meaningful. You should be able to understand their characteristics and their differences, and they should give you some knowledge about the data. When you use clustering in order to gain understanding of the data and the phenomenon reflected in the data, your main result is not perfect clusters but the knowledge you obtain during the process of clustering and interpretation. In this process, every run of the clustering algorithm can give you a new piece of knowledge. Hence, you need to perform clustering several times with different parameter settings until you feel that your understand the data and the corresponding aspect of the phenomenon well enough.

Let us consider partition-based clustering (which is the most frequently used type of clustering) using the running example with the London wards that we had before (Figures 4.15 to 4.18). The result of the partitioning into 4 clusters told us that the categories representing lower qualification levels (1 and 2), absence of qualification, and apprenticeship are interrelated in the sense that the proportions of these categories are often either simultaneously low or simultaneously high. The highest qualification level (4) is a kind of opposite from all others: its proportion is high where the proportions of the other levels are low, and vice versa. We have learned that there is a group of wards where the number of people having the highest qualification is much above the average, and these wards are mostly in the city centre. We have also discovered that quite many wards have higher than average proportions of people with lower qualification levels and without qualification, and these wards are mostly located on the eastern and southeastern periphery of the city. We can also notice that the wards with high proportions of highly qualified people have low proportions of schoolchildren and students 16-17 years old, and we notice the existence of wards distinguished by high proportions of people with "other qualifications" (this category includes qualifications that have no official levels in the UK, in particular, qualifications obtained in foreign countries).

Let us see what we can gain from increasing the number of clusters from 4 to 5. The bar charts in Fig. 4.20, top, confirm the knowledge we have got earlier. Additionally, we discover the existence of spatially scattered wards with very high (in relation to the average) proportions of the level 3 qualifications and students 18 and over years old. However, this group of wards is quite small (23 wards), as can be seen from the legend in Fig. 4.20, bottom. A large group of wards (a cluster with very pale, almost white colour) has quite average proportions of all qualification and study categories.

Increasing the cluster number to 6 reveals a rather small group of wards, mostly located compactly in the centre, with high proportions of level 4 qualifications and students aged 18 and more years and also relatively high proportions of level 3 and other qualifications (Fig. 4.21). Further increase of the cluster number reveals only quite small variations of the main profiles we have discovered earlier; so, we see no reason to continue the process.

If we compare the clustering results in terms of the average silhouette scores, we shall see that the average scores for 4, 5, and 6 clusters are 0.2788, 0.2916, and 0.2522, respectively; hence, the result with 5 clusters appears slightly better than the others. The average silhouette scores are computed from the individual silhouette scores of the cluster members, which may range from -1 to 1. Positive values indicate that a data item is closer to the members of its own cluster (regarding the mean distance) than to the members of any other cluster, and a negative value means that a data item is closer to members of some other cluster. The closer the individual

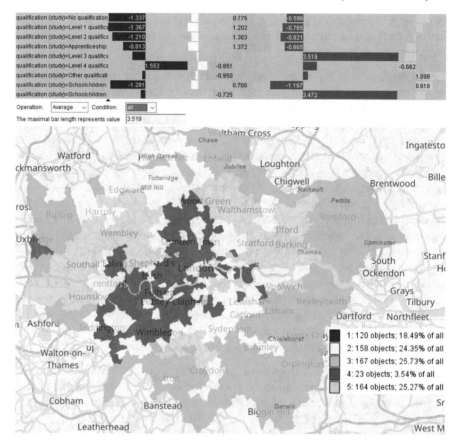

Fig. 4.20: The same dataset as in Figs. 4.15 to 4.18 has been partitioned into 5 clusters. Top: the cluster summaries in terms of the average z-scores of the attribute values. Bottom: the spatial distribution of the clusters.

value is to 1, the better, and higher average scores for clusters are meant to indicate better clustering results.

In our case, the result with 5 clusters is better than with 4 and 6; however, the average score for cluster 4 with 23 members (bright pink in Fig. 4.20) is very low, only 0.1679. The other results also include clusters with very low scores; thus, cluster 2 (yellow) in the result with 4 clusters has the average score of 0.1716, and cluster 4 (dark purple) in the result with 6 clusters has the average score of 0.0964 only. Increasing the number of clusters further does not improve the quality in terms of the silhouette scores; however, when we decrease the number of clusters to 3, the overall average silhouette score increases to 0.3369, with the minimum of the cluster averages being 0.3128. Hence, according to this measure of the cluster quality, the result with 3 clusters is the best.

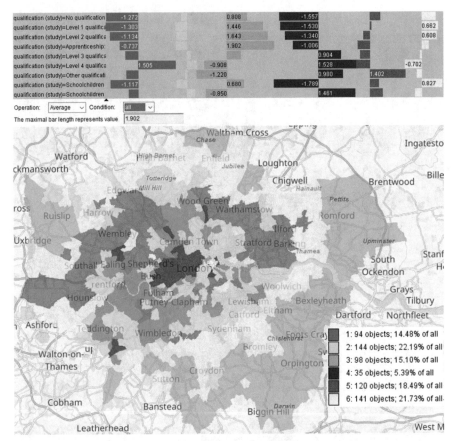

Fig. 4.21: The same dataset as in the previous figures has been partitioned into 6 clusters. Top: the cluster summaries in terms of the average z-scores of the attribute values. Bottom: the spatial distribution of the clusters.

Apart from the average scores, we can also compare the distributions of the individual scores using histograms, as in Fig. 4.22. Indeed, the scores for 3 clusters look better: there are more high values and fewer negative values.

The better silhouette scores for the result with 3 clusters do not mean, however, that we should ignore all other results. First, multiple results allow us to check the consistency of our findings. Thus, the result with 3 clusters (Fig. 4.23) tells us that the proportions of lower qualification categories, along with no qualification and younger students, are interrelated and opposite to level 4 qualifications. When we increase the number of clusters, we still see the same relationships, which increases our confidence in this finding. Second, increasing the number of clusters reveals subsets of the data whose profiles differ in some ways from the others. When these subsets are big enough, seeing their differences increases and refines our under-

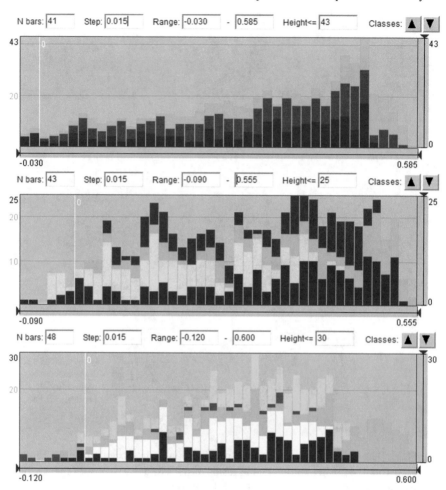

Fig. 4.22: Frequency histograms of the individual silhouette scores of the data items for the results with 3, 4, and 5 clusters.

standing of the data and the phenomenon. Thus, we have observed the variety of qualification profiles lying between the extreme high and low qualification profiles and found that the profiles with all proportions close to the average occur in a large part of the wards.

What we wanted to demonstrate with this example is that the main role of clustering in studying a phenomenon is not to merely produce good clusters but to provide valuable knowledge. In this respect, important criteria of cluster goodness are how meaningful and informative they are. Therefore, you should not rely only on numeric measures of cluster quality. On the other hand, you should not neglect such measures either, as they can tell you whether you can treat cluster summaries as

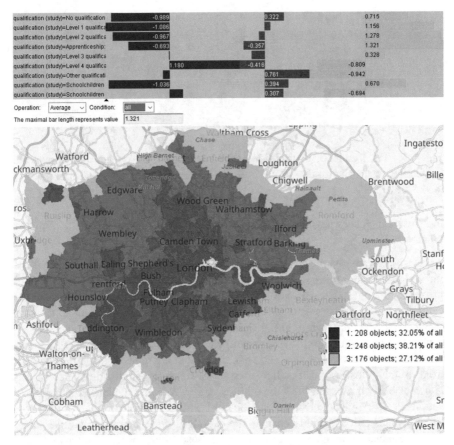

Fig. 4.23: The same dataset as in the previous figures has been partitioned into 3 clusters. This result is the best of all in terms of the average silhouette scores

valid and significant. For example, the lowest histogram of the silhouette scores in Fig. 4.22 tells us that the bright pink cluster has many (relative to its size) members with negative scores. These members are closer to members of other clusters than to the co-members in their cluster, which means that the summary profile of this cluster in Fig. 4.20 should not be trusted.

As we explained in Section 4.1.2, the role of clustering in visual analytics work-flows is to enable effective visual representation of large and/or complex data by organising the data into groups, which are then characterised by their statistical summaries. In our example, we characterised the clusters by the profiles consist-ing of the mean values of the attributes, which is a typical approach. Such profiles can be seen as *cluster centres*, usually called *centroids* in statistics and data mining. When the nature of the data does not permit computing some kind of "average ob-ject" from cluster members, clusters can be represented by their *medoids*. These are

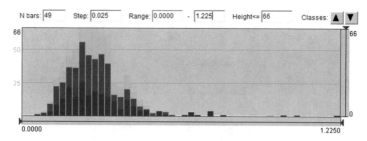

Fig. 4.24: The frequency distribution of the distances of the cluster members to the respective cluster centroids for the result with three clusters presented in Fig. 4.21.

the cluster members having the smallest sums of the distances to the other cluster members.

Apart from summarising cluster members, it is also appropriate to gain awareness of the *within-cluster variation*. When cluster centroids can be obtained, a possible approach is to look at the distribution of the distances from the cluster members to the centroids (or medoids, when centroids cannot be obtained). For example, the segmented frequency histogram in Fig. 4.24 shows the distribution of the member-to-centroid distances for the clustering result with three clusters. The histogram reveals the existence of several outliers, i.e., data items that are quite distant from the centroids of their clusters. Such outliers may affect the cluster summaries. Therefore, it is useful In such cases to filter the outliers out, re-compute the cluster summaries, and check if they have changed significantly. In our example, while the mean attribute values indeed slightly change, the overall profile patterns remain the same. If your analysis aims not only at finding the most typical patterns but also detecting and inspecting anomalies, the items distant from the cluster centroids deserve a closer look.

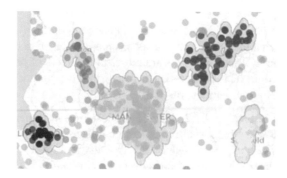

Fig. 4.25: An enlarged fragment of the map in Fig. 4.13, right demonstrates that density-based clusters may have complex shapes.

So far, our discussion referred to partition-based clustering. For results of density-based clustering, summarising clusters by averaging over the cluster members is, generally, inappropriate, and it is also meaningless to analyse the silhouette scores and distances to cluster centroids. As demonstrated in Fig. 4.25, density-based clusters may have arbitrary, sometimes very complex shapes. Distances between any two members of density-based clusters may be quite large, because each member is only required to be close to its neighbours. In a cluster having a complex, non-convex shape, there is no reasonable way to define the centre or centroid, and also the medoid cannot be considered as an adequate representative of the cluster.

When density-based clustering is applied to spatial or spatio-temporal objects, which are grouped according to the distances between them in space or space and time, the clusters can be summarised by computing their boundaries in space or space-time. These boundaries can be shown on a map, as we did in Figs. 4.13 and 4.25, and in a space-time cube, as will be seen in Fig. 10.19. Spatial or spatio-temporal clusters can also be presented visually by showing their members, as in Fig. 10.34. It is appropriate to clean the display from the noise and to differentiate the clusters by colours (however, as we noted earlier, the number of clusters may exceed the number of colours that can be easily distinguished by human eyes).

A more general approach to representing results of density-based clustering for interpretation and assessment is to, first, remove the noise, second, *partition* the clusters into sub-clusters with simple, convex shapes, and, third, summarise the sub-clusters in the same ways as applied to partition-based clusters. A possible approach to partitioning is to use the *x-means* [110] or another algorithm capable of finding an optimal number of clusters within a user-specified range. This algorithm is applied to the data after excluding the noise, i.e., to the members of all density-based clusters taken together. The number of the density-based clusters provides the lower bound for the number of the partition-based clusters. You can take some factor of this number, such as 1.5 or 2, as the upper bound, and it is appropriate to check the effect of this parameter on the resulting number of partition-based clusters and their summaries.

Another approach to summarising density-based clusters, which is suitable when averaging cannot be applied to the data, is to represent each cluster by as few as possible members such that the distance of any other member to one of these representatives is within a chosen distance threshold [15].

Like partition-based, density-based clustering also needs to be run several times, and the results of different runs need to be compared. Before taking efforts to obtain cluster summaries, which may be tricky, it is reasonable to compare the clustering results in terms of the amounts of the noise and numbers, sizes of the clusters (numbers of the members), and extents of the clusters, which can be estimated from the maximal distances between cluster members. Usually, when you do density-based clustering for gaining an overview of the data, a result with a large fraction of the data being categorised as noise is not very useful for you, because you will miss information about these data. On the other hand, having very large and extended clusters is not helpful as well, because the members of such clusters vary greatly and have nothing in common.

When you are not happy with the current results, you run the clustering algorithm another time with different parameter settings. To reduce the fraction of noise, you loosen the density conditions by either increasing the distance threshold (neighbourhood radius) or decreasing the required minimal number of neighbours (it is better to avoid changing both parameters simultaneously). To decrease the internal variance within clusters, you tighten the density conditions by making opposite changes of the parameter values.

However, it may happen that most of the clusters you have obtained are nice, while there are just a few clusters you are not happy with. Re-running the clustering with other settings may ruin the good clusters. To preserve the good clusters and improve the bad ones, you can do *progressive clustering*: you separate the good clusters from the rest of the data and apply clustering with tighter parameter settings to the rest.

Generally speaking, progressive clustering is application of clustering to a subset of data that has been chosen from the result of previous clustering. The purpose is to obtain a refined representation and, hence, a refined understanding of this data subset. For this purpose, the subset clustering is done with different parameter settings or even with a different distance function than the whole set. The use of different distance functions may be especially helpful when you deal with complex objects characterised by heterogeneous properties, such as trajectories of moving objects [116].

Summarising this section, we emphasise again that clustering is an analytical process rather than a single application of a clustering algorithm to data. In this process, you do clustering multiple times with different parameter settings, different data subsets, different distance functions, and, perhaps, different clustering algorithms (e.g., apply partition-based clustering to density-based clusters). You need to create appropriate visual representations of cluster summaries and properties for interpreting and comparing clustering results and gradually building up your knowledge about the data.

4.5.4 Assigning colours to clusters along the clustering process

As we discussed and demonstrated in Sections 4.5.2 and 4.5.3, visual exploration of clusters is greatly supported by representing clusters by distinct colours, and we advocated the use of data embedding as a meaningful, data-driven approach to assigning colours to clusters (Fig. 4.19). When you perform clustering as an analytical process involving multiple applications of a clustering algorithm and thus obtaining multiple sets of clusters, you are interested in keeping the cluster colours consistent between the results of the different runs, that is, similarity or dissimilarity of colours corresponds to similarity or dissimilarity of clusters across the different clustering results. How can this be achieved?

Most of the data embedding algorithms are non-deterministic with respect to the absolute positions given to data items in the embedding space. If you apply an embedding algorithm several times to the same data with the same parameter settings, the results will, most probably, look differently because the data will be put in different positions. However, the relative placements of the data items with respect to other data items will be consistent between the results. In fact, one result can be transformed to another one through rotation and/or flipping, plus, possibly, a little bit of scaling.

When you have two sets of clusters (obtained from different runs of a clustering algorithm) such that most of the clusters from one set have corresponding identical or very similar clusters in the other set, you can, in principle, find such transformations of the embedding of one set of clusters to that the clusters that have matches in the other set get the positions as close as possible to the positions of their matches in the embedding of the second set. However, we do not know any method to find such transformations automatically. What can be done instead is to let the embedding tool run many times producing many variants of embedding. This does not take much time because the number of clusters is usually quite small. Having many variants, you can choose the one that is the most similar to the embedding you used for assigning colours to the earlier obtained clusters. The degree of similarity can be expressed numerically as the sum of weighted distances between the positions of the matching clusters, the distances being weighted by the cluster sizes [13].

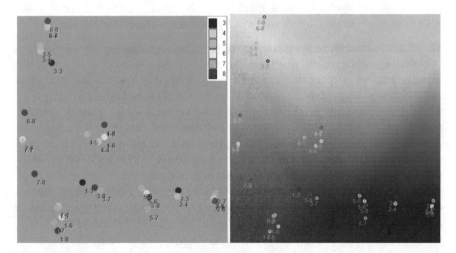

Fig. 4.26: Application of data embedding to centroids of clusters resulting from 6 runs of partition-based clustering with the number of clusters from 3 to 8.

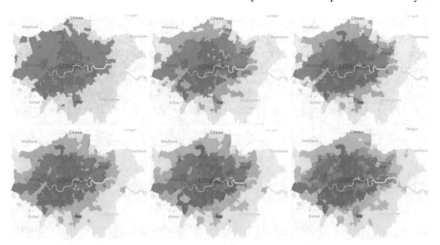

Fig. 4.27: Cluster colours obtained through the joint embedding of the cluster centroids (Fig. 4.26) are used for painting London wards on a map.

Even a better idea may be to apply embedding to the centroids (or summaries, or other representatives) of the clusters resulting both from the previous run and from the new run, i.e., two sets of clusters are taken together. In this case, there is no need to match clusters from two runs; similar clusters will be placed together in the embedding. The embedding tool can be run multiple times to produce embedding variants, from which the variant where the positions of the earlier obtained clusters are the closest to the positions they had before the last run of clustering is chosen. The positions (and, respectively, the colours) will not be exactly the same as before, but the changes are likely to be small. You will, however, need to re-assign the new colours to the previous clusters and adjust your mind to the colour change.

The idea of joint embedding can be applied not only to the results of the two last runs of clustering, but the results of all previous runs can be involved as well. Apart from consistent colour assignment, the benefit of joint embedding is showing similarities and differences between clusters from different runs. This is illustrated in Fig. 4.26, where data embedding has been applied to the centroids of the clusters resulting from 6 runs of partition-based clustering that produced from 3 to 8 clusters. The clustering was done on the same data referring to the London wards that we used in the previous illustrations. On the left, each cluster is represented by a dot in the embedding space coloured according to the clustering run the cluster results from: the colours from dark blue to red correspond to the runs with 3 to 8 resulting clusters. On the right, a continuous colour scale is spread over the embedding space. It can be used for assigning consistent, similarity-based colours to the clusters. In Fig.4.27, these colours are used for painting the wards on six maps representing results of the six clustering runs. The maps are easy to compare due to the high consistency of the colours.

4.5.5 Generalisation to other computational methods

The extensive discussion of the use of clustering presented in this section should be considered more generally. In analytical workflows, almost any method for computer analysis needs to be applied in an iterative way with varying parameter settings and/or modifying the input data. The purpose of running a method repeatedly is not only (and not so much) to obtain the best possible results but to incrementally build up human's knowledge about the data and the phenomenon reflected in the data. Results of different runs need to be interpreted, assessed, and compared. All these operations, which require involvement of human's cognitive abilities, need to be supported by visualisation. In this section, we showed and discussed how visualisation can support the process of using clustering for data analysis. This provides an example for the use of other kinds of computational methods. Perhaps, the techniques that we used for supporting clustering cannot be directly applied to results of another kind of computational method. However, this section demonstrates what needs to be cared about in the process of analysis with the use of computational techniques.

4.6 Topic modelling

We have mentioned in Section 4.4.3 that the methods known as "topic modelling", which have been developed for text analysis, are, in fact, methods of data embedding; more specifically, they belong to the category of dimensions-transforming methods. The new dimensions generated by the topic modelling methods are called "topics". Each "topic" is defined as a composition of weights of the original attributes. Usually, the number of "topics" to be generated is specified as a parameter of the method. Each data item is described by a combination of weights of the "topics". In classical applications of topic modelling methods, the original attributes are frequencies of various keywords occurring in texts. The "topics" are thus combinations of weights, or probabilities, of the keywords. A topic can be represented by a list of keywords having high probabilities.

Beyond the classical applications, topic modelling methods can be applied to other data that can be treated as "texts" in a vary general sense, i.e., as sequences or combinations of items taken from a finite set, which can be treated as a "vocabulary". For example, the "vocabulary" can consist of the names of the streets in a city, and trajectories of vehicles can be represented as texts consisting of the names of the streets passed on the way [41]. This approach, however, ignores the direction of the movement, treating street segments going from A to B and from B to A as equivalent, unless segments in opposite directions are encoded differently. Another example is a "vocabulary" consisting of possible actions, such as operations performed by users of a computer system, and "texts" consisting of action sequences [40]. In this case, application of topic modelling will ignore the order in which the actions were performed and take into account only the co-occurrence of different actions in one sequence.

Moreover, it is possible, in principle, to apply topic modelling to any data with a large number of positively-valued attributes having compatible meanings. For example, these may be data describing purchases of various kinds of products by customers of a e-shop, or combinations of educational modules chosen by students. In these examples, the products or modules play the role of "words", and their combinations can be treated as "texts".

4.6.1 General ideas and properties of methods

Since the methods for topic modelling have been originally developed for text documents, they are usually described using text-specific terminology. A more precise term to refer to this class of methods is *probabilistic topic modelling* [32], because these methods define a topic as a probability distribution over a given vocabulary, i.e., a set of keywords. In application to texts, the keywords with high probabilities represent the semantic contents of the topics. Apart from the *topic − keywords* distributions, topic modelling methods also produce *document − topics* distributions consisting of the probabilities of each document to be related to each topic.

It needs to be borne in mind that topics are defined based on frequent co-occurrences of keywords in texts. For example, a topic represented by the keywords "dog, tail, bark" can be derived only when these keywords frequently co-occur in same text documents. This means that topic modelling methods may fail to derive meaningful topics from collections of very short texts [155], such as messages consisting of one or a few sentences posted in online social media. A single message may contain only one or two informative keywords; hence, the total number of co-occurrences may be insufficient for deriving valid relationships between the words. To deal with this problem, short texts are aggregated into larger pseudo-documents [150]. This can be done based on the temporal proximity of the creation of the texts, commonality of message tags, and/or authorship of the texts. When geographic locations of the text creation are known, as in the case of the georeferenced tweets in the introductory example (Chapter 1), the texts can also be aggregated based on the proximity of their locations.

Let us extrapolate this problem and the solution to another application. Sets of products purchased simultaneously by individual customers may be too small for identifying groups of related products based on separate purchases. It may be reasonable to aggregate multiple purchases of different customers based on their temporal proximity, occurrence of products from the same category, and locations of the customers.

One of the most popular algorithms used for topic modelling is Latent Dirichlet Allocation (LDA) [33]. It is implemented in multiple open software libraries, computationally efficient, and known as more general than some other methods, e.g., Latent Semantic Analysis (LSA) [48] and Probabilistic Semantic Analysis (PLSA) [69]. A shortcoming of the method is its inconsistency across multiple runs. Being probabilistic, the algorithm does not guarantee the same results to be produced for the same input data and parameters. Nonnegative Matrix Factorization (NMF) [108] is a deterministic method, which generates consistent results in multiple runs when applied to same data. The drawback is the polynomial complexity, which limits the applicability of the method to large data sets.

4.6.2 How many topics?

Most of the topic modelling methods require the number of topics to derive to be specified by the user, for whom it may be hard to estimate how many meaningful and distinct topics exist. This is similar to specifying the number of clusters in partition-based clustering, and the same approach is taken to deal with the problem: run topic modelling several times and compare the results.

Another problem is that many topic modelling algorithms are non-deterministic and produce slightly different results in each run even if the number of topics is the same [43]. Therefore, better understanding of the contents of a text corpus can be gained by examining and comparing results of multiple runs of topic modelling, or, in other words, an ensemble of topic models [40].

Topics generated in different runs of an algorithm can be compared with the help of joint embedding, as, for example, shown in Fig. 4.28. We have earlier applied the same approach to results of several runs of partition-based clustering (Fig. 4.26). In the embedding space, similar topics are located close to each other and dissimilar ones are distant from each other. A cluster of spatially close topic indicates the existence of an archetype topic that is discovered by multiple runs of a topic modelling method. In other words, it is the same topic represented slightly differently in different models. The fact that it was found more than once indicates its significance and trustworthiness. Topics that are scattered over a wide area in the plot may be computational artefacts rather than representations of really existing topics. If there are a few topics that are distant from others, the analyst can check their trustworthiness through additional runs of the algorithm with incrementing the topic number in each run. An occurrence of similar topics in the additionally obtained topic models is an evidence of topic significance. Topics that remain far from all others can be ignored. In this iterative and interactive way, the analyst can gain understanding of what and how many topics exist.

Fig. 4.28: 2D space embedding (projection) of a combined set of topics resulting from multiple runs of a topic modelling algorithm with differing parameter settings. The topics are represented by pie charts, the sector sizes being proportional to the probabilities of the terms. Source: [40].

To obtain a good set of valid and distinct topics suitable for further analysis or for reporting, the analyst can select a representative topic from each group of close topics. The analyst can define the groups using the topic embedding display, e.g., by encircling clusters of dots. A suitable representative of a group is the medoid, i.e., the topic having the smallest sum of distances to all other topics in the group [40]. Here we mean not the distances between the dots in the embedding space but the distances in the multidimensional data space, in which the topics are represented by weights or probabilities of multiple components of the original data.

In principle, a suitable number of topics can be determined automatically based on some quality measures [27]. However, what is optimal in terms of statistical indicators in not necessarily the most meaningful and useful to a human. An analyst can gain better understanding and knowledge of the data from seeing and exploring the "topic space" with dense and sparse areas, clusters, and outliers.

4.6.3 Topic modelling versus clustering

The most significant difference between topic modelling and clustering is that the common methods for clustering assign each data item to a single cluster whereas topic modelling can associate a data item with several relevant topics. A basic premise of topic modelling is that any document may concern several topics, possibly, in different proportions; i.e., each document is modelled as a topic mixture. Topics are identified based on term co-occurrences. Even if there are no documents solely concerned with some topic A but there are documents where A is combined with various other topics $B, C, D...$, a topic modelling algorithm is quite likely to find topic A, because the co-occurrence frequencies of the terms pertinent to A will be high while their co-occurrences with the terms specific to $B, C, D...$ will be relatively less frequent.

Another difference is that clustering works based on distances between data items defined by some distance function. In cases of multidimensional data, the "curse of dimensionality" 4.2.1 makes division into clusters quite arbitrary and unstable regarding slight changes of parameters. Unlike clustering, topic modelling techniques are designed to deal with high-dimensional and sparse data, being based on term co-occurrences within data items (documents) rather than distances between the data items.

4.6.4 Use of topic modelling in data analysis

Topic modelling produces two kinds of outputs: first, the topics as such and, second, the profiles of the input data items in terms of the topics. Each of these outputs is valuable for data analysis.

Topics are defined by weights, or probabilities, of the input data dimensions, that is, keywords in text analysis or, more generally, attributes with non-negative numeric values in other applications. A topic is represented by the combination of keywords or attributes having high weights. This means that there are quite many data items in which these keywords co-occur or these attributes simultaneously have high values. This, in turn, means that these keywords or these attributes are semantically related. Hence, one possible purpose of using topic modelling is revealing semantic relatedness of data dimensions.

As we explained earlier, topic modelling is a specific sort of dimensionality reduction. It transforms highly multidimensional data into profiles consisting of topic weights. Since there are much fewer topics as the original data dimensions, the dimensionality is thus significantly reduced. A nice property of the transformed data is their intrepretability: knowing the meanings of the topics, it is not difficult to understand the profile of each item. Like results of any dimensionality reduction methods, results of topic modelling can be used for clustering or data embedding. The use of 2D data embedding is the most common approach to visualisation of results of topic modelling. The data items (particularly, documents) are represented by points in the embedding space arranged according to the similarities of their profiles in terms of the topics.

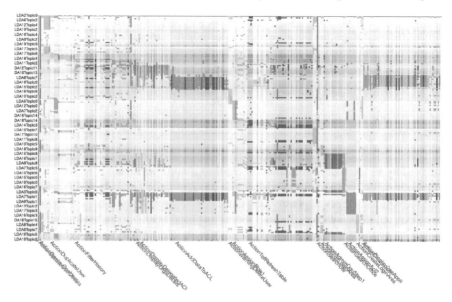

Fig. 4.29: A matrix display supporting interpretation of topics derived from sequences of actions. The rows correspond to the topics and the columns to the actions. The colour hues correspond to different categories of the actions. The saturation of the colours in the matrix cells encodes the weights of the actions for the topics. Source: [40].

The use of topic modelling in data analysis requires understanding of the meanings of the topics. In case of topics derived from texts, the meanings can be adequately represented by combinations of a few keywords having high probabilities, such as "environment, air, pollution" and "space, shuttle, launch". It is usually not difficult for a human analyst to guess the meanings of so represented topics. The same applies to other cases where the meanings of the data dimensions are well known to the analyst. Examples are types of products sold or names of educational modules. However, this approach may not work well when the meanings of the dimensions are not immediately clear and/or more than two or three dimensions can have high weights in the topic definitions. In such cases, interpretation of the topics requires visualisation support. Since topic definitions are multidimensional numeric data, the visualisation methods designed for this kind of data can be used for this purpose. These methods are presented in Chapter 6. An example of a visual display supporting topic interpretation is shown in Fig. 4.29.

4.7 Conclusion

There are lots of techniques for computational processing and analysis of data of different types. This chapter discusses only a few classes of such techniques that are most often used in combination with interactive visualisation in visual analytics activities, as represented schematically in Fig. 1.16. All computational techniques produce some derived data, which need to be interpreted by the analyst and used in the following analysis and reasoning. To enable perception and understanding by the human, the derived data need to be represented visually. Obviously, the choice of computational techniques depends on the type and characteristics of the original data, and the choice of the visualisation techniques for representing the outputs of the computational processing depends on the type and characteristics of the derived data.

In the following chapters, we shall discuss and exemplify applications of computational techniques (together with visualisations) to different types of data. Apart from the common classes of techniques discussed in this chapter, there will be examples of applying more specific techniques suitable for particular data types.

Part II

Visual Analytics along the Data Science Workflow

In Chapter 2 (Section 2.1), we argued that different stages of a typical data science workflow require understanding of three different subjects: the data under analysis, the real-world phenomenon reflected in the data, and the model that is being built. Visual analytics approaches, which are designed to support human reasoning and enable understanding of various kinds of subjects, can thus be very helpful at all stages of the data science workflow. In the following chapters, we shall consider how visual analytics can help in exploring, understanding, and preparing data (Chapter 5), using data to understand phenomena that involve different kinds of components and relationships (Chapters 6 to 12), and using the acquired understanding of the data and phenomena in building mathematical or computer models (Chapter 13).

Chapter 5
Visual Analytics for Investigating and Processing Data

Abstract In this chapter, we discuss how visual analytics techniques can support you in investigating and understanding the properties of your data and in conducting common data processing tasks. We consider several examples of possible problems in data and how they may manifest in visual representations, discuss where and why data quality issues can appear, and introduce a number of elementary data processing operations and computational techniques that help, in combination with visualisation, understand data characteristics and detect abnormalities.

5.1 Examples of data properties that may affect data analysis

In a perfectly organised world, all data that land on the desk of a data analyst, are collected and verified carefully, documented thoroughly, and cleaned from any occasional problems. The reality often differs from this description, unfortunately. We have seen many data sets that were collected by different people and organisations using different equipment, methods, and protocols. The data are often represented in different formats using different notations, making their fusion a challenging task. Erroneous and missing values are very usual in any data.

In this chapter, we shall write about data sets in general, irrespective of their specifics and representation. The following chapters will address different types of data in detail. So, we shall talk here about a general *dataset* consisting of multiple *data items*. Data items are composed of *fields*, which may contain values of attributes, references to entities, places, or times, or to items in another dataset. All data items in a dataset have homogeneous structure, i.e., consist of the same number of fields having the same meaning and containing the same kind of information. The fields usually have names. The contents of the fields are called the *values* of these fields. Some fields in a dataset may be empty. The absence of a value in a field may

N. Andrienko et al., *Visual Analytics for Data Scientists*,
https://doi.org/10.1007/978-3-030-56146-8_5

151

have different meanings: either no value exists, or some value exists in principle but could not be determined. Knowing the meaning of the field can help understand what the absence of a value means. When this is not clear and not described in metadata, it is necessary to obtain additional information about the dataset, e.g., by contacting the data collector.

In many data sets, dummy values, like 999 or −1, have special meanings, such as "missing value", "anything else" (e.g., when values in a field are categories or classes), or "error". Sometimes, different dummy values have the same meaning in a single data set, if it was prepared by different people or at different times. For example, both "n/a" and "-" may mean "not applicable". It may also happen that the same dummy value has different meanings. A series of data records may lack consistency in measurement units (e.g. metres or kilometres), formatting (decimal dot or decimal comma), and representation of dates and times. If a data set has been collected over a long period of time, consistency may be lacking due to changes in equipment, policies, daily routines, or personal habits and preferences. Data items representing times may be inconsistent due to wrong time zones (e.g. if a tourist did not set a correct time zone in the photo camera after moving to a different continent) or ignoring switches to/from daylight saving time. Textual data components may be misspelled, contain abbreviations and jargon, texts in different languages, etc. The same meanings may be expressed using synonyms, thus complicating processing and analysis.

During processing, field values may lose their *precision*. It may be a result of insufficiently careful transformation of the data format (e.g., from a spreadsheet to a text file) or an attempt to decrease the size of a file with data. Figure 5.1 shows the impact of rounding up geographic coordinates from 5 to 2 decimals. The dots are rendered with 70% transparency for enabling the assessment of the densities. By comparing two maps, it is visible that, as a consequence of the displacement of the points, some real patterns disappear while artefacts, or fake patterns, emerge.

Fig. 5.1: The same set of 3,083 points (Twitter messages posted in London) is displayed with precision of 5 and 2 decimals (left and right).

Another possible reason for precision loss is striving to protect sensitive data, for example, to hide exact locations visited by individuals and preclude in this way identification of the individuals and their activities. Location data may also be inaccurate due to the data collection procedures employed. For example, if points describing locations of mobile phones originate from a mobile phone network, their coordinates may represent the positions of the cell tower antennas rather than actual positions of the phones.

Missing data, i.e., absence of valid values in fields, may be occasional (e.g. no value for a specific age category, no measurement in a single location) or systematic (for example, no observation during weekends due to absent staff). Sometimes, missing values may indicate absence of something (e.g., absence of crime incidents in a given place), which can be appropriately represented by the value 0.

Opposite to missing values, sometimes data sets contain *duplicates*. For example, if it is expected to have a single data record per minute, two or more records with the same temporal reference are duplicates. Duplicates may have the same field contents, but they may also differ. In such cases, it is necessary to determine which of the records has higher validity.

The notes and examples from this section indicate the need to investigate and understand data properties before starting any analysis. As explained in Section 2.1 and schematically illustrated in Fig. 2.1, investigation of data properties and preparation of the data to the analysis is an essential stage of the data science process.

We refer an interested reader to the paper [62] that introduces a general taxonomy of problems in data. Although it specifically focuses on temporal data, a great part of the material is also valid for other kinds of data. A recent paper [91] considers possible data problems for the most commonly used data types, such as graphs, images and videos, trajectories of moving objects, etc. In this book, we give specific recommendations for different types of data in the further chapters. The following section considers data quality problems at a more general level and provides generic recommendations for dealing with outliers and missing data irrespective of data types.

There are a number of visual and computational tools that help us to investigate of new datasets that we encounter. We have discussed these tools in Chapters 2 and 3. In this chapter, we are revisiting some of them in the context of conducting an exploration into data properties.

5.2 Investigating data properties

The first thing you will need to do when starting a data science project is to, as most data scientists would say, "play with the data". Through this interaction with the data, you get familiarised with them and get some understanding of how the fields represent the phenomenon you want to study. You need to question critically whether the data provide an effective and suitable representation of the phenomenon and determine if you need to get additional data from other sources or perform some data transformations. You also investigate whether the data have sufficient quality and, if not, what needs to be done to bring them to a state that is suitable for any further analysis. Data investigation, assessment, and preparation to the analysis (often referred to as *data wrangling*[1]) is a phase of utmost value, and like it or not, one of the most time consuming, labour intensive parts of the data science process. One of the interviewees from Kandel et al.'s study that looked into data analysis practices at enterprises [77] expresses this view very nicely: "*I spend more than half of the (my) time integrating, cleansing and transforming data without doing any actual analysis. Most of the time I'm lucky if I get to do any analysis...*"

An important point to make clear here is that your emphasis in this phase of the analysis process is on the data itself, i.e., the data is the subject of the analysis (Fig. 2.1 and Section 2.1). Of course, you always need to keep the underlying phenomena in your mind, as certain decisions are only viable when you put them in the context what the data is referring to. For instance, a suspiciously high value within rainfall measurements might mean a malfunctioning device but could also be the result of a rare weather event that is of utmost interest.

5.2.1 Overall view of a distribution

It is typical to begin the investigation of the data properties with considering the **descriptive statistics**[2]. Unlike inferential statistics, which is designed for inferring properties of a population by testing hypotheses and computing estimates[3], descriptive statistics measures are calculated based on the given sample of data and reflect properties of this particular set. The most common measures are those that estimate the *central tendency*, i.e., where the central, average value of observations fall, and *dispersion*, i.e., how much the values vary. Common statistics for central tendency are the *mean* – the sum of all the values divided by the number of data items, *median* – the middle value that separates the half with higher values from the other half with lower values in the dataset, *mode* – the most frequent value within the records, especially useful when the data are categorical. The dispersion is estimated using the *standard deviation*, the *interquartile range (IQR)*, which is the difference between the values of the upper and lower quartiles, or simply the difference between the maximum and minimum.

---

[1] https://en.wikipedia.org/wiki/Data_wrangling

[2] https://en.wikipedia.org/wiki/Descriptive_statistics

[3] https://en.wikipedia.org/wiki/Statistical_inference

Introduced in the John Tukey's seminal book on exploratory data analysis [138], the box plot technique (Sections 3.2.1 and 3.2.4, Figures 3.2 and 3.3) visually displays most of the descriptive statistics. However, as noted in Section 3.2.7, both the box plot and the descriptive statistics it portrays may be insufficient and even inappropriate for characterising a set of numeric values when the distribution is not unimodal. While the statistics and the box plot provide a compact summary, it is necessary to see the data in more detail. Thus, we have started Chapter 3 with an example of the "Anscombe's Quartet": a series of small data sets with very different patterns, but exactly the same statistical indicators. Another example of this kind is the "datasaurus dozen" (Fig. 5.2).

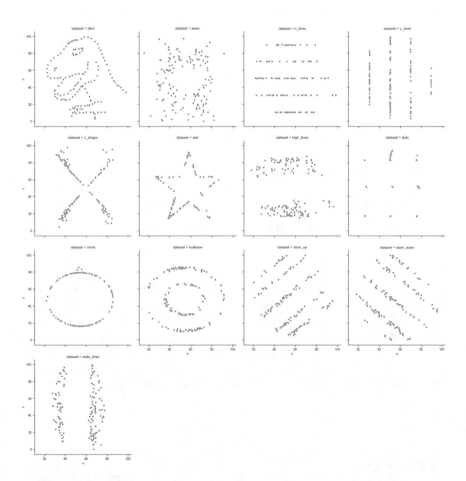

Fig. 5.2: The datasaurus dozen [97] – a good reminder of why you always need to inspect your data visually before performing any further analysis. In this set of scatterplots representing pairwise relations with the same x-mean, y-mean, x-standard deviation, y-standard deviation and correlation, the visualisations show us a completely different story.

Possible ways of combining high-level statistical summaries portrayed by a box plot with more detailed representations of the data distribution have been demonstrated in Fig. 3.2 and 3.3. Figure 5.3 shows how box plots combined with dot plots can be used for exploring and comparing distributions of multiple variables. Thus, we see that "attribute-8" has a remarkably higher variation than the other attributes, with some data records having extreme values, i.e., outliers. Upon encountering such cases, you need to investigate what this attribute is and what the underlying reasons for the high variation and outliers might be.

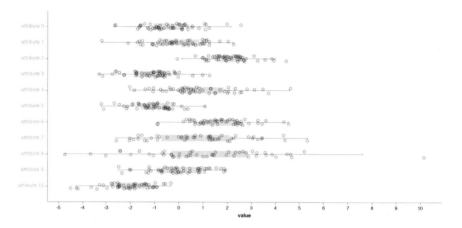

Fig. 5.3: Box plots combined with dot plots portray value distributions of 11 numeric attributes, we spot *attribute-8* as standing out and requiring further investigation.

Box plots combined with dots are suitable only if the amount of data is rather small. For large data volumes, it is more appropriate to investigate distributions of attribute values using frequency histograms (see Section 2.3.3). Different distribution properties exhibit themselves in particular histogram shapes and features. Various histogram shapes and possible interpretations of their meanings are discussed extensively in statistical literature. Many examples of possible meanings of the shapes are explained in the book [130]; we reproduce and summarise them in Table 5.1.

5.2.2 Outliers

Outliers[4] are data items that differ significantly from others. In Section 2.3.2, we said that outliers do not join patterns formed by groups of other items because they

---

[4] https://en.wikipedia.org/wiki/Outlier

Shape	Interpretation
	Normal. A common pattern is the bell–shaped curve known as the "normal distribution". In a normal distribution, points are as likely to occur on one side of the average as on the other. Be aware, however, that other distributions look similar to the normal distribution. Statistical calculations must be used to prove a normal distribution. Don't let the name "normal" confuse you. The outputs of many processes—perhaps even a majority of them—do not form normal distributions, but that does not mean anything is wrong with those processes. For example, many processes have a natural limit on one side and will produce skewed distributions. This is normal — meaning typical — for those processes, even if the distribution isn't called "normal"!
	Skewed. The skewed distribution is asymmetrical because a natural limit prevents outcomes on one side. The distribution's peak is off center toward the limit and a tail stretches away from it. For example, a distribution of analyses of a very pure product would be skewed, because the product cannot be more than 100 percent pure. Other examples of natural limits are holes that cannot be smaller than the diameter of the drill bit or call-handling times that cannot be less than zero. These distributions are called right – or left–skewed according to the direction of the tail.
	Double-peaked or bimodal. The bimodal distribution looks like the back of a two-humped camel. The outcomes of two processes with different distributions are combined in one set of data. For example, a distribution of production data from a two-shift operation might be bimodal, if each shift produces a different distribution of results. Stratification often reveals this problem.
	Plateau. The plateau might be called a "multimodal distribution". Several processes with normal distributions are combined. Because there are many peaks close together, the top of the distribution resembles a plateau.
	Edge peak. The edge peak distribution looks like the normal distribution except that it has a large peak at one tail. Usually this is caused by faulty construction of the histogram, with data lumped together into a group labeled "greater than..."
	Comb. In a comb distribution, the bars are alternately tall and short. This distribution often results from rounded-off data and/or an incorrectly constructed histogram. For example, temperature data rounded off to the nearest 0.2 degree would show a comb shape if the bar width for the histogram were 0.1 degree.
	Truncated or heart-cut. The truncated distribution looks like a normal distribution with the tails cut off. The supplier might be producing a normal distribution of material and then relying on inspection to separate what is within specification limits from what is out of spec. The resulting shipments to the customer from inside the specifications are the heart cut.
	Dog food. The dog food distribution is missing something—results near the average. If a customer receives this kind of distribution, someone else is receiving a heart cut, and the customer is left with the "dog food," the odds and ends left over after the master's meal. Even though what the customer receives is within specifications, the product falls into two clusters: one near the upper specification limit and one near the lower specification limit. This variation often causes problems in the customer's process.

Table 5.1: Examples of histogram shapes and what they mean;
source: http://asq.org/learn-about-quality/data-collection-analysis-tools/overview/histogram2.html

are dissimilar to everything else. Particularly, outliers among numeric values are manifested in histograms as isolated bars located far away from others and in dot plots and scatterplots as isolated dots distant from the bulk of the dots. As noted in Section 2.3.2 it is necessary to understand whether outliers are errors in the data that need to be removed, or they represent exceptional but real cases that require focused investigation.

For numeric attributes, statisticians have developed a number of computational techniques that, assuming a given distribution of the values, indicate if some values are outliers. The type of the value distribution may be understood from the shape of a frequency histogram (see Table 5.1) For example, if a distribution is *normal*[5] (exhibited by a histogram having a bell shape, as in the top row of Table 5.1), the values can be checked for outlierness by computing their distances to the quartiles of the value set. If $Q1$ and $Q3$ denote respectively the lower (first) and upper (third) quartiles and $IQR = Q3 - Q1$ is the interquartile range, then an outlier can be defined as any observation outside the range $[Q1 - k \cdot IQR, Q3 + k \cdot IQR)]$ for some constant $k > 1$. According to John Tukey who proposed this test, $k = 1.5$ indicates an "outlier", and $k = 3$ indicates that a value is "far out".

A data item consisting of values of multiple attributes can be an outlier although the value of each individual attribute may not be an outlier with respect to the other values of this attribute. Such multidimensional outliers can be detected using a suitable distance function (Section 4.2) taking care of data normalisation and standardisation when necessary (Section 4.2.9). The outlierness can be judged based on the distribution of the distances of the data items to the average or central item, such as the centroid or medoid of the dataset.

After outliers are identified, they need to be examined by a human analyst for making an informed decision whether to treat them as mistakes or as correct data signifying interesting properties of the phenomenon reflected in the data. Suspicious data may originate from procedures of data collection (e.g. by selection of an inappropriate sample of a population), from hardware or software tools used for data collection (incorrect measurements, wrong units etc.), or from mistakes inevitably made by humans in the course of manual recording. The ways to understand whether data are plausible or suspicious are usually domain- and application-specific and require expert knowledge. It is important that a knowledgeable person looks at different data components from different perspectives and in different combinations, using a variety of potentially useful visual representations and corresponding interactive controls from those described in Section 3.

Outliers regarded as errors need to be removed from the data or corrected, if such a possibility exists. It should be remembered that removal or correction of outliers may change the descriptive statistics substantially; therefore, they need to be recalculated. Moreover, visual representations of the value distributions, such as frequency histograms and scatterplots, may also change. It can happen that some data

---

[5] https://en.wikipedia.org/wiki/Normal_distribution

items that were earlier not identified as outliers, either statistically or visually, will now stand far apart from the others. Hence, the procedure of outlier detection needs to be repeated each time after some outliers have been detected and removed or corrected.

All considerations so far referred to **global outliers** that differ from all other items in a data set. In a distribution of some values or value combinations over a base with distances (see Section 2.3.1), such as space or time, a **local outlier** is an element of the overlay (i.e., a value or combination) that is substantially dissimilar from the overlay elements in its neighbourhood. An example is shown in Fig. 5.4, where the value in one of the London wards is much higher than everywhere else in the surrounding area. A map representation of values in comparison to a reference value (e.g. the average) enables identification of local outliers.

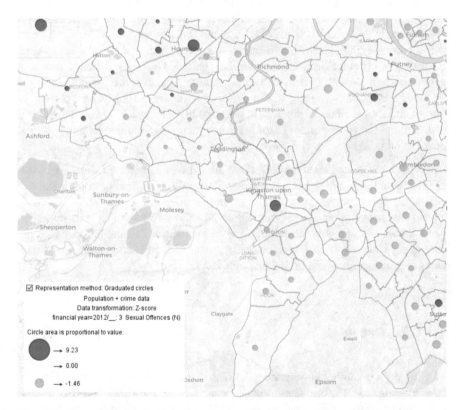

Fig. 5.4: An example of a local outlier in geographic space: a ward in Kingston is characterised by a high number of sexual offence accidents in 2012, but is surrounded by much safer areas.

The concept of neighbourhood can be defined not only for distribution bases with inherently existing distances between the elements (positions) but also in any set of items for which a suitable distance function can be specified (see Section 4.2). The neighbourhood of an item positioned in a base, in terms of the specified distance function, is defined as all positions in the base whose distances from the position of the given item are below a certain chosen threshold. The distance function can be used for data spatialisation by means of embedding (Section 4.4). The distribution of the data items in the embedding space can be represented by marks positioned in a projection plot, similarly to the circle marks on the map in Fig. 5.4. Local outliers will be manifested by marks dissimilar from the neighbouring marks in the embedding space.

When a data set is large, it may be difficult to detect local outliers in a distribution by purely visual inspection. To support the detection computationally, the differences of the data items – elements of the distribution overlay – from their neighbours in the base need to be expressed numerically. Again, a suitable distance function is needed for this purpose, that is, the function must be applicable to the elements of the distribution overlay. It may be the same function as is used for the base of the distribution, if the elements of the base and the overlay are of the same kind; otherwise, a different distance function is required. Using the distance function, the dissimilarities of all overlay elements to their neighbours are computed and the frequency distribution of these dissimilarities is examined. High-valued outliers of this frequency distribution potentially correspond to local outliers of the original distribution. The corresponding data items and their neighbourhoods in the original distribution need to be specially inspected.

Apart from global and local outliers, there may also be *unexpected values*, i.e., values that usually do not occur in a given place and/or time and/or combination of conditions. For example, a zero count of mobile phone calls in a business area on a weekday afternoon appears suspicious and needs to be investigated. It may happen due to a problem with the cell network infrastructure, or a public holiday, or an emergency situation, or be just an error in the data. Detection and inspection of such cases requires involvement of domain knowledge.

5.2.3 Missing data

Missing data is another important aspect of data quality that always requires a thorough investigation. Unfortunately, many visual representations, by their very design, cannot indicate that some data are missing; therefore, an analyst has no chance to detect the problem. Imagine a scatterplot displaying 1,000 data items with attributes a_1 and a_2. If 20 records have missing values of a_1, 10 records miss values of a_2, and 5 of them miss both values, only $1000 - 25 = 975$ dots will be plotted. When you see this plot, you cannot guess that 25 dots are missing. Like this scatterplot, other common visualisations cannot help you to detect that some data are missing, because they show the data that are present but do not show what is absent.

Apart from the existence of data items that miss values in some fields, it may happen that entire data items (records) that are supposed to be present in a dataset are absent. Missing records can occur in data that are collected and recorded in some regular way, for example, at equally spaced time moments or in nodes of a spatial grid. When some recordings have been skipped, e.g., due to a failure of the data recording equipment, the dataset will miss respective data items.

It is usually not a big problem for analysis when a few data items miss some field values or a few records are missing in a dataset. A problematic situation arises when a large portion of the data misses values in the fields that are important for the analysis and understanding of the subject, or when a large fraction of the supposed-to-be-there items are absent. In such cases, it is necessary to evaluate whether the remaining data are acceptable for the analysis. For the data to be acceptable, the following conditions must fulfil:

• The amount of the data must be sufficient for the analysis and modelling, i.e., contain sufficient number of instances for making valid inferences and conclusions.

• The coverage of the data must be sufficient for representing the analysis subject. This requirement may concern the area in space for which the data are available, the period in time, the age range of the individuals described by the data, and the like.

• The missing data must be randomly distributed over the supposed full dataset, so that you can treat the missing data as a randomly chosen sample from the whole data. Otherwise, the available data will be biased and thus not suitable for a valid analysis.

It is easy to count how many data records miss values in analysis-relevant fields and, hence, how many records remain for the analysis. This can be done, in particular, using functions of a database. It is also not difficult to obtain the ranges of field values, or value lists for categorical fields, for assessing the coverage. The spatial coverage can be examined using a map showing the bounding box of the available data items. A more accurate judgement can be supported by dividing the bounding box by a grid and representing the counts of the data items in the grid cells.

It may seem quite tricky to check whether the missing data constitute a random sample of the supposed whole dataset, which does not exist in reality. However, the requirement of the missing data to be random is equivalent to the requirement that the distributions of the missing and available data over analysis-relevant bases are similar, i.e., contain the same patterns. This, in turn, means that the *proportion of the missing data* is uniform over the base, i.e., is nearly constant in any part of the base. Thus, when data include references to time moments or intervals, the proportions of missing data should be approximately the same in different time intervals. The same requirement applies to different locations or areas in space when data are space-referenced, to different groups of entities when data refer to entities, and to different intervals of numeric attribute values when relationships of this attribute to other data components need to be studied.

Hence, investigation of the data suitability is done by comparing distributions of the available and missing data, or by investigating the distributions of the proportions of the missing data over different bases. This is where visualisation is of great help, because, as we explained in Chapter 2, the key value of visualisation in data analysis arise from its ability to display various kinds of distributions. The general approach is to partition the base of a distribution into subsets, such as intervals, compartments, groups of entities, etc., and obtain counts of the available data items for each partition, as well as for combinations of partitions from different bases.

Let us consider an example of a dataset describing mobile phone calls in Ivory Coast provided by the telecommunications company Orange for the "Data for Development" challenge [35]. The dataset includes records about mobile phone calls made during 20 weeks (140 days) in 2011-2012. Each record includes the date and time of the call and the position of the caller specified as a reference to one of 1231 cell phone antennas in the country. It is supposed that the dataset contains data about all calls performed within the given period.

In this dataset, the absence of records about phone calls can be detected by considering the counts of the recorded calls by the network cells and time intervals. We therefore aggregate the data into daily counts by the cells. Each cell receives a time series of the count values. Zero or very low values in this time series may indicate that all or a large part of data records for this cell and this day are missing, because our common sense tells us that people make phone calls every day. However, it is daunting to look at the individual time series of each of the 1231 cells. We shall instead look at the overall temporal patterns made by all time series.

In the upper time graph in Fig. 5.5, the individual time series are represented by thin grey lines; the thick black line connects the mean daily values. There were several days with unusually high counts of calls in many locations, especially on January 1, 2012. These peaks correspond to major holidays and are plausible therefore. At the right edge of the time graph, we see three drops of the mean line, which correspond to 3 days in April, specifically, 10th, 15th, and 19th. In these days, the call counts were everywhere substantially lower than usual, which is an indicator of missing data. When we look at the mean line more attentively, we notice that similar drops occurred also at other times. The time graph in the middle of Fig. 5.5 shows the individual time series of a few randomly selected cells. It can be seen that there were multiple days with no records. Moreover, there were periods of missing data extended over multiple days or even weeks.

To investigate the temporal distribution of the missing records more thoroughly, we generate a segmented time histogram, as shown at the bottom of Fig. 5.5. Each bar corresponds to one day, and the full bar length represents all 1231 cells. The bar segments represents the numbers of the cells whose daily counts of the phone calls fall into the class intervals 0-5, 5-10, 10-20, 20-50, 50-100, 100-200, 200-500,

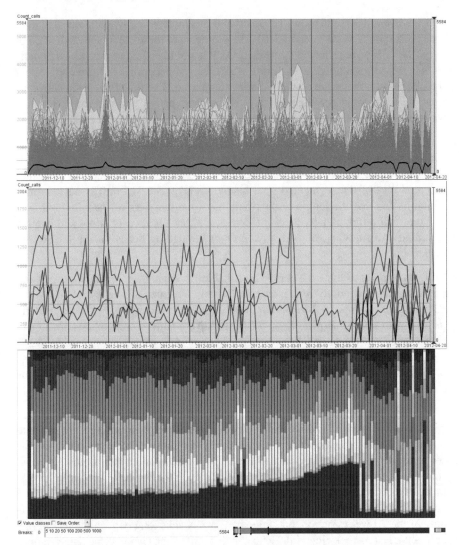

Fig. 5.5: The time graph at the top represents the time series of the daily call counts for 1231 cells. The time graph in the middle portrays the time series for 4 selected cells. The time histogram at the bottom shows the proportions of the cells with the daily call counts falling into given class intervals. Missing data are manifested as zero values of the call counts.

500-1000, and 1000 and more. The interval 0-5 is represented by dark blue and light blue corresponds to the interval 5-10. The segments of these colours mean that no or very few calls were reflected in the data. As it is not likely that nobody made phone calls in the corresponding cells during whole days, most probably, the records about the phone calls that were actually made are missing. The time histogram shows us multiple irregular occurrences of days when very many cells had no or very few records, and we also see an increasing trend in the number of the cells with no records starting from the beginning of the period until the last three weeks.

The temporal patterns we have observed signify that the lacunae in the recordings of the calls are not randomly scattered over the dataset. The height of the blue-coloured part of the segmented time histogram (Fig. 5.5, bottom) is not uniform, which means that the temporal distributions of the numbers of the cells with and without data differ. Hence, it can be concluded that the dataset is not suitable for revealing and studying long-term patterns of the calling behaviour on the overall territory.

Still, some parts of the dataset may be suitable for smaller-scale analyses focusing on particular areas and/or time intervals. To find such parts, it is necessary to investigate the spatio-temporal distribution of the gaps in the recordings. This can be done using a chart map as shown in Fig. 5.6. The map contains 2D time charts positioned at the centres of the phone network cells. The columns of the charts correspond to the 7 days of the week and the rows to the sequence of the weeks. The daily counts of the recorded phone calls are represented by colours. The dark blue colour means absence of recordings.

Obviously, the map with 1231 diagrams on it suffers from overplotting and occlusions. While it is easy to notice many dark blue rectangles signifying absence of data in the respective places, it is necessary to apply zooming and panning operations to see more details for smaller areas. In Fig. 5.7, there is an enlarged map fragment showing a business area of one of the major cities in the country, Abidjan. The majority of the phone cells in this area miss recorded data for the initial 17 weeks, and only the last 3 weeks of data are present. Some other cells have data since the beginning of the 20-weeks period, but there are 2 or 3 weeks long gaps in the recordings before the last 3-weeks interval. Hence, in this part of the territory, the data can only be suitable for analysing the phone calling activities of the people during the last 3 weeks. The consideration of the whole time period of 20 weeks is only possible for the relatively small area on the southwest of the country (Fig. 5.6), where days with missing recordings seem to occur sporadically and infrequently.

This example shows that missing data may occur systematically rather than occasionally. The presence of non-random patterns in the distribution of missing data signifies that the available dataset in its full extent is not acceptable for analysis, but it may contain parts suitable for analysis tasks with smaller scope. Such parts, if they exist, can be found by investigating the distributions of the missing and available data. Our last example demonstrated visual exploration of the spatial and spatio-temporal distributions. Many examples of approaches to investigation of other kinds

Fig. 5.6: A chart map representing the spatio-temporal distribution of the gaps in the data recording. The 2D time charts positioned at the centres of the phone network cells show the local temporal distributions of the phone call counts. The counts are represented by colours; dark blue means absence of recordings.

of distributions occur throughout the book, and many of these approaches can be used for investigating various distributions of missing data.

When cases of missing values or records occur occasionally and rarely, it is sometimes possible to estimate the likely values based on some assumed *model* of the data distribution. For example, assuming a smooth continuous distribution of attribute values over a base with distances, such as time or space, estimates for missing values can be derived (*imputed*) from known values in the neighbourhood of the missing values. Imputation of missing values in a numeric time series can also be done based on a periodic pattern observed in the distribution of the available values [61]. Assuming a certain interrelation between two or more numeric attributes,

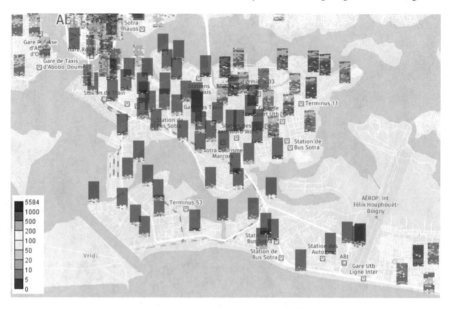

Fig. 5.7: An enlarged fragment of the map from Fig. 5.6 showing a part of the country's largest city Abidjan.

it may be possible to impute a missing value of one of them from known values of the other attributes. Imputation of missing values usually requires creation of a mathematical or computer model representing the distribution or interrelation. The model is built using the available data.

5.2.4 Impacts of data collection and integration procedures

The ways in which data are collected and processed may have a large impact on the patterns emerging in data distributions. Let us consider the example demonstrated in Fig. 5.8. The time histogram portrays data that are supposed to reflect the spread of an epidemic in some country. The bars in the histogram represent the daily numbers of the registered disease cases. We want to analyse how the epidemic was evolving and expect that the data tell us how many people got sick each day. However, the shape of the histogram indicates that the data, possibly, do not conform to our expectations. The strangest feature is the drop to a negative number in the last day. Quite suspicious are also two very high peaks and a day with no registered cases.

To understand what these features mean, we need to investigate how the data were collected. We find relevant information and learn that the data come from two groups of clinics: state clinics, which report the disease cases every day, and private clinics,

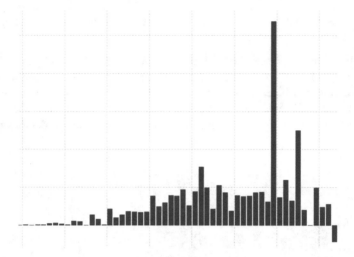

Fig. 5.8: The abrupt peaks and drops in a time histogram showing the number of reported cases of an epidemic disease emerge due to specifics of the data collection and integration.

which report their data from time to time. The two high peaks in the daily counts emerged due to arrival of two batches of data from the private clinics. The records from these batches were not assigned to the days in which the cases were registered but to the days in which the data arrived. The gap with the zero daily count near the right edge of the histogram corresponds to delayed reporting rather than absence of disease cases. The bar next to the gap is approximately twice longer than the bar before the gap and the following two bars because it represents the registered cases from two days. The negative count at the end reflects a correction of the data: some of the earlier registered disease cases turned out to be wrongly classified, i.e., the people did not have this particular disease.

Hence, the temporal distribution of the disease cases is distorted in the data by artefacts caused by data collection specifics. These data thus cannot be immediately used for the analysis. If we cannot get better data, we need to make corrections in these data taking some assumptions. One assumption we can take is that the numbers of the disease cases identified in the private clinics are distributed over time in the same way as the disease cases identified in the state clinics. Under this assumption, the data records from the two batches that came from the private clinics can be spread over the previous days proportionally to the counts of the cases in the state clinics. Another assumption is that the fraction of the wrongly classified cases is approximately the same in each day. Bases on this assumption, we can compute the ratio of the absolute value of the negative count to the total number of disease cases and decrease each daily count by the resulting fraction. The gap with zero records can be filled by dividing the next day's count into two days. After all these corrections, we can hope that the data give an approximate account of how the epidemic was developing.

Generally, it happens quite often that problems in data are caused by insufficiently careful integration of data coming from different sources. Thus, when data come from spatially distributed collecting units (such as clinics, radars, or other kinds of sensors) that observe certain areas around them, an integrated dataset may have spatial gaps due to missing data from some of the units, as demonstrated in Fig. 5.9.

Fig. 5.9: Spatial coverage gaps in a dataset obtained by integration of records from multiple radars.

When the "areas of responsibility" of different collecting units overlap, it may happen that the same information is reflected two or more times in the data. It is not always easy to understand that two data items are meant to represent the same, because these items are not necessarily identical. There may be discrepancies due to measurement errors or differences in the times when the measurements were taken or recorded. Specific problems arise when the data refer to some entities represented by identifiers. Sometimes, different collecting units may use the same identifiers for referring to different entities. It is also possible that different identifiers are used for referring to the same entity. Detecting such problems usually requires domain knowledge: you need to know the nature of the things or phenomena reflected in the data in order to understand how good data *should* look like and be able to uncover abnormalities.

Let us consider the example shown in Fig. 5.10. Here, spatial positions of two moving objects (namely, two cars) were recorded independently by two positioning devices. When the data from multiple positioning devices are put together in a single

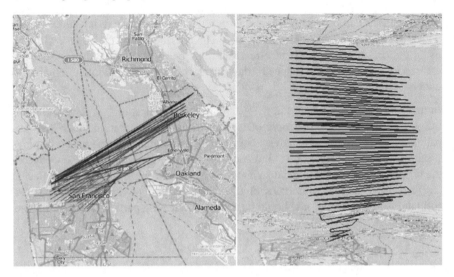

Fig. 5.10: When two simultaneously moving objects are denoted in data by the same identifier, connecting the positions in the temporal order produces a false trajectory with a zigzagged shape and unrealistically distant moves between consecutive time steps. The two images show the appearance of such a "trajectory" on a map (left) and in a space-time cube (right).

dataset, the outputs of different devices are distinguished by attaching unique identifiers to their records. It somehow happened that identical identifiers were attached to the records coming from these two devices. Trajectories of moving objects are reconstructed by arranging records with coinciding identifiers in the chronological order. When this procedure was applied to the records having the same identifiers but originating from distinct devices, the positions of the different moving objects were mixed, which produced a false trajectory with an unrealistic zigzagged shape. We can easily recognise the abnormality of this shape because we have background knowledge of how cars can move.

When you have a very large dataset with positions of many moving objects, it may be hard to detect such abnormalities from a visual display showing all trajectories. Fortunately, we also have other background knowledge about car movement; particularly, we know what speed values can be reached by a car. We can do a simple computation of the speed in each position by dividing the spatial distance to the next position by the length of the time interval between the positions. In a false trajectory composed of positions of two or more cars, temporally consecutive points are distant in space but close in time; therefore, the computed speed values will be unrealistically high. High values of the computed speed can also result from occasional erroneously measured positions. To distinguish trajectories with occasional errors from trajectories composed of positions of multiple objects, we can inspect the distribution of the counts of the positions with too high speed values per trajectory.

Such positions can be selected by a query in which we specify a threshold beyond which the speed is considered unrealistic. The distribution of the counts will give us an indication of the kinds of errors that may be present in the dataset, and we can separately look at the shapes of the trajectories with high counts to obtain full understanding.

In this subsection, we wanted to demonstrate that visualisation can reveal artefacts introduced in data by procedures of data collection or integration. Such artefacts can be detected by human analysts because their appearance in the visual displays disagrees with analysts' knowledge and expectations. The analysts' knowledge of the phenomena reflected in the data and the ways in which the data can be collected suggests them what distributions should be inspected and, possibly, what data transformations and/or selections should be made for detecting particular kinds of errors and artefacts.

5.3 Processing data

After data properties are investigated and understood, some kind of data processing may be required to make the data suitable for further analysis. In this section, we give a brief overview of the major data processing operations. Not all of them can benefit from using visualisations; however, there are operations in which a human analyst needs to make informed decisions, and visual displays are required for conveying relevant information to the human. We shall consider such operations in more detail.

5.3.1 Data cleaning

During the recent decade, a number of commercial and free tools for data cleaning have been developed. Most of them apply a combination of visual and computational techniques for identifying data problems and fixing them. Representative examples of such tools are Trifacta data wrangler[6] and OpenRefine (formerly Google Refine)[7]. These tools focus mostly on tabular data, supporting such operations as data cleaning in columns containing numeric values (detecting and replacing dummy values, ensuring formatting consistency, detecting outliers), dates and times (fixing formatting issues, adjusting time zones, changing temporal resolution), and texts (fixing misspelled words, abbreviations, etc.). Table columns are usually considered independently. Histograms are commonly used for assessing distributions of

---

[6] https://www.trifacta.com/
[7] https://openrefine.org/

numeric attributes. Interactive query tools are used for inspecting outliers and fixing errors.

5.3.2 Modelling for data preparation

In discussing the problem of missing data (Section 5.2.3), we mentioned that it may be possible to impute missing values based on some assumed model of the data distribution. Data modelling may be helpful not only for reconstruction of missing data but also for detection and correction of errors in data and for adaptation of the data sampling to the requirements of the following analysis.

As an example, let us consider *map matching*[8] of GPS-tracked mobility data. GPS positioning results may contain errors and/or temporal gaps between recorded positions (i.e., missing data), especially when the visibility of satellites is restricted. For supporting transportation applications, it is often necessary to snap GPS coordinates to streets or roads. A straightforward geometry-based approach with finding the nearest street segment for each point in the data may not work well. For example, an erroneous position may be located closer to a street lying aside of the movement path than to the street where a vehicle was actually moving. A more appropriate but computationally costly approach is to compute a plausible route through the street network that matches the majority of the tracked positions, and then snap the positions to the segments of this route.

This transformation provides several benefits. First, erroneously recorded positions lying beyond streets of on wrong streets are replaced by plausible corrected position. Second, after linking the resulting positions by lines, it becomes possible to determine the likely intermediate positions between the recorded positions. This allows one to increase the temporal and spatial resolution of the data (i.e., generate more data records with smaller temporal and spatial distances between them) and/or to create a dataset with regular temporal sampling (i.e., a constant time interval between consecutive records).

This example demonstrates the following general idea: from a set of discrete data records, derive a mathematical or computer model representing the variation or distribution of the data as a *continuous function*, and use this model for (a) filling gaps in the data, i.e., reconstructing missing values, (b) replacing outlying values (possible errors) by values given by the model, and (c) creation of a dataset with regular spacing (distances) between data items within some distribution base with distances, such as time or space.

---

[8] https://en.wikipedia.org/wiki/Map_matching

The process of construction of a continuous function representing some distribution, particularly, temporal or spatial, is called *smoothing*[9]. The process of creating additional data items positioned in a distribution base between existing data items is called *interpolation*[10]. The process of creating a dataset with a regular (equidistant) distribution of the data items over a base with distances is called *resampling*[11]. A need for data resampling may arise because many methods for time series analysis and for spatial analysis require all data items to be uniformly distributed and equidistant in time and space, respectively, while real data do not always satisfy this requirement.

In terms of the concepts introduced in Section 2.3.1, smoothing and interpolation complement the overlay of a distribution from a discrete set of sampled data items to a set covering the complete base of the distribution. Resampling picks a subset of equidistant elements from the complete base of the distribution; for these elements, the corresponding elements of the reconstructed complete overlay are taken.

Figure 5.11 demonstrates how the general idea of data modelling applies to a numeric time series. Methods for time series smoothing will be considered in Chapter 8.

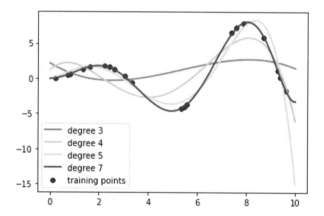

Fig. 5.11: Approximation of a set of observations by a smooth function.

Obviously, domain knowledge and experience of a human expert need to be involved in the process of building a data model. Visually-driven approaches are suitable for this purpose.

---

[9] https://en.wikipedia.org/wiki/Smoothing

[10] https://en.wikipedia.org/wiki/Interpolation

[11] https://en.wikipedia.org/wiki/Resampling_(statistics)

5.3.3 Transformation of data elements

After cleaning data from errors and outliers and/or reconstructing missing values, field values may still need to be improved for facilitating analysis. If data items include values of multiple attributes, it may be useful to bring them to comparable ranges. For this purpose, *normalisation* and *scaling* techniques can be applied, see Section 4.2.9.

Another potentially useful operation is binning, or *discretisation*, that transforms a numeric attribute with a continuous value domain into an attribute with ordered categorical values. We have mentioned the concept of discretisation in Chapter 3 (particularly, in Section 3.3.3) in relation to defining classes of values of a numeric attribute for visual encoding in choropleth maps (Fig. 3.9) and 2D time charts (Fig. 3.12). There exist a wide range of approaches to discretisation. The most popular are *equal intervals* (having the same widths of the bins but a varying number of data items in each bin) and *equal quantiles*, or equal size (having the same amount of data items in each bin, but varying bin widths). Discretisation can be applied either to raw data or to results of normalisation. For example, some analysis tasks may benefit from equal interval binning with the interval length (i.e., the bin width) equal to the attribute's standard deviation.

It should always be kept in mind that the way in which data are discretised affects the distribution of the resulting values, Figures 2.5 and 3.13 demonstrate how the choice of the bin widths in a frequency histogram can affect the visible patterns of a frequency distribution. Figure 3.12 demonstrates how the pattern of a temporal distribution can depend on the definition of attribute value classes. Discretisation can also affect spatial distribution patterns, particularly, patterns visible on a choropleth map.

5.3.4 Synthesis of data components

When data consist of multiple components, it may be useful or necessary for analysis to create additional components by synthesising the existing ones. A simple example is assessing the overall magnitudes of numeric time series based on the maximal, average, or total value. This creates a new data component, namely, a numeric attribute, such that each time series is characterised by a single value of this attribute derived from the multiple values originally present in this time series. Even in this simple example, it can be useful to involve human reasoning supported by visual analytics approaches, so that only relevant components are taken into account in the synthesis process. Thus, an analyst may want to assess time series of traffic counts according to the values attained in rush hours of working days. This requires careful selection of days that were working (excluding holidays, special events etc.)

and hours of intensive traffic (taking into account seasonal changes, daylight saving time, particular traffic patterns on Fridays, and other possible sources of variation).

Another example is determining the dominant attribute (i.e., the one with the highest value) among multiple attributes. Obviously, the original attributes need to be comparable either by their nature (e.g. proportions of different ethnic groups in the total population of geographic areas) or as a result of data standardisation, such as transformation to z-scores, see Section 4.2.9. Even a simple operation of finding the attribute with the highest value in real-world data may require a human expert to make informed decisions: What to do when all values are almost equal? What amount of difference between the highest value and the next one is sufficient? How to handle cases when all attributes have very small values? To make such decisions and obtain meaningful results of the synthesis, the analyst needs to analyse the distributions of the attribute values and their differences and to see the effects of possible decisions on the outcomes. This activity requires the use of visual displays and interaction operations [8].

A need in synthesising multiple attributes can arise in the context of *multi-criteria decision analysis* [95, 75], where *decision options* are described by a number of attributes used as *criteria*. Criteria whose low values are preferred to higher values are called *cost criteria*, and those whose high values are preferred are called *benefit criteria*. There may be also criteria whose average values are desirable. More generally, one can define a certain *utility function* assigning degrees of utility to different values of a criterion. By means of this function, the original attributes describing the options are transformed into a group of comparable attributes whose values are the utility degrees. These derived attributes are then used for comparing the options and choosing the best of them.

It happens very rarely that there exists a decision option with the highest possible utility degrees for all criteria. Usually, each option has its advantages and disadvantages, and it is necessary to select such an option that the advantages overweight the disadvantages. A typical approach is to combine the values of the transformed criteria into a single utility value and then select one or a few options with the highest combined values. The criteria may differ in their importance; these differences are represented by assigning numeric weights to the criteria. Having the criteria-wise utility degrees and the weights of the criteria, the overall utility score of each option can be calculated as a *weighted linear combination* of the individual utility degrees for the criteria. There are also more sophisticated methods for combining multiple utility degrees, but all of them use numeric weights, which are supposed to be assigned to the criteria by a human expert.

A problem hides here: it is usually hard for a human expert to express the relative importance of criteria in exact numbers. Should the weight of this criterion be 0.3 or 0.33? It may be tricky to decide, whereas a slight change of the criteria weights may have a significant impact on the scores and ranks of the options. To make a reasonable final choice, the analyst needs to be aware of the consequences of small changes of the criteria weights. Such awareness can be gained by performing *sensitivity analysis* of the option ranks with respect to the criteria weights. As any analysis process where human reasoning and judgement is critical, it should be supported by visual analytics techniques, as, for example, proposed in paper [16]. The analyst can interactively change any of the weights and immediately observe the changes of the option scores and ranks, which are shown in a visual display. There is also a more automated and thus more scalable way of the analysis. The analyst specifies the variation ranges of the criteria weights and the steps by which to change the weights. In response, a computational tool calculates the option scores and ranks for all possible combinations of the weights taken from the specified ranges. The summary statistics of the resulting scores and ranks (minimum, maximum, mean, median, quartiles) are then represented visually, so that the analyst can find one or more options having robustly high scores. Instead of the summary statistics, frequency histograms could be used. Their advantage is a more detailed representation of the distributions. Thus, a bimodal distribution (which would be very strange to encounter in this context) can be recognised from a histogram but not from the summary statistics.

Synthesis of multiple attributes into integrated scores or ranks is quite a widespread procedure, which is applied in evaluating students, universities, countries, living conditions, and many other things. In all cases, weighting is applied for expressing the relative importance of different attributes. It is often useful to investigate how the final scores depend on these weights.

To summarise, synthesis of multiple data components often requires taking certain decisions and analysing the consequences of different possible decisions. This requires involvement of human reasoning and judgement; therefore, the use of visual analytics techniques is highly appropriate. Their primary role is to allow the analyst to make changes in the settings and inputs of the synthesis method and observe the effects of these changes on the synthesis results. In complex cases, when there are too many possible settings to be tested, the analysis of the sensitivity of the results to the settings is supported by combining automated calculations of a large number of possible outcomes with visualisation of the distributions of these outcomes.

5.3.5 Data integration

Data required for analysis may be originally available in several parts contained in different sources. These parts need to be integrated. For example, one database table describes salaries of individuals over time, and another table lists employees per department. If a goal of analysis is to study the dynamics of salaries per department, it is necessary to integrate these two tables. Standard database operations[12][13] can be

---

[12] https://en.wikipedia.org/wiki/Join_(SQL)

[13] https://en.wikipedia.org/wiki/Merge_(SQL)

applied for this purpose. Before linking the data, it is necessary to check whether there is a correspondence between the identities of the individuals specified in the two tables.

There may be more complex cases when data integration is not limited to simple key-based join. For example, sophisticated spatial queries need to be used for combining population data with land use information in order to assess more accurately the distribution of the population density. The integration enables exclusion of unpopulated areas (e.g. water bodies, agricultural lands etc.) from the areas for which population counts are specified. This approach is called *dasymetric mapping*[14])

Data integration includes not only operations on pre-existing database tables or data files but may involve the use of external services for obtaining additional data. Data wrangling software (e.g. OpenRefine) can automate the process of obtaining geographic coordinates based on addresses by means of online services for geo-coding. Similarly, named entity recognition services can be used for extracting names of places, persons, organisations, etc. from texts. Language detection tools may help to determine the languages of given texts. Sentiment analysis can be applied for assessing emotions in texts.

Open linked data[15] technologies can support data integration technically. All these tools extend available data by deriving additional attributes for the existing data items. However, such new attributes need to be checked for possible problems similarly to the original data.

5.3.6 Transformation of data structure

An example of data structure transformation appeared in Section 5.2.3. The original data describing individual phone call events were transformed into spatially referenced time series of call counts by means of spatio-temporal aggregation. In relational databases, such transformations are done by means of GROUP BY queries, which characterise groups of table rows by the counts of the rows and various statistical summaries of attribute values, such as the sum, average, minimum, maximum, etc. In spreadsheets, similar functionality is available via *pivoting*. The languages for statistical processing and data science R and Python include a DataFrame structure that enables very sophisticated and flexible pivoting operations[16].

Possible transformations are not limited to generic operations on rows and columns of flat tables. There are transformations specific to data of a certain nature. Imagine

---

[14] https://en.wikipedia.org/wiki/Dasymetric_map

[15] https://en.wikipedia.org/wiki/Linked_data

[16]   https://pandas.pydata.org/pandas-docs/stable/reference/api/pandas.DataFrame.pivot_table.html

a data set consisting of time-referenced geographic positions of vehicles tracked over some time period. The chronological sequence of positions of each vehicle can be treated as a *trajectory* of this vehicle. Having a trajectory, it is possible to extract *stops*, i.e., events when the position of the vehicle did not change during sufficiently long time. For this transformation, it is necessary to define the minimal significant duration of staying in the same position, and it may also be necessary to specify a tolerance threshold for ignoring minor variations of the recorded spatial positions, which may happen due to measurement errors. Taking decisions concerning the temporal and spatial threshold values and assessing the results of these decisions require visualisation support.

A further transformation can be applied to the extracted stop events: by means of density-based clustering (Section 4.5.1), areas of repeated stops can be identified. For these areas, the stop data can be aggregated into time series of stop counts. As explained in Section 4.5, the process of clustering requires visualisation support. Further examples of transformations of spatio-temporal data will be considered in Chapter 10, and transformations of temporal data are discussed in Chapter 8.

Doing data transformations usually requires a number of decisions to be made by the analyst. Thus, the most popular type of transformation, aggregation, requires defining, first, the groups of data items or the bins in the numeric, temporal, or spatial domains and, second, the kind(s) of summary statistics to be computed for the groups or bins. Not only the immediate result of the transformation will depend on these decisions but also what can be derived from the transformed data in the following process of analysis and, eventually, model building. There are plentiful examples of how statistics may become misleading or even lying, either occasionally or intentionally. Let us consider a simple example of two-level aggregation of voting results for two candidates A and B with a simple majority voting at each level. Imagine 5 districts with the distributions of the votes 2:1, 0:3, 2:1, 0:3 and 2:1. This gives preferences to A, B, A, B, and A, respectively. The overall winner is A, because the final score derived from the first-level preferences is 3:2 in favour of A. However, if the data were aggregated by counting all primary votes, the score would be 6:9 in favour of B. Figure 5.12 illustrates this example. Interestingly, this toy example reflects the reality of presidential elections in some countries[17], with more details explained elsewhere[18].

We point interested audience to excellent books demonstrating how statistics and data aggregation can lie [72, 100]. To avoid fooling yourself by inappropriately transformed data, you should be very careful in making your decisions. For this purpose, you need to explore the properties and distributions of the data with the help of informative visualisations.

---

[17]   https://en.wikipedia.org/wiki/United_States_presidential_
elections_in_which_the_winner_lost_the_popular_vote
[18] https://en.wikipedia.org/wiki/United_States_Electoral_College#
Contemporary_issues

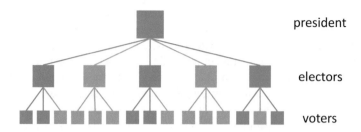

Fig. 5.12: A two-level aggregation procedure applied to primary votes for two candidates can give the victory to a candidate supported by fewer voters than the other candidate.

5.3.7 Data reduction and selection

There may be many possible reasons for data reduction and selection:

- some available data are not relevant;
- some available data are outdated;
- data are too numerous to be handled by the analysis tools;
- data are excessively detailed;
- data are excessively precise, while a lower precision is sufficient for analysis.

Data filtering is the tool for removing irrelevant or outdated data from the dataset to be used in the analysis. As described in Section 3.3.3, filtering can remove data items with unsuitable attribute values, or data referring to times or spatial locations beyond the period or area of interest, or aged data that may not be valid anymore. Filtering can also be used for selecting manageable portions of data in order to test whether the planned approach to the analysis will work as expected, or to do the analysis portion-wise. In this case, it is necessary to take care that the selection is reasonable according to the domain specifics as well as common sense. For example, when data portions are selected based on time references, it may not be very meaningful simply to take continuous time intervals containing equal numbers of records. It may be more appropriate to take time intervals including the same number of full days or weeks, or to do the selection based on the positions in the time cycles (e.g., analyse separately mornings and evenings). When the data represent some activities, such as journeys, or events that happen during some time, such as public events, it is reasonable to ensure that the selection does not contain unfinished activities or events. Sensible decisions on how to select or divide data need to be taken by a human expert, who needs to be supported by visualisations showing the consequences of different possible decisions.

Opposite to deterministic filtering, **data sampling** takes a random subset of data items of a requested size. When performing sampling, it is necessary to ensure that the sample is representative for data and task, i.e., the patterns of the task-relevant distribution are the same as in the full dataset. Let us consider an example of sampling movement data consisting of position records of different vehicles over time. Such data set can be sampled on the level of (a) recorded positions, (b) complete trips between stops, (c) all trajectories of a sample of vehicles, (d) trajectories of selected types of vehicles, etc. The approach to sampling is, obviously, chosen by a human analyst. Irrespective of the chosen approach, it is useful to overview the data distributions in several samples and check if the data properties and patterns are consistent.

A general approach to reducing excessively detailed data is **aggregation**, in which groups of data items are replaced by single items. In this process, it is necessary, first, to select a suitable method of grouping, so that the items in a group are similar in terms of attributes and/or close in space and/or in time. Second, it is necessary to select an appropriate aggregation operator: whether it should be the mean, or median, or minimum, or maximum, or a certain percentile. Again, this requires human reasoning and informed decision making, which needs to be supported by visual analytics techniques.

For certain kinds of data, reducing excessive details can be achieved by means of **modelling** and **resampling**, as discussed in Section 5.3.2, or using specific **simplification** techniques, for example, simplification of shapes of lines and polygons for representing them by fewer points. In such cases, the degree of simplification and the appropriateness of the results needs to be controlled by a human expert through method parameters. As in other cases, the impacts of different parameter settings need to be visualised and examined by the expert.

When reducing data precision, it is necessary to ensure that the data remain suitable for analysis. Figure 5.1 presents an example of the negative effect of precision loss in geographic coordinates. Another example is a comb-shaped frequency histogram caused by reduced precision of data values, see Table 5.1 for the example and its explanation. It is always necessary to control changes in data precision through visual analytics tools.

5.4 Concluding remarks

In this chapter, we argued that data properties need to be studied and understood with the help of visual displays showing distributions of data items with respect to different components: value domains of attributes, time line and time cycles, space, and sets of entities. Human analysts use these displays to check if the distributions appear consistent with their background knowledge and, if inconsistencies are noticed, understand their reasons. We demonstrated how visual displays can be helpful in detecting different problems in data quality. Visualisation of distributions can reveal outliers, unrealistic items and combinations of items, gaps in coverage of a distribution base, biases, and wrong references to entities, time moments, or spatial locations. Such problems may sometimes make the data unsuitable for a valid analysis, but even when they are not that severe, they need to be taken into account in performing the analysis.

Data suitability for intended analysis depends not only on the quality but also other features, such as regularity, resolution, comparability of field values, structure, and presence of required kinds of information. To become suited for the analysis, data may need to undergo preparatory operations, some of which may require involvement of human knowledge, interpretation, and reasoned decision making. In all such cases, visual displays are necessary to inform the humans, support their cognitive activities, and evaluate and compare the outcomes of possible alternative decisions.

In the following chapters, we shall mainly focus on the use of visualisations for studying and understanding various kinds of real-world phenomena, but we shall also discuss the structure and properties of data reflecting these kinds of phenomena and mention possible specific problems in such data.

5.5 Questions and exercises

- Does a box plot convey sufficient information about properties of a set of numeric values?

- What features in histogram shapes require special attention in the context of exploring data quality?

- What is a local outlier in a numeric time series and how can it be detected?

- When some data are missing in a dataset, what needs to be checked for judging whether the available data are suitable for a valid analysis? How can this be checked?

- What data transformations can alter properties of data distributions? How to avoid such alterations?

Chapter 6

Visual Analytics for Understanding Multiple Attributes

Abstract One very common challenge that every data scientists has to deal with is to make sense of data sets with many attributes, where "many" can sometimes be tens, sometimes hundreds, and even thousands. Whether your goal is to do exploratory analysis on the relationships between the attributes, or to build models of the underlying phenomena, working with many dimensions is not trivial. The high number of attributes is a barrier against using some of the standard visual representations: just try to imagine a scatterplot matrix where you want to look at the pairwise distributions combinations of 100 variables. Moreover, any computational method that you apply produces results that are challenging to interpret. Even linear regression, one of the easiest models to understand, becomes quite complex if you need to investigate the interactions between hundreds of variables. This chapter will discuss how not to get lost in these high-dimensional spaces and how visual analytics techniques can help you navigate your way through.

6.1 Motivating example

You might think that you know your city well and you have a good idea which neighbourhoods are similar and which are different. Let's assume that you live in London, in a neighbourhood called Hackney, which is full of bars and coffee shops and trendy looking youngsters who are likely early in their careers or still studying. You could probably tell how Hackney is different to Westminster, which is full of governmental buildings and businesses and people who live there are pretty wealthy. Here, you are distinguishing these two boroughs by the dominant kinds of businesses, and the age and average (likely) income of its residents. These are relevant observations, but these aspects are only a small portion of different characteristics one can think of. What if I ask you to think of 72 different characteristics of the London boroughs all at the same time, including the "number of cars per household

© Springer Nature Switzerland AG 2020
N. Andrienko et al., *Visual Analytics for Data Scientists*,
https://doi.org/10.1007/978-3-030-56146-8_6

of residents", "turnout at 2014 local elections", "ambulance incidents per hundred population" and so on? How would you then define similarity and find those that are similar in some way? Which characteristics will you "value" more in describing the boroughs? Are all 72 characteristics equally informative or descriptive? These are some of the questions that are representative of the kinds of challenges one faces when working with data sets with many attributes. And the answer to such "multidimensional" questions is never straightforward and requires us to make use of computational and visual methods hand in hand.

Let's take the London boroughs example and look how we can approach such a scenario. We use a publicly available dataset from the London Data Store[1]. The data include 72 unique attributes describing characteristics of the 33 London boroughs as we mentioned above. For simplicity, we only consider the numeric attributes, 69 of them. In our exploration, we will make use of data embedding (Section 4.4) performed by means of a non-linear dimension reduction algorithm called Sammon mapping, or Sammon projection [120]. We feed the whole data to the projection algorithm, which produces the result shown in Fig. 6.1. We can see how certain boroughs stand out and how some are in close proximity in the embedding space. *City of London* stands far out from the rest, while *Newham* lies in the opposite corner, which, probably, means that it is the most dissimilar to the City of London.

What we can't understand from the above result is in what ways these boroughs are similar or different. Moreover, since we used all the attributes in the embedding method, our understanding of the result is complicated even further. This is because the attributes used are a mixture of social, economic, demographic, and other characteristics, and it is hard to tell which factors are more influential. We will discuss some visual analytics techniques later to see how we can approach this, but for now, let's take a simpler step. Seeing that the data set contains information on the "emotional" state of the residents (the data were gathered through the Annual Population Survey by Office for National Statistics, UK[2]) through variables such as "Life Satisfaction Score", "Feeling Worthwhileness Score" or "Feeling Anxious Score", we produce a new Sammon projection with this selected "emotion" subset; see Fig. 6.2.

We are seeing quite a different picture now. *City of London* is now standing even farther apart from the rest. *Kensington and Chelsea* together with *Barnet* are on the opposite side of the plot. *Westminster* was previously separated from the others, but now it is in the core of the dense cluster, which means that it is very similar to many others. *Newham* and *Tower Hamlets* are not opposite to City of London anymore and also join the dense cluster. To unravel, to a certain extent, the meaning of the coordinates in the projection space, we colour the dots according to the values of the "Happiness Score" attribute, so that dark green means high reported happiness and paler greens are indicative of a grimmer status. Now we see that *City of London* has

---

[1] https://data.london.gov.uk/dataset/london-borough-profiles
[2]   http://data.london.gov.uk/dataset/subjective-personal-well-being-borough

Fig. 6.1: Sammon mapping applied to the 33 London boroughs using data that describe the boroughs in terms of 69 numerical features.

the lowest happiness score while the differences among the remaining boroughs are not very high. *Kensington and Chelsea*, which lies opposite to City of London, has the highest happiness score, and *Hackney*, lying on the other side of the cluster, has the second lowest score after City of London but not as much different from those of the remaining boroughs. Generally, we can see that the relative arrangement of the boroughs in the projection plot is consistent with the values of the happiness score. This may mean that either this attribute has high impact on the result or that it is correlated with the other "emotional" attributes.

This example only scratches the surface of what we could be doing in such an analysis. It would be appropriate to investigate whether the attributes are interrelated and whether some of them are non-informative or more informative. We have seen that, depending on which aspects you consider (e.g., our manual selection of "emotional" attributes), you get very different observations. Think of all possible combinations that can be taken from the 69 attributes – the full space of possibilities is practically intractable. The question is which combinations are meaningful and worth considering. This section now looks at visual analytics techniques to help us work with such high-dimensional data sets and enable us to answer a wider variety of questions, and eventually reach deeper insights.

Fig. 6.2: Sammon mapping applied to the 33 London boroughs using only 4 attributes that relate to the emotional state of the residents.

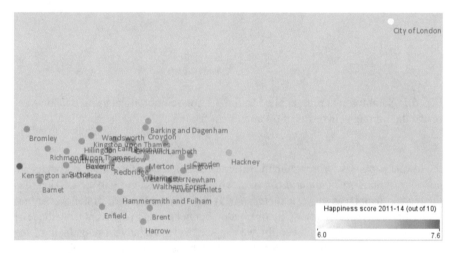

Fig. 6.3: The same result as in Figure 6.2 with the colour indicating the "Happiness Score" values, darker green meaning higher reported happiness.

6.2 Specifics of multivariate data

The data sets that we deal with here are those sets where for each data record we have several attributes. Depending on the context, the number of attributes can nowadays easily reach hundreds and even thousands. Most of the methods available to data scientists are not geared for such high numbers of attributes, e.g., even simple linear regression models become hard to interpret once you try to build it with hundreds of independent variables. Moreover, statistical analysis tools often come with many assumptions on the distributional characteristics of the attributes, which are rarely met in practice. Visualisation methods alone are also hardly helpful here, since the visual representation techniques are mostly designed for data with low dimensionality. For instance, scatterplot matrices are used widely when investigating the relations within pairs of a dozen or at most a few dozens of attributes but would lack screen space if the number of attributes is larger.

Due to these inherent characteristics, neither the visual nor the computational approaches alone are sufficient for the effective analysis of high dimensional data sets, and iterative visual analytics approaches combining visualisations with computations are needed. Here we list a number of specific challenges that are common for this kind of data:

• **High amount of attributes:** The large number of attributes makes the analysis of data challenging in many ways. Most of the visual representations are designed with fewer dimensions in mind, and visual representations do not scale well to capture such complexity. Most often, a computational method is needed to reduce the dimensionality; such methods have been described in Section 4.4. However, the use of this approach entails further challenges as discussed below.

• **Heterogeneous characteristics of attributes:** The space of the attributes is often heterogeneous, with individual attributes carrying distinctive characteristics: some could be continuous, others could be categorical; numeric attributes may greatly differ in their value ranges and distribution properties. Many algorithms make assumptions regarding the underlying characteristics of the data, e.g., PCA assumes that the attributes are normally distributed. These algorithms will not be able to produce satisfactory results if/when categorical attributes are fed into them or if the assumptions concerning the distributions are not met.

• **Local structures and interdependencies:** It is often the case that there are correlations and dependencies between the attributes. These relationships skew the results of the computational methods and decrease their validity. Besides, without carefully considering the interdependencies, you will be dealing with significant redundancies within your set of attributes. Not only the excessive amount of attributes is a problem but also the risk that multiple attributes representing, in essence, the same aspect of a phenomenon will dominate and prevent other aspects from being properly taken into account. This problem has been discussed in Section 4.3.

6.3 Analytical Goals and Tasks

The typical high level analytical goals referring to high dimensional data analysis include:

- **Investigating relationships:** As discussed above, there often are inherent relationships between the attributes. For instance, consider the variables we used in the example in Fig.6.1. It is not unusual to expect some degree of dependency between the survey responses for the "happiness" and "life satisfaction" scores. One fundamental task therefore is to discover and study the relationships between the attributes, which is important for understanding and modelling the phenomenon these attributes describe.

- **Feature selection:** One natural follow-up task to the above is to filter the attributes for removing redundancies and selecting what is important. The selection requires ranking the features according to different criteria.

- **Formulating and modelling multi-variate relationships:** It may be necessary not only to discover and understand the existing relationships between the attributes but also to formally represent and model them. This can be done by building mathematical or computer models, such as linear regression or support vector machines. Such a model can eventually serve an analyst as a predictor, classifier, or an explainer.

- **Studying data distribution in the multidimensional space:** Basically, the goal is to understand the variation of the multi-attribute characteristics (i.e., combinations of attribute values) over a set of items described by these attributes. This goal can be re-phrased in the following way: imagining a multidimensional space of all possible value combinations, the goal is to understand which parts of the space are more populated than others and which are empty, whether there are groupings of close items in this space and isolated items distant from all others (outliers), whether there exists a concentration of items in some part of the space and a decreasing density trend outwards from this part. Basically, analysts may be interested in finding and studying the same kinds of distribution patterns as in a usual two-dimensional space; see Section 2.3.3 and examples of spatial distribution patterns in Fig. 2.7.

6.4 Visual Analytics Techniques

Here, we organise the visual analysis techniques according to the high-level capabilities they provide within the course of an analysis process.

6.4.1 Analysing characteristics of multiple attributes

Small multiples of attribute profiles: When you have data with multiple attributes, it often makes sense to begin with considering the value frequency distribution of each of them. This can be done by creating multiple frequency histograms, as in the example shown in Fig. 2.11. That example demonstrates that it may be possible not only to examine the value distribution of each individual attribute but also, by means of interaction, explore the interrelations between the attributes. The interaction techniques needed for supporting this exploration are selection and highlighting of data subsets, as described in Section 3.3.4. The idea is that differently coloured segments of the histogram bars show the value distribution in the selected data subset. Another example of the use of this approach can be found in the paper by Krause et al [85].

Attribute Spaces: One can easily imagine that even using the most space-efficient visual representation for each dimension would not scale (in terms of the screen space or perceptually) once we start working with very high numbers of dimensions. A promising approach is creation of so-called attribute spaces [140]. The main idea is that the value distributions of the attributes are characterised by a number of numeric features, such as skewness, kurtosis, and other descriptive statistics. This creates, additionally to the data space, where the data items are represented by points positioned according to their values of the attributes, an attribute space, where the attributes are represented by points positioned according to their distributional features. For this attribute space, one can use the same visualisation techniques as for the data space. For example, one can create a scatterplot where two attribute features form the axes and the attributes are represented by dots positioned according to the values of these features.

In the example in Fig. 6.4a, we see a visualisation of two dimensions of the attribute space (left, yellow background) with more than 300 attributes. The data we are looking at here is from a longitudinal study of cognitive ageing that involved 82 participants with 373 numeric attributes derived from the imaging of the brain using various modalities and from the results of cognitive capability tests [140]. The axes of the scatterplot correspond to the *skewness* and *kurtosis*. For instance, the attributes that exhibit high values in both skewness and kurtosis are in the top right quadrant. The scatterplot gives you a mechanism to select a subset of attributes with particular values of these distributional features and visualise the data with regard to these attributes or perform further analysis on this selected subset.

Another key application using attribute spaces is to interactively investigate the differences within subsets of the data records across all the attributes. To do this, one can select a subset of records, calculate all the descriptors for the subset, for instance looking at the standard deviation of the data within the selected sample, and then compare the re-computed metrics to the metrics computed using all the available data records. Once the difference information is visualised, it reveals in what ways the attribute value distributions differ when a subset is considered. Figure 6.4c

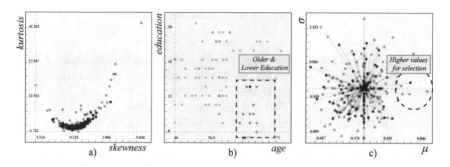

Fig. 6.4: Dual analysis framework, where visualisations of data items and attributes have blue and yellow backgrounds, respectively. a) A scatterplot of the attributes regarding their skewness and kurtosis values. b) A scatterplot of the data items (which describe people) regarding the attributes 'Age' and 'Education'. A group of older people with lower educational levels is selected. c) A deviation plot shows how the mean and standard deviation values of the different attributes change when the selection in (b) is made. Source: [140]

demonstrates how the values of the mean $\bar{\mu}$ and standard deviation $\bar{\sigma}$ change when a subset of the participants who are older and have lower level of education is selected, as shown in Fig. 6.4b. The plot in Fig. 6.4c can be called a *deviation plot*. Here, the centre refers to the zero deviation in $\bar{\mu}$ and zero deviation in $\bar{\sigma}$. We observe, for instance, certain subsets of attributes that have remarkably high values for the selected subset. These attributes (which are marked in Fig. 6.4-c) are potential candidates for being discriminatory features for the older and lower educated sub-cohort.

6.4.2 Analysing Multivariate Relations

Interactively ranked scatterplot matrices: What we have looked at so far were mostly one-dimensional characteristics of the attributes, i.e., each attribute has been considered on its own, irrespective of the interrelations between the attributes. It is natural to start an investigation of the interrelationships with considering the bivariate (i.e., pairwise) relationships. An often employed technique here is to use a scatteplot matrix that we introduced in Section 3.2. This is a viable approach when you have a moderate number of attributes, but once the number gets high, you need filtering capabilities to select from all possible attribute pairs those that are potentially interesting and relevant to your goals. The rank-by-feature framework [122] is an approach where all possible pairwise combinations of attributes are evaluated and ranked according to some criteria that is deemed interesting by the analyst, as can be seen in Fig.6.5. Examples of potentially useful ranking criteria are the *correlation coefficient*, the *number of outliers*, according to statistical measures of the outlierness, or the *uniformity of the point distribution* over the whole 2D space of the plot. These characteristics of the joint 2D distributions can be combined with the distributional characteristics of the individual attributes. An analyst can sort the pairs according to some of the characteristics, choose those that are of potential interest, and focus the analysis on those. Hence, ranking of attributes and attribute pairs by various distributional features can help you to navigate the huge attribute space more effectively.

Fig. 6.5: Rank-by-feature framework interface for ranking histograms and scatterplots. Source: [122].

Scagnostics: While discussing the rank-by-feature framework above, we mentioned a few metrics to characterise and rank bivariate relationships. Many of these features tend to be parametric. Visualisation researchers have been devising methods to capture more complex relation types. Scagnostics, an abbreviation of **sca**tterplot dia**gnostics**, is a set of measures describing a distribution of points in a scatterplot. Scagnostic measures were first proposed by John and Paul Tukey [139] and were later elaborated through a graph-theoretic approach by Wilkinson et al. [152]. These metrics, as seen in Fig.6.6, try to capture "visible" characteristics of the spatial distributions of the dots in the scatterplot space, such as how "skinny" a distribution is, or how clumpy, or how straight. Many further measures have been proposed since the introduction of scagnostics [44].

ScagExplorer [47] utilises these measures to guide the visual data analysis process. Multiple scatterplots are described by a vector of scagnostic characteristics, which then paves the way for further analysis of a huge number of bivariate relationships. Thus, it enables finding clusters of similarly looking scatterplots and finding the ones that are most dissimilar to the others.

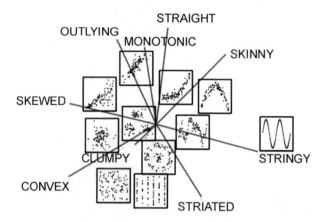

Fig. 6.6: Various scagnostic measures to characterise spatial patterns of dot distribution observed in 2D scatterplots [152].

6.4.3 Analysing higher-order relations and local structures

Correlation networks: The methods so far investigated one-dimensional distribution features and pairwise relationships between attributes. However, there are also higher-order relationships involving more than two attributes. It is not unusual to find "clusters" of attributes that are highly interrelated or those that are distinctively different and independent from others. An effective visual analytics method for the discovery of such structures is the use of correlation networks. The core idea in this technique is to construct a network (we talk more about networks in the next chapter) where attributes are the nodes, and the edges between them are the correlation strengths [157]. Once such a network is constructed, a suitable layout algorithm could be used, for instance, the force-directed layout[3], which puts strongly linked nodes close to each other in the display space. This reveals clusters of related attributes. Manual selections can be made in such a display by means of interaction operations. What is even more interesting is the ability to use graph-theoretic measures in application to this network. For instance, one can find the most "central" dimensions or look for "bridging" attributes that relate strongly to different subsets of attributes.

To illustrate this approach, let's have a quick look at the correlation network for the attributes describing the London boroughs in the dataset that we investigated in

---

[3] https://en.wikipedia.org/wiki/Force-directed_graph_drawing

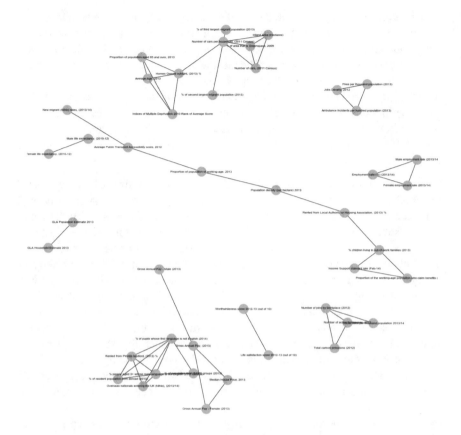

Fig. 6.7: A correlation network constructed with the 69 numeric attributes characterising the London boroughs that we have in the motivating example in Section 6.1. Each node represents one of the attributes. The edges represent strong correlations between attributes, i.e., where $\rho > 0.75$.

Section 6.1. We calculate the Pearson correlation coefficient ρ between all attribute pairs, construct a network with 69 nodes corresponding to the 69 numeric attributes and connecting edges between those nodes where there is a strong positive correlation between the attributes; specifically, we take $\rho > 0.75$ in this example. In the result, we get is a network with unconnected sub-components, as seen in Fig.6.7. Each component represents a small group of attributes with strong interrelationships among them.

Employing machine learning methods: The machine learning literature offers several effective tools to work with high-dimensional data sets, from reducing the dimensionality (as considered in Section 4.4) to multivariate modelling. The survey by

Endert et al. [51] is a good source to read about the integration of machine learning techniques into visual analytics approaches. Particularly, dimensions-transforming data embedding techniques (discussed in Section 4.4.3) can be helpful for revealing relationships among attributes. The most widely adopted method is the Principal Component Analysis (PCA), which represents variation in data in terms of a smaller number of artificial attributes, called principal components, derived as weighted linear combinations of the original attributes. The original attributes that have high weights in one and the same principal component can be considered as somehow related. It is a bit like the clusters of attributes we saw in the correlation network in the previous section.

A problem with using PCA or similar techniques for high-dimensional data is that the number of principal components to consider may become too large. A possible approach to deal with this problem is to use *subspace clustering* [86]. The main idea of subspace clustering is to find subsets of data that form dense clusters in spaces constructed from subsets of the whole set of dimensions. For example, a dense cluster of data points can exist in a space of k particular attributes taken from a larger set of n attributes, but these points may be scattered with respect to any of the remaining $n - k$ dimensions[4]. In other words, there may be a subset of data items that are similar in terms of particular attributes but not similar in terms of the other attributes. The goal of the subspace clustering algorithms, which are developed in machine learning, is to find subsets of attributes such that there are dense and sufficiently big clusters of data items with similar combinations of values of these attributes. The resulting data clusters may be existing in distinct subspaces. The groups of attributes forming these subspaces are locally related with regard to certain subsets of data. These local structures can be explored to gain some understanding of the data distribution and the relationships among the attributes. To generate a visual representation of the subspaces and the respective data distributions, one can apply PCA or another dimensionality reduction technique.

An example is shown in Fig. 6.8, where the display contains multiple projection plots representing different subspaces. The coloured bars along the x- and y-axes of the projection plots indicate the relevance of different dimensions for the corresponding principal components used to create the plots. The projection plots are laid out in the display space according to the similarities between the subsets of the attributes forming the represented subspaces. This can be achieved by means of a data embedding method, such as MDS, which is applied in Fig. 6.8.

---

[4] https://en.wikipedia.org/wiki/Clustering_high-dimensional_data#Subspace_clustering

Sight map with sight glyphs of sub-space projections   Control panel

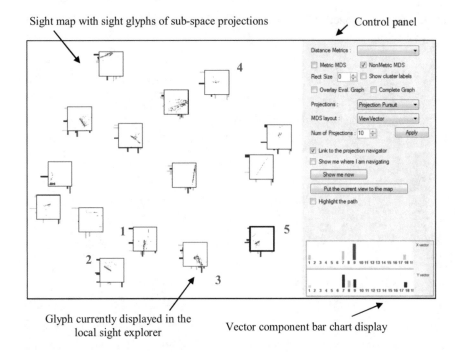

Glyph currently displayed in the
local sight explorer   Vector component bar chart display

Fig. 6.8: Multiple projection plots representing different subspaces of a high-dimensional attribute space are laid out in the display space according to their similarities in terms of the relevant attributes. The degrees of attribute relevance are indicated by the sizes of the bars drawn along the x- and y-axes of the plots. Source: [104]

6.5 Further Examples

6.5.1 Exploring projections through interactive probing

The paper by Stahnke et al. [127] was mentioned multiple times in Section 4.4, where we discussed data embedding. It provides good examples of how to visualise and explore distortions of between-item distances in a projection (Figs. 4.6 and 4.7) and how to explore and compare multi-attribute profiles of selected subsets of data items (Fig. 4.9). In addition to the illustrations provided in Section 4.4, Figure 6.9 demonstrates interactive selection of a subset of dots in a projection plot and obtaining information about the subset of data they represent in the form of smoothed frequency histograms corresponding to multiple attributes. The histograms show the distributions of the attribute values in the subset in comparison to the distributions in the whole set. Hence, it is possible to identify attributes that are distinctive for the selected subset.

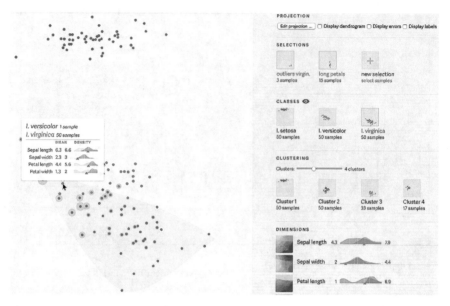

Fig. 6.9: The "Probing Projections" approach enables selection of a subset of data items and viewing the distributions of the values of multiple attributes over the selected subset in comparison to their distributions in the whole dataset. Source: [127]

Another useful technique, attribute distribution heatmaps (Fig. 6.10), is motivated by the problem that the dimensions of the data embedding space have no intuitively understandable meaning, as we mentioned in Sections 4.4.4 and 6.1. For individual attributes, heatmaps show the variation of the values over the projection space. In the example in Fig. 6.10, an MDS projection has been built from multi-attribute data describing a set of countries. The large heatmap on the left shows the value variation of one of the attributes, namely, 'Educational Attainment'; darker areas correspond to higher values. We can see that countries such as Japan and Korea are in the "darker" areas whereas Brazil, Mexico and Turkey are on the "paler" side of this heatmap. The small heatmaps on the right show the value variations of the other attributes. Any of them can be expanded for a detailed examination.

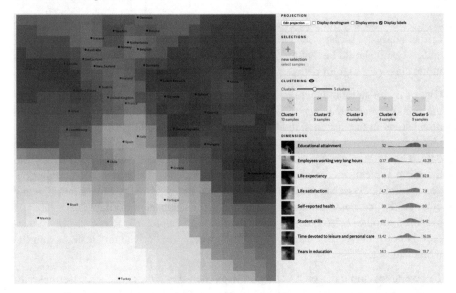

Fig. 6.10: The distributions of the values of individual attributes over the projection space are visualised in the form of heatmaps. Source: [127].

6.5.2 Manually crafting projections through "Explainers"

An automatic dimensions-transforming method of data embedding, such as PCA, strives to create artificial dimensions such that they capture the bulk of data variation, i.e., the existing data items should be differentiated sufficiently well in terms of these dimensions. There is a potential problem here: the distinctions that have made the greatest impact on the algorithm result are not necessarily those that you care about, whereas other distinctions that are more important or interesting for you may be buried among derived dimensions with low importance, according to the algorithm's criteria. For some analyses, you would like to have the power to construct artificial dimensions that differentiate particular data items from the others. These dimensions can be then used for analysing data distributions, in particular, by constructing projection plots.

A derived dimension is a function involving some of the original dimensions (attributes) as variables. Most typically, it is a weighted linear combination of the original attributes, such that the weights indicate the importance of the attributes. The task of differentiating a group of data items (also referred to as a "class") from others based on attribute values is known as *classification* task in machine learning, and a function that can do the differentiation is called a *classifier*. There are many machine learning methods that build classifiers. Hence, it is possible to use these methods for constructing artificial dimensions, and also for revealing the most important differences of a group of items from the others. Not all kinds of classifiers are suitable for the purpose of constructing interpretable dimensions to be used in further analysis. A classifier should be an explicit function of the original attributes, and its result should be a numeric value expressing the probability of an item being in the class or a kind of degree of item relevance to the class.

One of many suitable machine learning methods that can build such a classifier is the support-vector machine (SVM)[5]. This method is used for "crafting" projection dimensions in the system called Explainers [54]. The authors of the approach emphasise that classifiers are meant here for describing and explaining the existing data. The authors call these functions "explainers", to distinguish this usage from the more typical case when classifiers are created for the purpose of prediction, i.e., assigning classes to previously unseen data items. Usually, a classifier, like any other kind of computer model, is generated using a subset of available data data and tested on another subset, as will be described in Chapter 13. However, this procedure is not followed in creating explainers, which are derived and tested using the whole dataset.

To give an example of constructing an explainer, the authors use a dataset describing the livability (i.e., the living conditions) in 140 cities of the world in terms of 40 attributes. In this example, the analyst wishes to construct a dimension which could differentiate the American cities from all others. The analyst selects the group of American cities, and the system constructs an SVM classifier for differentiating these cities from the others. The function is: $x_1 - x_2 - x_3$, where x_1 is an attribute reporting the availability of public health care, x_2 is the availability of sporting events, and x_3 is a measure of healthcare quality (lower numbers are better). This artificial dimension of the data can be called "American-ness".

This function, however, does not make a clear-cut separation between the American and not American cities. The distribution of the function values is shown in Fig. 6.11. The bluish colouring of the histogram segments and the lower box plot corresponds to the group of American cities and the greenish colouring to the remaining cities. Most of the highest values of the function represented by the rightmost bar of the histogram correspond to American cities, but some non-American cities also have similar values. The other bars of the histogram mostly contain non-American cities, but some American cities occur there as well. The separation quality can additionally be assessed by considering the box plots.

The function we are discussing is not the only function that can be built for separating American cities from others. Typically, there are many possible classification functions that differ in their complexity and differentiating quality. It is quite usual that higher classification accuracy is achieved at the cost of high complexity, which diminishes the interpretability of the function. Low interpretability is an undesired property for a function that is intended to be used in analytical reasoning. Hence, an analyst would strive for a suitable trade-off between higher accuracy and lower complexity. The Explainers system propose the user several candidate functions for

---

[5] https://en.wikipedia.org/wiki/Support-vector_machine

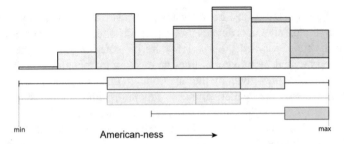

Fig. 6.11: The distribution of values of a classification function "American-ness" is shown by a frequency histogram and box plots. The bluish colouring of the histogram segments and the lower box plot corresponds to the target group of data items (namely, American cities) and the greenish colouring to the remaining items. The upper box plot coloured in light grey represents the function value distribution in the whole dataset, This representation shows how the function differentiates the American cities from the other cities of the world. Source: [54].

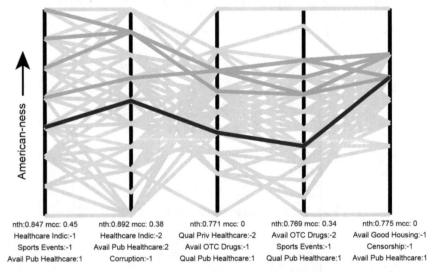

Fig. 6.12: The axes of the parallel coordinates plot correspond to 5 distinct explainer functions generated for the same target group of data items. The lines connecting the axes represent different cities. The lines of the American cities are coloured in blue. Source: [54].

consideration and selection of the most appropriate one. The candidate functions are generated so that they are diverse in terms of involving distinct subsets of the original attributes, and functions with integer weights of the attributes are preferred for their simplicity.

Figure 6.12 demonstrates how an analyst can compare several candidate functions. Each axis in the parallel coordinates plot represents one function. The attributes involved in the functions and their weights are indicated below the axes. The data items are represented by polygonal lines connecting the positions on the axes corresponding to the scores given by the functions. In this example, the axes represent different variants of the "American-ness" function, some emphasising healthcare while other emphasising housing or other criteria. The lines in blue correspond to the American cities. The dark blue line corresponds to the city San Jose, which is given the lowest scores by all functions except the last one. This city appears to be the least American among all American cities. All five functions fail to completely separate the American cities from not American. The advantage of the first function is that it gives the highest possible score to American cities while the others to non-American. This is the function whose value distribution is seen in Fig. 6.11.

An explainer can also be constructed with the purpose to separate a single item from all other items, i.e., the function is expected to give the highest possible score to the selected item and lower scores to all others. Such a function characterises what is unique about the selected item compared to the others. For example, one can build a function describing and explaining the uniqueness of Paris.

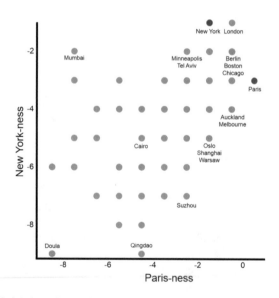

Fig. 6.13: Artificial data dimensions "Paris-ness" and "New York-ness" constructed for capturing the distinctive characteristics of Paris and New York are used as axes of a projection plot, in which all cities are positioned according to their scores in terms of "Paris-ness" and "New York-ness". Source: [54].

As we have mentioned, explainers can be used as dimensions of a projection space in which all data items are positioned according to their scores. An example is shown in

Fig. 6.13, where the axes of the projection plot correspond to the explainers "Paris-ness" and "New York-ness". The red dots represent Paris and New York, which have the highest values of the respective functions. The positions of the other dots show how similar or different to Paris and New York the respective cities are. Thus, London is very New York-like, whereas Mumbai is quite similar to New York but is opposite to Paris. Of course, the similarities and differences refer to specific attributes that are involved in the definitions of the functions "Paris-ness" and "New York-ness".

The following takeaways can be drawn from this example:

- In reducing data dimensionality and creating data embeddings, an analyst may wish to obtain artificial dimensions that capture distinctive features of particular items or groups of items. Hence, the analyst needs to have control on the process of dimension construction, which is not possible when automatic methods are used.

- Human-controlled creation of dimensions with desired properties can be supported by employing machine learning methods that build classification functions with numeric outputs.

- The purpose of using the classification functions is data description and explanation rather than prediction. Hence, interpretability of the functions is crucial.

- For gaining better, more comprehensive understanding of the analysis subject, it is beneficial to consider several distinct variants of classification functions that involve different groups of attributes.

- The functions constructed under the analyst's supervision can be used in further analysis, particularly, as dimensions of a data embedding space.

- To select one of multiple possible functions for using in further analysis, the analyst seeks a suitable trade-off between the classification accuracy and complexity.

6.6 Concluding remarks

The complexity of high-dimensional data necessitates the use of computational techniques together with visualisations that enable human interpretation, reasoning, decisions concerning further steps in the analysis, and control over the computer operations. Machine learning offers a wide spectrum of computational methods designed to deal with such kind of data. Several examples of combining visualisations with different types of machine learning techniques, including data embedding, clustering, and classification, have been described and discussed in this chapter as well as in Chapter 4. Of course, these examples do not cover all possibilities of how visualisations and computations can complement each other, but they demonstrate a few general ideas that can be re-used in other analytical scenarios:

- Support interpreting computation results by visualising value distributions of original and derived attributes.

- Perform multiple computations and use visualisations to support comparison of the results.

- Direct the work of the computational methods through visually supported inter-active selection of subsets of attributes and/or data items.

6.7 Questions and exercises

Use the publicly available dataset describing the London boroughs that is referred to in Section 6.1, or take a similar dataset where you can define several groups of semantically related attributes. What approach(es) would you try for comparing the boroughs in terms of each group of attributes? What techniques can help you to identify the unique and mainstream items with regard to each attribute group? Experiment with applying data embedding and clustering to different subsets of the attributes and different subsets of the data items.

Chapter 7

Visual Analytics for Understanding Relationships between Entities

Abstract A graph is a mathematical model for representing a system of pairwise relationships between entities. The term "graph" or "graph data" is quite often used to refer, actually, to a system of relationships, which can be represented as a graph, rather than to the mathematical model itself. In line with this practice, the term "graph" is used in this chapter as a synonym to "system of relationships". Graph data have high importance in many application domains, such as social network analysis, transport network analysis, systems biology, transaction analysis, to name a few. The relational nature of graph data sets presents unique challenges that open up space for distinctive and innovative visual analytics methods. Graphs often contain inherent hierarchies, and the analysis outcomes vary significantly depending on the *scale* (i.e., aggregation level) at which the relationships are considered. Identification, analysis, and interpretation of inherent structures, such as tightly connected parts of the network, requires a combination of algorithmic and interactive methods. Analysis of graph structure may require involvement of different graph theoretic properties of nodes and links, consideration of paths, and extraction of recurring connectivity patterns. All of these analyses are taken to a different level of complexity when these graphs are multivariate, i.e., where each node carries multiple attributes, and dynamic, i.e., when the structure of the graph changes with time. In this chapter, we discuss these challenges in depth and present examples of employing visual analytics techniques for extracting valuable findings from this highly rich and interesting data type.

7.1 Motivating example

Our example is based on a paper presenting a visualisation technique called Contingency Wheel [5]. However, our focus will be not on the technique but on exploration of relationships between groups of entities. We shall consider a process of analysing a data set with ratings of movies by users, that is, by people who viewed the movies. The data set is publicly available and known as MovieLens[1]. Each record of the data includes information about the user (identifier, sex, age, and occupation), about some movie (title, release year, and genre), and the rating given by the user to the movie. In the analysis we are going to discuss, not the ratings are of interest but the users' choices of the movies to view. The analyst wants to know how characteristics of the users (their sex, age, and occupation) are related to characteristics of the movies they choose (year and genre). For this purpose, the analyst will perform two kinds of operations: first, define groups of users and groups of movies based on the attributes of the former and latter and, second, extract and explore the following relationships:

• associations between the groups of the users and groups of the movies;

• similarities between the user groups in terms of movie choices;

• links between the movie groups due to shared audiences.

7.1.1 Extracting relationships

For extracting the associations between groups of users, on the one side, and the movies or groups of movies, on the other side, the analyst transforms the original data into a so called *contingency table*. It is a matrix whose rows and columns correspond to values of two attributes and the cells contain frequencies of the occurrence of the value pairs in the data. Usually, contingency tables are generated for attributes with categorical (qualitative) values. Figure 7.1 shows an example of a contingency table where the rows correspond to the movies and the columns to the user groups by occupations. Similar tables can be built for movie groups based on the genre or for user groups based on the sex. To define groups of users or groups of movies based on numeric attributes, such as 'age' and 'release year', the analyst needs to divide the ranges of the numeric values into meaningful intervals. For example, the analyst may define the following age groups of the users: below 18, from 18 to 25, from 25 to 35, and so on.

Tools for creation of contingency tables are commonly available in various data analysis systems, libraries, and frameworks, such as R, KNIME, WEKA, and Python. Having a contingency table with frequencies of pairs of some items, it is possible to assess the strengths of the associations between the items. If a pair occurs more frequently than it can be expected, it is said that the items are positively associated. If a pair occurs less frequently than expected, it is said that the items are negatively associated. If the frequency is close to what is expected, the items are not associated. The strength of a positive or negative association can be measured based on

---

[1] https://grouplens.org/datasets/movielens/

occupations / movies	K-12 student	self-employed	scientist	executive	writer	homemaker	academic/educator	programmer	technician/eng.	other	clerical/admin	sales/marketing	college/grad stud.	lawyer	farmer	unemployed	artist	tradesman	customer service	retired	doctor/health care	f_{i+}
1 One Flew Over the Cuckoo	25	79	47	196	96	23	191	107	109	195	59	81	200	39	3	20	86	22	33	42	72	1725
2 James and the Giant Peach	29	19	10	42	39	10	35	29	33	69	19	20	75	9	1	7	35	4	10	7	23	525
3 My Fair Lady (1964)	19	30	15	62	43	17	81	32	34	75	27	25	66	14	0	8	38	1	8	15	26	636
4 Erin Brockovich (2000)	41	57	26	165	59	18	126	75	101	131	38	81	188	30	1	13	49	9	18	39	50	1315
5 Bug's Life, A (1998)	72	73	49	159	80	32	103	117	145	215	52	94	247	23	4	25	78	18	33	19	65	1703
⋮	⋮	⋮	⋮	⋮	⋮	⋮	⋮	⋮	⋮	⋮	⋮	⋮	⋮	⋮	⋮	⋮	⋮	⋮	⋮	⋮	⋮	⋮
3703 Broken Vessels (1998)	0	0	0	0	0	0	0	0	0	0	0	1	0	0	0	0	0	0	0	0	0	1
3704 White Boys (1999)	0	0	0	0	0	0	0	0	1	0	0	0	0	0	0	0	0	0	0	0	0	1
3705 One Little Indian (1973)	0	0	0	0	1	0	0	0	0	0	0	0	0	0	0	0	0	0	0	0	0	1
3706 Five Wives, Three Secretar	0	0	0	0	0	0	1	0	0	0	0	0	0	0	0	0	0	0	0	0	0	1
f_{+j}	23290	46021	22951	105425	60397	11345	85351	57214	72816	130499	31623	49109	131032	20563	2706	14904	50068	12086	21850	13754	37205	f_{++}

Fig. 7.1: A contingency table containing the frequencies of all pairs movie + viewer's occupation. Source: [5]

the difference between the actual and expected frequencies. The expected frequencies are estimated based on the null hypothesis assuming absence of association. Let $i+$ and $+j$ be two items corresponding, respectively, to the row i and to the column j of a contingency table. Let f_{i+} and f_{+j} be the overall frequencies of the items $i+$ and $+j$ in the data, and let f_{++} denote the sum of all frequencies contained in the table; see Fig. 7.1. Then, the expected frequency $\hat{e}_{ij}$ can be estimated as the product of the overall frequencies of the items divided by the sum of all frequencies: $\hat{e}_{ij} = f_{i+} \cdot f_{+j}/f_{++}$. If the actual frequency of the item pair f_{ij} is close to $\hat{e}_{ij}$, it means that the null hypothesis holds, that is, the items are not associated; otherwise, they are associated, either positively, when $f_{ij} > \hat{e}_{ij}$, or negatively, when $f_{ij} < \hat{e}_{ij}$.

The significance of the difference between the actual and expected frequency values can be estimated by means of the standardised, or adjusted, Pearson residual:

$$r_{pearson_{ij}} = \frac{f_{ij} - \hat{e}_{ij}}{\sqrt{\hat{e}_{ij} \cdot (1 - f_{i+}/f_{++}) \cdot (1 - f_{+j}/f_{++})}}$$

The values of $r_{pearson_{ij}}$ have a normal distribution with the mean of 0 and the standard deviation of 1. It is convenient to re-scale these values to the range $[-1, 1]$, so that -1 corresponds to the strongest negative association (it would mean that the items never occurred together, i.e., $f_{ij} = 0$) and 1 to the strongest positive association (it would mean that the items occurred only together with each other and never with any other items, i.e., $f_{ij} = f_{i+} = f_{+j}$). To obtain the scaling factor, the maximal theoretically possible absolute value of $r_{pearson_{ij}}$ is calculated based on the contingency table. For obtaining the maximal possible negative association value, the frequency of the pair $(i+, +j)$ is set to 0, and the aggregate frequencies f_{i+}, f_{+j}, and f_{++} are updated accordingly. For obtaining the maximal possible positive association value, the frequencies of all pairs in the ith row and in the jth column, except for the pair $(i+, +j)$, are set to 0, and the aggregate frequencies f_{i+}, f_{+j}, and f_{++} are decreased accordingly.

7.1.2 Visualising relationships

To represent the associations visually, the authors of the paper [5] propose a design demonstrated in Figs. 7.2 and 7.3. The display has an appearance of a wheel and is called Contingency Wheel. The wheel is divided into segments corresponding to columns or rows of a contingency table. Thus, the wheel segments in Fig. 7.2 correspond to groups of users according to the occupation, and in Fig 7.3 to groups of movies according to the genre. The widths of the segments may differ proportionally to the frequencies of the respective items, as in Figs. 7.2. Inside the segments, the associations with the items corresponding to the other dimension of the table are represented in a detailed mode, as in Fig. 7.2, or in an aggregated way, as in Fig 7.3. In the example we consider, the analyst is only interested in the positive associations between the users and the movies; therefore, the negative associations, if any, are not shown in the display.

In Fig. 7.2, each movie is represented by a dot whose distance from the inner side of the segment is proportional to the strength of the association between the movie and the user group represented by the segment. When there are many movies with the same association strength, the corresponding dots may overlap, and the quantities will be hard to estimate. In such cases, it is better to represent the associations using histograms, as in Fig 7.3 and Fig 7.4, instead of dots. In these figures, the histograms are drawn in a peculiar way corresponding to the wheel-like design of the display. Such a fancy design is not essential; it would also be OK to use standard histograms.

The use of histograms, either curved or standard, creates additional opportunities for the exploration of the associations. The bars of the histograms can be divided into segments corresponding to groups of the associated items. Thus, in Fig. 7.4, the segments correspond to groups of movies according to the release year (top) and according to the genre (middle). The legends at the bottom explain the encoding of the years and genres by colours of the bar segments and also show the frequency distributions of the values of these attributes. Such segmented histograms are good for revealing associations between groups of items. The histograms in Fig. 7.4 reveal the movie choice preferences of some occupation-based user groups. We can see that the group "K-12 Student" (i.e., from kindergarten to the 12th school grade) has strong associations with recent movies while the group "Retired" is stronger associated with older movies. Concerning the movie genres, the group "K-12 Student" has very strong associations with movies for children and with comedies,

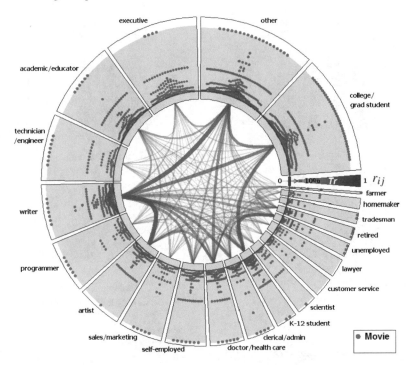

Fig. 7.2: A Contingency Wheel display showing the links between the viewers' occupations based on the associations with the same movies. Source: [5]

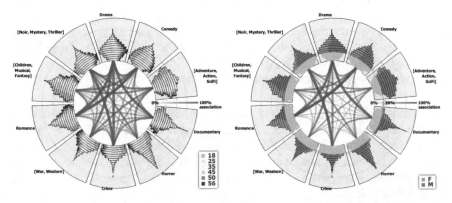

Fig. 7.3: A Contingency Wheel display showing the links between the genres of the movies based on the associations with the same viewers. The numbers of the associated viewers are represented by the lengths of the curved histogram bars in the wheel sections corresponding to the movie genres. The bars are divided into coloured segments corresponding to the age groups (left) and to the gender (right) of the viewers. Source: [5]

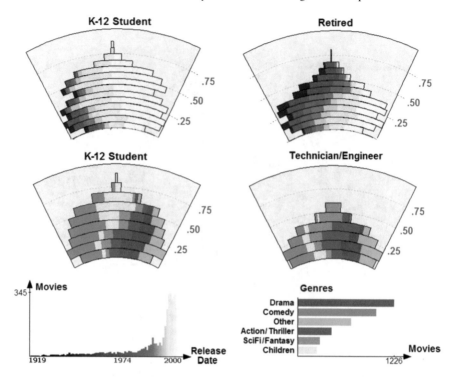

Fig. 7.4: Exploration of the associations between the occupations of the movie viewers and the movies regarding their release times (top) and genres (middle). The numbers of the associated movies are represented by the lengths of the curved histogram bars. The bars are divided into coloured segments corresponding to the release years (top) and genres (middle) of the movies. The vertical positions of the bars correspond to the strengths of the associations. Source: [5]

whereas the group "Technician/Engineer" has moderately strong associations with SciFi/Fantasy and Action/Thriller. The histograms in the wheels in Fig. 7.3, where the wheel segments correspond to the movie genres, show the associations between the movie genres and the user groups by age (left) and sex (right). Particularly, we can notice that some movie genres have much stronger associations with male users than with female.

As we said at the beginning of the section, apart from the associations between the users and movies, there are also relationships between the users based on similar choices of the movies to view and relationships between the movies based on shared audiences. The strength of the similarity relationship between two users can be measured by the number of the movies that were viewed by both users. The strength of the similarity relationship between two user groups is measured by the number of the movies having sufficiently strong associations with both user groups, that is, the association strength is higher than a chosen threshold. The strength of the relationship between two movies can be measured by the number of the users who viewed both movies. The strength of the relationship between two groups of movies can be measured by the number of the users having sufficiently strong (i.e., above a threshold) associations with both groups. In the analysis scenario described in the paper [5], the analyst used the threshold 0.4.

The Contingency Wheel display is designed to represent simultaneously the associations between the columns and rows of a contingency table (by the histograms in the wheel segments) and the similarity relationships between the columns or between the rows (by the arcs linking the wheel segments). However, as we noted earlier, it is not essential to use this particular design. It would be quite suitable to represent two kinds of relationships separately. Usual histograms can represent the column-row associations. The similarity links can be visualised by means of standard graph drawing techniques, which are available in many software systems and data analysis frameworks, including Python and R.

7.1.3 Exploring relationships

Let us return to the analysis of the relationships hidden in the movie rating data set. The analyst first creates a contingency table of the user's occupations and the movies as shown in Fig. 7.1. The analyst explores the relationships between the user groups based on the associations with the same movies. She finds strong relationships between the groups "K-12 student" and "college/grad student" and between "programmer" and "technician/engineer". More surprisingly, there is also quite a strong relationship between "academic/educator" and "retired". To investigate this relationship, the analyst extracts and examines the subset of the movies associated with both these groups. She finds that the subset mostly consists of old movies; only a few of the shared movies were released in the 1990-ies and all others are older. Based on the noticed similarities, the analyst merges "K-12 student" with "college/grad student", "programmer" with "technician/engineer", and "academic/educator" with "retired". The further analysis is done based on the redefined user groups.

Having learned that the group [academic/educator, retired] has a strong inclination to viewing old movies, the analyst compares this group to the other user groups with regard to the distribution of the release years of the associated movies. She finds that the distribution for the group [academic/educator, retired] is quite different from the distributions for the other user groups. In particular, the group [college/grad student, K-12 student] has opposite preferences (Fig. 7.4).

To explore the relationships between the movie genres, the analyst creates a contingency table with the columns corresponding to the genres and the rows to the users. After visualising the relationships, the analyst finds high overlaps of the audiences of "Musical" and "Children", "Action" and "Adventure", and "War" and "Western". Contrary to her expectations, she does not see much overlap between the audiences of the genres "War" or "Western" and "Crime". Like with the user groups, the analyst redefines the groups of the movies according to the observed commonalities of the audiences. Then the analyst explores the age and gender differences between the audiences associated with the redefined movie groups using segmented histograms, as in Fig. 7.3. She finds that the user age distributions for the movie groups look quite similar. Concerning the gender, she sees that, as could be expected, the most strongly associated users of the movie group "Romance" are female, but, surprisingly, the audience of "Horor" does not include fewer women than the other movie groups. The analyst also investigates the audience compositions of the different movie groups in more detail using appropriate histogram displays. Particularly, for the movie group "Children", she finds that there is an almost equal distribution between male and female viewers, and that most viewers are in the age group of 18–24 year, while the analyst expected the prevalence of younger users. The movies of the groups "War" and "Western" are viewed more by men, which are often executives and in the age group of 35-44 years for "War" and older age groups for "Western".

The analysis was done for a movie rental service company and allowed them to better understand their customers. Based on the insights gained, the company can simplify and improve their movie recommendation system.

7.1.4 Main takeaways from the example

The example demonstrates the following things:

- Pairwise relationships between entities or categories are not always explicitly specified in data, but they can be extracted from data. In particular:
  - Associations between entities or categories from two distinct sets can be extracted from the frequencies of the co-occurrences of the set elements in the data.
  - Similarity relationships between entities or categories from the same set can be defined based on their common associations with elements of another set.
- There may be numeric measures expressing the strengths of relationships. When relationships are extracted from data, it is often possible to estimate also their strengths.
- Relationships between elements from two distinct sets and relationships between elements within the same set require different techniques for visualisation and analysis.

- Between-set relationships can often be analysed using frequency histograms or equivalent displays, which show for each element or group of elements of one set the frequency distribution of the elements or groups of elements of the other set.

- Within-set relationships are usually represented by means of graph visualisation techniques, such as node-link diagrams. Thus, a Contingency Wheel is a node-link diagram, where the wheel segments are the nodes, which in this particular visualisation have a complex internal structure.

- Elements of sets under analysis may be put into groups according to various criteria. Grouping helps to deal with large sets of entities, and it supports abstraction, disregarding minor details, and gaining an overall picture. In the process of analysis, it may be reasonable to redefine the groups by merging, splitting, or taking a different grouping criterion.

7.2 Graphs as a mathematical concept

7.2.1 Definition

A *graph G* consists of a set of *vertices V*, which are also called *nodes*, and a set of *edges E*, which are also called *links*: $G = (V, E)$. A vertex represents an individual entity and an edge indicates the existence of a relationship between two entities; hence, each edge refers to some pair of vertices. For instance, in a graph representing a friend network on a social media site, a node would be an individual user, for instance *Mary* or *Bob*, and an edge between Mary and Bob would mean that they are friends on this site and thus *related*. In our motivating example, there were relationships of shared audience between groups of movies and relationships of common movie choices between groups of viewers. The groups of movies or the groups of users could be represented as vertices in two graphs, and the relationships between the groups would be represented by graph edges.

Graphs can be *directed*, which means that there is a direction associated with each edge to specify which vertex is the origin and which is the destination of the relation. A typical example could be a graph representing message communications or money transfers between people or organisations, where each edge would be directional and indicate a sender and a receiver.

Often in real-world applications vertices and edges in graphs come with multiple attributes associated with them. These attributes can be any form of data from numeric values, to textual information to multimedia data. In our motivating example, the movies and the users had specific attributes, and the groups were described by distributions of values of these attributes.

A special numeric measure, which is treated as the strength or *weight* of the relationship, can be associated with the edges of a graph. In such a case, the graph is called *weighted*. In our motivating example, there were numeric measures of the strengths of the relationships between the movies or movie groups (based on the number of common users) and between the users or user groups (based on the number of the same viewed movies). These measures could be represented as the weights of the graph edges. In the internal circular area of a Contingency Wheel, the arcs between the wheel segments represent visually the graph edges, and the widths of the arcs are proportional to the weights of the edges. In a graph representing money transfers, the edge weights could represent the amounts of the transferred money.

In various applied disciplines, including machine learning, data mining, data science, transportation, biology, and others, the term *network* is used as a synonym for *graph*.

7.2.2 *Graph-theoretic metrics*

For identifying important, or interesting vertices and edges of graphs, a number of graph centrality measures are proposed [38]. Frequently used measures include:

- The *degree centrality* of a vertex is defined by the number of its connecting edges. For a directed weighted graph, it is necessary to consider four measures of degree centrality: (1) *in-degree* and (2) *out-degree* are the numbers of the edges directed into and out of a vertex respectively, and (3) *weighted in-degree* and (4) *weighted out-degree* are the sums of the weights the edges directed into and out of a vertex.

- The *closeness centrality* is based on the distance, measured as the length of the shortest path, of a vertex to all other vertices. Closeness centrality of vertex v is defined as the inverse of the total distance of v to all other vertices.

- The *betweenness centrality* is based on the number of the shortest paths connecting different pairs of vertices that pass through a given vertex v. It is defined as the proportion of the shortest paths that pass through v to the total number of the shortest paths in the graph.

- The *clustering coefficient* represents the likelihood that two neighbours of a vertex v are connected, which indicates, in a sense, the importance of the vertex in its immediate neighbourhood. The clustering coefficient of vertex v is defined as the fraction of v's neighbours that are also neighbours of each other.

In application of the centrality measures to weighted graphs, it is important to take into account the meaning of the weights. In some cases weights can be treated as "costs", such that lower values are better (e.g., travel time or distance), while in other cases higher weights are considered to be better, i.e., the weights represent a kind of "benefits" (e.g., friendship duration or frequency of communication). The algorithms computing the shortest paths in weighted graphs typically assume that the edge weights represent some costs that need to be minimised; hence, the shortest path is the one with the minimal total cost, which is the sum of the edge weights along the path. Consequently, the centrality measures that involve the shortest path calculation (i.e., the closeness and betweenness) inherit the assumption of the cost character of the weights. If the weights in your graph express benefits while you want to calculate shortest paths and related measures, you need to reverse the weights, for example, replace each weight w by $1/w$ or $1/(w+1)$, if zero weights occur in the graph.

The *global clustering coefficient* of a graph is computed based on the number of closed and open node triplets. A closed triplet is a group of three vertices where any two vertices are connected by an edge. An open triplet is a group of three vertices where there are two edges and one edge is missing. The global clustering coefficient is the ratio of the number of closed triplets existing in a graph to the total number of all existing triplets, both closed and open.

7.3 Specifics of this kind of phenomena/data

Before analysing data sets representing relationships, it is necessary to ensure that the data correspond to expectations and don't have obvious errors and omissions. Paper [91] considers, among other data types, possible problems in graph data and approaches to deal with them.

Specifically for graphs / networks, it is necessary to check their topological structure. It is reasonable to begin with calculating and inspecting overall characteristics of the graph: density, diameter, global clustering coefficient, connectivity (existence and number of disconnected parts), and average vertex degree. Analogously to detection of missing data in tables, one may need to check correspondence between vertices and edges, detect disconnected vertices and duplicated edges. Topology-based outlier detection in graphs is based on analysing the frequency distributions of the values of graph-specific indicators, such as centrality measures.

Depending on an application, it may be needed to test whether a given graph is a tree (i.e., each vertex has at most one incoming edge), a planar graph (i.e., can be drawn in the plane without edge crossings), or a directed acyclic graph (i.e., has no closed paths). One may need to analyse topology-based constraints: a tree has exactly $n-1$ edges, a planar graph can have at most $3*n-6$ edges, where n is the number of vertices. Further topology-based checks may include finding frequent patterns, such as motifs (small subgraph consisting of a few vertices), small cycle patterns, such as triangles, and other subgraphs with specific topological features, such as paths, trees, stars, and complete subgraphs.

For weighted graphs, it is necessary to check also the properties of attributes associated with the vertices and edges. Specifically, missing data, outliers, unusual combinations of values of multiple attributes require special attention, among other potentially suspicious patterns. Such patterns may indicate errors in data that need to be fixed before the analysis.

7.4 Graph/network visualisation techniques

We shall illustrate graph visualization approaches using data from VAST Challenge 2015[2]. The data set consists of simulated trajectories of 11,374 individuals who spent 3 days in an imaginary amusement park, repeatedly visiting 73 places of interest (POIs) belonging to 11 distinct categories, including park entry, information, food, beer gardens, shopping, restrooms, shows, and different categories of rides (thrill rides, rides for everyone, and kiddie rides). Figure 7.5 shows the spatial layout of the park, the locations of the POIs, and the flows of the visitors. The POIs and the flows between them make a directed weighted graph where the POIs are the nodes and the flows are the edges. This is a specific kind of graph: its nodes have fixed spatial locations. To have a more generic kind of graph for further illustrations, we shall make some transformation of the data. Specifically, we aggregate the POIs into groups, represent the groups by graph nodes, aggregate the between-POI flows into the between-group flows, and obtain in this way a *state transition graph* providing an aggregated representation of the collective movement behaviour [19].

A meaningful way for grouping locations is by their types or categories. In our example, we make groups and create graph nodes based on the POI categories. However, taking into account the specifics of the challenge, there are a few places requiring separate consideration; we thus represent each of them by a separate graph node. After aggregating the flows, we obtain a directed weighted graph, where the weights of the nodes are the counts of the visits to the POIs of the respective categories, and the weights of the links correspond to the counts of the moves between the POIs of the categories represented by the nodes. The graph is not fully connected, as a few of potentially possible transitions did not ever occur in the data.

There are two major approaches to graph visualisation [90]: a graph can be represented by an adjacency matrix and by a node-link diagram. Figure 7.6 demonstrates two variants of a matrix representation of the state transition graph constructed from the amusement part data. The variants differ in the ordering of the rows and columns. Meaningful ordering may enable identification of structures and connectivity patterns and help in data interpretation. Jacques Bertin was the first to introduce matrix reordering as an approach to finding patterns [31]. Bertin suggested that the reordering should aim at "diagonalisation" of the matrix, so that most of the filled cells are

---

[2] http://vacommunity.org/VAST+Challenge+2015

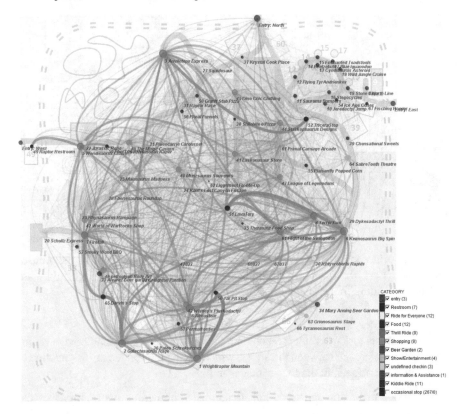

Fig. 7.5: Flows of visitors in an amusement park. Labelled dots represent the POIs. The dots are coloured according to the place categories, and the sizes are proportional to the counts of the visits. The magnitudes of the flows between the POIs are represented by the line widths.

arranged along the matrix diagonal. Since the pioneering work of Bertin, numerous matrix reordering algorithms have appeared [30].

Figure 7.7 represents our state transition graph in the form of node-link diagrams in different automatically produced layouts. Some layouts position the nodes on the plane irrespective of the edge weights while others (e.g., the so-called force-directed layout) treat the edge weights as forces in a virtual system of springs: repulsive forces exist between nodes that are not linked by edges and attractive forces exist between linked nodes. The algorithm strives to balance the repulsive and attractive forces. A survey of graph layout techniques can be found in [53].

In a node-link diagram, edges are represented by straight or curved lines connecting nodes. When the edges are directed, the direction is indicated by an arrow or by a specific line shape, such as decreasing width. In our examples, the direction is represented by the increasing line curvature. In a directed graph, it may be suitable for

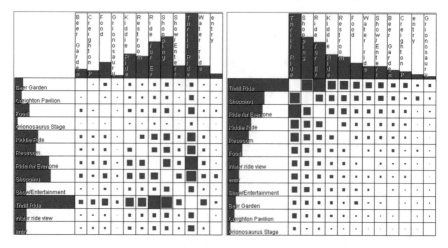

Fig. 7.6: Adjacency matrices representing the collective movements in the amusement park. Left: the rows and columns are ordered alphabetically; right: they are ordered according to the visit counts.

certain analysis tasks to replace each pair of opposite edges by single edge representing the difference between the edges, which can be interpreted as the "net flow". Examples can be seen in the lower images in Fig. 7.11.

A big problem in drawing node-link diagrams is numerous intersections of the edges. This problem can be dealt with by using edge bundling [71], in which multiple edges that have close positions and similar directions are united in a single thick line representing a "bundle" of edges. However, such edge bundles may be misleading due to representing artificial patterns that do not really exist in the data. This relates especially to geographically-referenced graphs.

Sometimes it may be useful for analysis to create specialised data-driven layouts, utilising additional information beyond the weights of nodes and links. For example, Figure 7.8 shows the same state transition graph as before. For projecting the nodes to a 2D plane, a data embedding method (namely, Sammon mapping) was applied to the time series of the hourly counts of the place visitors. Two different variants of the projections were obtained using different variants of normalisation of the same time series of the counts.

There exist approaches that combine node-link diagrams and adjacency matrices in the same display. Matrices are used to represent strongly connected groups of nodes, thus decreasing the number of intersecting lines that would be drawn in a pure node-link diagram. Obviously, neither adjacent matrices nor node-link diagrams scale to very large graphs. Graph drawing community continuously develops new methods for layout optimisation, while information visualisation and visual analytics communities focus on devising approaches to visually-steered graph aggregation and abstraction [90].

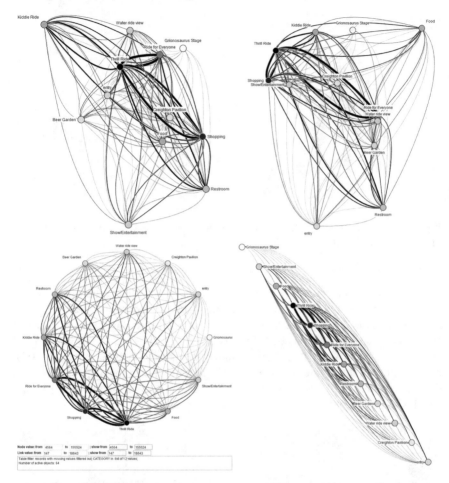

Fig. 7.7: Graph representing collective movements in the amusement park are drawn with different layouts: two variants of force-directed layouts (top), circular (bottom left) and diagonal (bottom right) layouts.

To some extent, scalability of graph visualisation can be improved by interactivity. Graphs can be filtered, omitting rendering of nodes and edges with low weights. This operation results in showing only the most important graph components, whereas potentially relevant contextual information may be lost.

A good instrument for analysing large graphs is abstraction through aggregation of nodes. The idea is that groups of nodes are replaced by single nodes, and the links connecting the original nodes from different groups are joint into links between the aggregate node. The nodes that have been merged in a single aggregate node and the links between them can be viewed by applying an interactive operation of zooming into a node. The grouping of nodes can be done according to different criteria. The

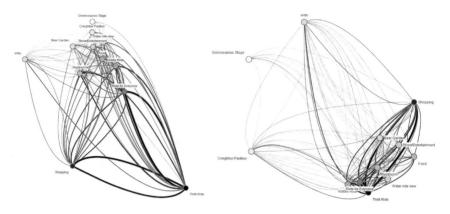

Fig. 7.8: Two variants of graph layout with node positions defined by a 2D embedding of time series of node attributes.

most obvious approach is to use domain-specific hierarchical relationships in which the entities represented by the graph nodes are involved, as demonstrated in Fig. 7.9. Another possible criterion for node grouping and following aggregation is strong connectivity among several nodes. This criterion is often combined with additional requirements, such as attribute-based similarity or spatial neighbourhood [143]. It can also be useful to aggregate parts of a graph according to the degree of interest: the parts that are currently in the focus of the analysis are shown in the necessary detail whereas the other parts are shown in an aggregated form [2].

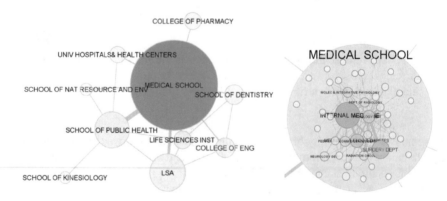

Fig. 7.9: Graph abstraction by aggregating nodes and edges. The original graph in this example represents collaborations (graph edges) between individual researchers (graph nodes). Left: The graph nodes have been aggregated according to the organisations in which the researchers work. Right: by zooming into an aggregated node, the corresponding sub-graph can be viewed at a lower level of aggregation (here: by departments within an organisation). Source: [57].

In analysis, it may be necessary to compare multiple graphs of similar structure, for example, graphs constructed from subsets of data. Figure 7.10 shows two graphs produced by querying the park attendance data by days, Friday on the left and Sunday on the right. Visual comparison of two images suggests high similarity of flows in these two days. It is very difficult to find differences between the two images. Representation of the net flows in the same two days (Fig. 7.11) helps to notice structural differences between the two graphs. Thus, Sunday differs from Friday in the existence of many out-flows from the "Thrill Rides" to the other categories of POI.

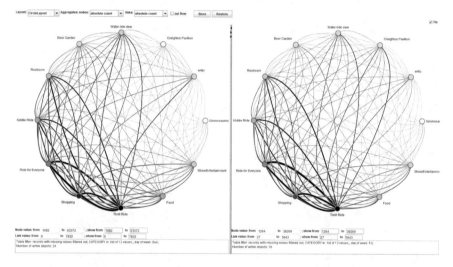

Fig. 7.10: Visual comparison of two graphs: data for Friday (left) and Sunday (right).

A useful operation supporting comparison is to compute difference graphs and show the differences explicitly. Different colour hues can be used for depicting positive and negative differences. For example, Figure 7.12-left shows the differences of the weights of the nodes and links on Sunday with respect to Friday, i.e., the Friday's weights are subtracted from the Sunday's weights. The positive differences are shown in red and negative in blue. The magnitude of the difference is encoded by the colour value for the nodes and by the line widths for the links. It is visible that almost all weights are substantially bigger on Sunday, with a few exceptions related to "Creighton Pavilion". To see structural rather than quantitative differences between these two graphs, it is necessary to normalise the weights to meaningfully comparable value ranges. For example, the weights of the nodes can be converted to their proportions in the total weight (i.e. divided by the sum of all weights), and the same operation can be applied to the links. Figure 7.12-right shows the differences between the normalised weights. Now the changes related to "Creighton Pavilion" are visible more clearly, as well as other changes, such as substantial decreases of the transitions from "thrill rides" to "shopping" and "ride for everyone".

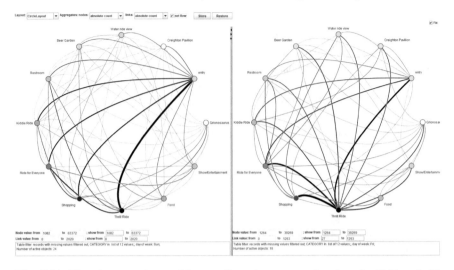

Fig. 7.11: Visual comparison of the net flows in two graphs: data for Friday (left) and Sunday (right).

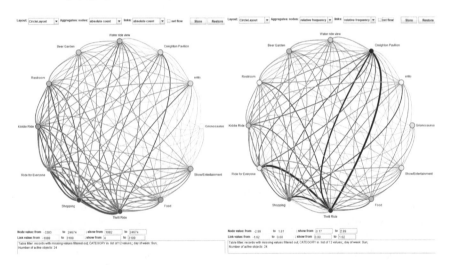

Fig. 7.12: Difference graphs, Sunday - Friday, based on absolute (left) and normalised (right) weights.

To summarise, interactive visualisation is an important instrument for exploratory analysis of graph data. Major display interaction and data transformation techniques include reordering and layout change, filtering of visual elements, creating graphs from different subsets of data, and computation of differences between graphs. However, applicability of interactive visualisations is limited to rather small graphs. Respectively, it is necessary to combine visualisations with computational methods of analysis. Possible computational approaches include clustering of graphs or their components. Grouping and aggregation of nodes allows dealing with much larger graphs, but calls for flexible approaches for multi-scale analysis.

7.5 Common tasks in graph/network analysis

One of common tasks in graph/network analysis is *community detection*. Community is a group of strongly interconnected homogeneous nodes. Communities often occur in social media (a group of users with common interests and opinions) and in bibliometrics (groups of authors with frequent cross-citations). Communities also occur in geographically referenced graphs, where neighbouring locations are often linked by massive commuter movements or frequent phone calls. Communities detected in a large graph enable multi-scale hierarchical analysis. There exists a number of community detection algorithms, which can be used in analytical workflows in combination with visualisations.

Many algorithms for graph analysis are being developed in data mining, where there is a special sub-area dedicated to graph mining. A general task in graph mining is to detect and extract patterns (sub-graphs) of interest. These may be sub-graphs having particular structure that may have special meaning to an analyst, or frequently occurring structural patterns, called motifs. Another general task is graph classification, when there are examples of graphs of different classes (e.g., graphs representing molecules of active and inactive chemical substances), and it is necessary to determine the class of a new graph. Both pattern detection and graph classification involve the basic task of *comparison of graph structures*, which is performed for whole graphs or parts of a graph. Automatic comparison is performed using appropriate distance functions (as mentioned in Section 4.2.7), whereas a human analyst needs visualisation support for understanding what is similar and what is different between graphs and what it means in regard to the phenomenon represented by the graphs.

An important problem is analysis of *dynamic graphs*, i.e., sequences of states of a graph where the changes from one state to the next may include adding and/or removing nodes and/or edges, changes of weights, and/or changes of other attributes. To represent all states of a dynamic graph in a uniform way, it is usually possible to construct a super-graph that includes all nodes and edges that ever existed. The changes of the graph can then be represented by time series of Boolean existence indicators for the nodes and edges and/or time series of the weights, where a zero weight indicates the absence of a node or edge. These time series can be analysed using established techniques for analysis of time-related data, see Chapter 8. Besides, graph states can be characterised by general features: counts of nodes and edges, statistics of weights (e.g. distribution histograms), selected graph-theoretic metrics and their distributions etc. Time series analysis is applicable also to time series of graph features.

Generally, there exist numerous implementations of graph-specifics computations, so finding a ready-to-use code may be quite easy for different development environments and application settings. However, combining computational methods with appropriate visualisation techniques in valid visual analytics workflows requires good understanding of data properties and semantics, engineering of relevant features, and selection of appropriate visualisation techniques for interpreting and evaluating computation results.

7.6 Further Examples

A variety of workflows for analysing graphs were proposed recently. These workflows combine decomposition of large graphs into components, feature engineering and selection, and the use of the so defined features for assessing the similarity of graphs and graph components by means of suitable distance functions. A frequently used operation is clustering of graphs or sub-graphs according to their similarity, which is determined by means of suitable structure-based or feature-based distance functions (see Section 4.2.7). In the following subsections, we shall consider two representative examples of visual analytics approaches to graph data analysis.

7.6.1 Analysis of graphs with multiple connected components

Paper [144] proposes an approach for identifying common structures in economic networks. A data set under investigation consists of financial and ownership data of German companies. Shareholding relationships between the companies are modelled as a directed graph. The companies are represented by nodes of this graph, while their shareholding relationships are reflected by directed (who-owns-whom) and weighted (size of the share) graph edges. The graph consists of 300,000 nodes.

It is well known to domain experts that the structure of shareholding among the German corporations contains several large groups of highly interconnected companies, which are not interesting for the purposes of analysis. Such groups are represented by strongly connected sub-graphs (communities), which can be easily identified by applying graph-theoretic measures (Section 7.2.2). In this way, the analysts detected two large communities, one with 115,000 nodes connected by 135,000 edges, and another with 20,000 nodes. The edges linking the entities within these two communities were excluded from the further analysis. The analysts and focused their attention on 40,000 much smaller sub-graphs with up to 110 nodes each. For characterising these smaller graphs, a library of graph-specific features has been created, including the following features:

- *general* features of the network, such as its size, degree of completeness measured as the ratio between the existent and potentially possible links, statistics of the edge weights, etc.;

- *reciprocity* features representing the distribution of bi-directional edges and their weights;

- *distance* features representing statistics of distances in a network, e.g., the diameter of a graph;

- *centrality* features characterising the distributions of different graph-theoretic measures of the nodes and edges, as described in Section 7.2.2;

- *motif-based* features representing frequencies of user-selected predefined structures (motifs, see examples in Fig. 7.13) that reflect particular domain semantics, such as out-star motifs for companies with many subsidiaries.

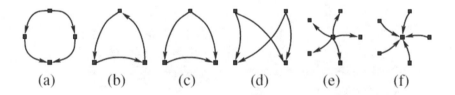

(a)   (b)   (c)   (d)   (e)   (f)

Fig. 7.13: Examples of graph motifs: a) Caro, b) Feedback, c) Feedforward, d) Bipartite, e) Out-star and f) In-star. Source: [144].

Computing the whole set of features for a large number of graphs is time consuming. However, this needs to be performed only once, and can be parallelised in a distributed computing environment. Since the computed features are numeric attributes, the characterisations of the graphs in terms of these features have the form of usual multidimensional data, which can be analysed using the whole range of methods designed for this kind of data; some of them have been discussed in Chapter 6. Feature selection methods (Section 4.3) can help to select and rank the features that are most important for a task at hand. Clustering and embedding operations (Sections 4.5 and 4.4) can be used for identifying groups of similar graphs.

In the analysis workflow we discuss here, embedding and spatialisation of multiple graphs is done by means of SOM. Examples of visualisations supporting data analysis with the use of SOM have been shown in Figs. 4.8 and 4.10. In those examples, the SOM results were visualised by means of grids with hexagonal cells. Some of the cells represented the nodes (neurons) of the SOM network while the cells between them where used to encode the distances between the data represented by the nodes; see the description of Fig. 4.8. In the work we discuss here, the result of SOM is visualised in the form of a rectangular matrix with the cells corresponding to the neurons, as shown in Fig. 7.14. The orange shading of the cells encodes the counts of the items (i.e., graphs) represented by the respective neurons. The distances between these groups of items (i.e., the U-matrix) are represented by grey shading of the spaces between the rectangular cells. The cells in the display include images of representative graphs for the respective groups of graphs. Each group can also be seen in detail, and the analyst can also examine the frequency distribution of the distances of the group members to the centroid, i.e., to the SOM neuron representing this group.

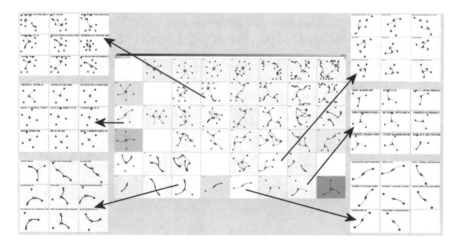

Fig. 7.14: A result of SOM embedding of a set of graphs is visualised in the form of a matrix with rectangular cells corresponding to the SOM nodes (neurons). Each cell contains an image of a representative graph, while the background colour indicates the size of the group of graphs represented by the node. The groups can be viewed in detail, as shown on the figure margins. Source: [144].

In addition to the matrix of the neurons, SOM component planes for the features used in the SOM derivation can be visualised as shown in Fig. 7.15) , which is similar to the earlier demonstrated Fig. 4.10 but uses a rectangular grid instead of hexagonal.

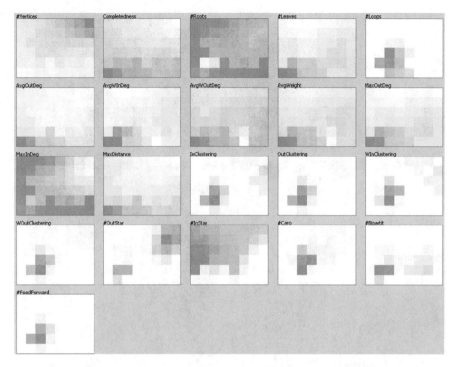

Fig. 7.15: Visualisation of the SOM component planes for the graph embedding. Source: [144].

The analyst can progressively refine large groups of graphs assigned to the same SOM node by selecting one or more cells and creating a new SOM for the selection, as demonstrated in Fig. 7.16.

The analysis reported in the paper [144] enabled discovery of several interesting patterns and supported understanding of the overall structure of the shareholding system. This was achieved through representation of the relationships in the form of a graph, detection and removal of parts that were not relevant to the goals of the study, efficient feature engineering, interactive progressive spatialisation and visual inspection of groups of similar graphs.

7.6.2 Analysis of dynamic graphs

Spatialisation by means of data embedding can also be employed for analysis of dynamic graphs. We have earlier (in Section 4.4.4) mentioned the Time Curve visualisation technique, which employs data embedding for representing dynamic phe-

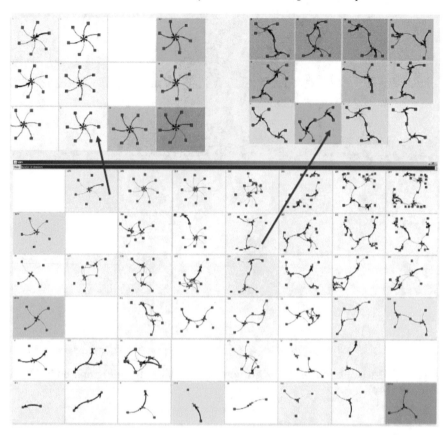

Fig. 7.16: Refinement of a SOM embedding. The upper two images show the SOM embeddings created from groups of graphs assigned to selected cells of the initial SOM. Source: [144].

nomena. The idea is to represent different states of a phenomenon by points in an embedding space and connect these points in the chronological order. This general idea can be applied, in particular, to dynamic graphs, as demonstrated in Fig. 7.17. The display shown in this figure was used in the analysis of the dynamics of a social network presented in paper [141]. More specifically, the data describe face-to-face contacts between individuals tracked by special sensors in the context of studying the mechanisms of spreading infectious diseases. Each state of a graph represents the contacts that occurred during one hour time interval within the time frame of the study. The graph nodes correspond to the individuals and the edges to the contacts.

To measure the dissimilarity between the states of the graph, the analysts transformed the dynamic graph into multidimensional time series, where each state was represented by a vector (i.e., an ordered set) of multiple numeric attributes. The vec-

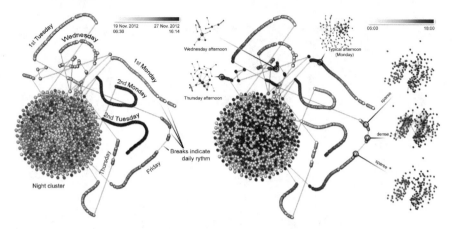

Fig. 7.17: States of a dynamic social network are represented by points in an embedding space. The points are connected by lines in the chronological order. On the left, the points are coloured according to the dates and times of the state occurrence; on the right, the same display is shown with the points coloured by the times of the day. The graph structures corresponding to several selected points are shown around the projection plot. Source: [141]

tors were obtained from the adjacency matrices representing the graph states by concatenating the matrix rows into sequences. After such a transformation, any distance function suitable for multiple numeric attributes, such as Euclidean distance, can be applied for assessing the dissimilarities between the graph states. Since the number of the resulting attributes is very high, it is reasonable to apply feature selection and/or dimensionality reduction techniques for avoiding the curse of dimensionality, as discussed in Chapter 4.

The analysts tried different data embedding methods for creating spatialisations of the dynamic network, in particular, PCA and t-SNE; the latter is used in Fig. 7.17. The methods provided somewhat different views of the network evolution, and it was useful to see both and compare them. Both methods generate embeddings with a large dense cluster of states corresponding to the night hours of all days, when there were no or almost no contacts. The t-SNE embedding shows more clearly that the days differed with regard to the sets of the contacts that occurred in these days, because the "trails" corresponding to the different days do not overlap or form joint clusters. The analysts have gained a number of interesting findings, which are reported in the paper [141].

The analysts also utilised the result of the PCA in a different way: they constructed a time graph where the x-axis corresponds to the time and the y-axis to the first principal component generated by PCA (Fig. 7.18). The time graph clearly shows the periodic repetition of the daily patterns of the social contacts.

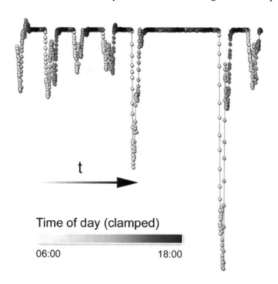

Fig. 7.18: Evolution of a dynamic social network is represented by a time graph where the vertical axis corresponds to the first principal component generated by the PCA method. Source: [141].

In analysing data describing players' movements and activities during a football (soccer) games, graph structures can be used for representing relationships between the players from the two opponent teams, such as the relationships of the pressure exerted by the defending players onto the attacking players [11]. In the analysis, a game is divided into a sequence of episodes, which can be defined according to various criteria. In particular, each episode can consist of a time interval of the continuous ball possession by one team followed by an interval of the ball possession by the other team. Each episode can be characterised by a bipartite graph [3] of the pressure relationships, where the edges are directed from the players of the defending team to the players of the ball-possessing team, and the weights of the edges represent the accumulated amounts of the pressure exerted during the episode. Figure 7.19 shows an example of such a pressure graph. Bipartite graphs have specific mathematical properties, which can be used to optimise the feature engineering for similarity search or clustering. Since the potential number of the edges in a pressure graph is limited to maximum $10 \cdot 10 = 100$, it is possible to represent any pressure graph by a vector of 100 edge weights. This representation is much more compact than the vectorised adjacency matrices in the work discussed previously.

---

[3] https://en.wikipedia.org/wiki/Bipartite_graph

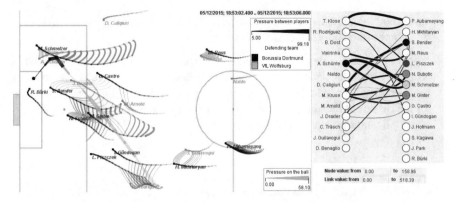

Fig. 7.19: Left: Player-to-player pressure forces during a game episode are visually represented on a map of the football pitch by curved linear symbols connecting the positions of the players. Right: For this game episode, a pressure graph shows the total amounts of the pressure from the players on the ball (by the darkness of the node shading) and on the opponents (by the line widths). Source: [11]

7.7 Concluding remarks

The main take-away messages from this chapter are the following:

- In many applications, a graph is a suitable structure for representing and analysing relationships between entities. The entities are represented by nodes of the graph and the relationships by edges. The strengths of the relationships can be expressed as weights of the edges.

- Relations between entities are not necessarily described in data explicitly, but they can be identified and extracted by means of computations chosen according to the domain semantics and analysis goals.

- There are several established techniques for visual representation of graphs. The most frequently used are matrix representations and node-link diagrams; both are not scalable to large graphs. Graph abstraction by aggregating nodes and links is a helpful approach to dealing with large graphs.

- Interactive operations may partly compensate for the limitations of the visualisations and facilitate analysis. The possible operations include filtering of the nodes and links in both representations, reordering of the rows and columns in a matrix, changing the layouts of node-link diagrams, and zooming of a graph represented in an aggregated form. For comparison of graphs with similar structure, visualisation of their differences and ratio may be useful.

- There exist an established set of graph-theoretic metrics and graph-specific computations, such as shortest path calculation, that can be used for characterising and comparing graphs, identifying their parts with special properties, decomposing graphs into sub-graphs, and other purposes.

- Some applications require considering multiple graphs and, often, time related graphs. Similarity search, embedding, and clustering are useful techniques in such contexts. The use of these techniques requires that an appropriate measure of the dissimilarity (distance) between graphs is chosen or defined according to the application semantics and analysis goals.

7.8 Questions and exercises

Consider typical activities that you perform during a usual week or a working day. This may include sleeping, preparing food, eating, Potentially interesting relationships are transitions between the activities: how often, when and where they happen, and how long are the time lapses between two consecutive activities. Think about a possible representation of your data, their expected properties and complexities, consider how to check data quality, formulate some analysis tasks, and design appropriate data analysis approaches and procedures. What kinds of data transformations, computational methods and visualisations are necessary / useful / nice to have?

You can extend this task by considering the associations between the activities and the locations in which they take place. In this case, you may need to represent your data by a bipartite graph, with one subset of nodes representing the locations and another the activities. Again, define the data collection and processing methods, analysis tasks, and corresponding analytical procedures.

Chapter 8
Visual Analytics for Understanding Temporal Distributions and Variations

Abstract There are two major types of temporal data, events and time series of attribute values, and there are methods for transforming one of them into the other. For events, a general analysis task is to understand how they are distributed in time. For time series, as well as for events of diverse kinds, a general task is to understand how the attribute values or the kinds of occurring events vary over time. In analysing temporal distributions and variations, it is essential to account for the specific features of time and temporal phenomena, particularly, recurring cycles and temporal dependence. To see patterns of temporal distribution or variation, people commonly apply visual displays where one dimension represents time and the other is used to show individual events, attribute values, or statistical summaries. To see the data in the context of temporal cycles, a common approach is to use a 2D display where one or two cycles are represented by display dimensions. For large and/or complex data, visual displays need to be combined with techniques for computational analysis, such as clustering, embedding, sequence mining, and motif discovery. We show and discuss examples of employing such combinations in application to different data and analysis tasks.

8.1 Motivating example

We have already dealt with temporal distributions when we investigated the epidemic outbreak in Vastopolis in Chapter 1, and we briefly discussed possible patterns in temporal distributions in Section 2.3.3. We have introduced the term *event* for discrete entities (things, phenomena, actions, ...) that appear or happen at some time moments and disappear immediately (instant events) or exist for limited time (durable events). When events are numerous, like the microblog posts in Chapter 1, we often want to investigate their *temporal frequency*, that is, when how many events happened and what is the pattern of the *change* of the event number over time. To

© Springer Nature Switzerland AG 2020
N. Andrienko et al., *Visual Analytics for Data Scientists*,
https://doi.org/10.1007/978-3-030-56146-8_8

see the variation of the event number over time, it is convenient to use temporal histograms like in Figs. 1.5 and 1.6. Figure 1.14 demonstrates that the bars in a time histogram can be divided into coloured segments to represent counts of events of different types, like the two distinct diseases in Vastopolis. Observing patterns in these time histograms was essential for our analytical reasoning concerning the time of the outbreak and its development trends.

Further examples of time histograms appear in Fig. 2.6. The upper image is similar to the histograms used in Chapter 1: the horizontal axis represents time, which is divided into intervals of equal length (bins), and the vertical dimension is used to represent the counts of the events in these intervals by the heights of the bars. Such a histogram is called linear: the time is represented as a line. However, the linear representation of time may be not ideal (or, rather, not sufficient) when the variation of the number of events (or any other kind of temporal variation) may be related to temporal cycles.

Thus, in the upper image of Fig. 2.6, we see high peaks in the event number that occur at approximately equal time intervals between them. This is not surprising because the histogram represents the events of taking photos of cherry blossoms. The underlying dataset consists of metadata records of geolocated photos taken on the territory of the North America in the years from 2007 till 2016 that were published online through the photo hosting service flickr[1] and have keywords related to cherry blossoming in the titles or tags attached. It is natural that these photo taking events are much more frequent in spring than in other seasons of a year, which means that the number of events varies according to the annual (seasonal) time cycle.

While the linear histogram informs us that the distribution is periodic, it does not explicitly show the relevant time cycles and is not convenient for checking if the peaks occur at the same times in each year. This information can be seen much better when we use a two-dimensional display, like the 2D time histogram in the lower image of Fig. 2.6, in which one dimension represents positions within a time cycle (such as days in a year) and the other dimension represents a sequence of consecutive cycles (years). We obtain a matrix the cells of which may contain certain symbols representing counts of events or values of another measure, for example, the number of distinct photographers that made the photos. In Fig. 2.6, bottom, the horizontal dimension (matrix columns) represents the days of a year grouped into 7-day bins and the vertical dimension (matrix rows) represents the sequence of 10 years from 2007 at the bottom to 2016 at the top. We see that the periods of mass photo taking activities related to cherry blossoming did not happen at the same times in different years but they happened earlier in the years 2011 and 2012 and later in the following three years.

Time histograms are of good service when we deal with large numbers of instant events and need to investigate the temporal variation of the frequencies of event occurrences. However, when we have durable (but not so numerous) events, we may

---

[1] https://flickr.com/

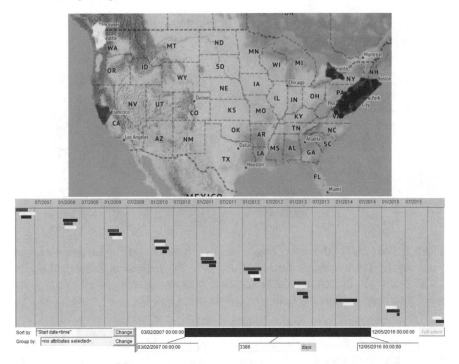

Fig. 8.1: The map shows the regions where the events of mass photo taking happened. In the timeline display below the map, the horizontal dimension represents the time, and the horizontal bars represent the periods when the events were happening. The bars are coloured according to the regions where the events took place.

need to investigate and compare the periods of event existence. Let us use the same example of taking photos of cherry blossoms, but now we want to consider not the occurrences of the elementary photo taking events but large spatio-temporal concentrations (clusters) of photo taking activities, when people make many photos in some area over multiple days. We have detected 34 clusters that include at least 50 elementary events (the sizes of the clusters obtained range from 57 to 711 members). We treat these clusters as events of mass photo taking. The map in Fig. 8.1, top, shows that the events mostly happened in four regions: Northeast, including Washington D.C., New York City, Philadelphia, and Boston (blue), area of Vancouver and Seattle (yellow), San Francisco and its neighbourhood (red), and Toronto (purple). There was also an occasional event in the area of Portland (green, south of the Vancouver-Seattle region). Unlike the individual photo takings, these events have duration. In Fig. 8.1, bottom, a timeline display (Section 3.2.4), represents the times when the events were happening by the horizontal positions and lengths of the horizontal bars, which are coloured according to the event regions. The bars are ordered according to the start times of the events and make compact groups corresponding to different years. This display allows us to compare the times of event happening in different regions in each year, however, it is not so easy to make comparisons between the years.

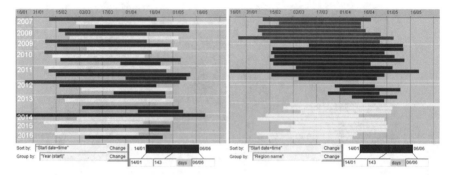

Fig. 8.2: The times of the mass photo taking events have been aligned relative to the yearly cycle. The horizontal dimension of the display represents the days of a year (in all years). On the left, the bars representing the events are grouped by the years when they happened, on the right – by the regions.

A technique that supports comparisons between cycles, like the yearly cycles in our case, is transformation of the event times from absolute to relative with respect to the cycle start. In our case, simply speaking, we ignore the differences in the years of the event happening and treat them as if they would happen in one year. We build a timeline display according to the relative dates of the events (Fig. 8.2). We can now group the bars by the years (Fig. 8.2, left) and compare the beginning and ending times of the events across the years focusing our attention on the bars of particular colours or on the whole bar groups corresponding to the different years. We can also group the bars according to the regions (Fig. 8.2, right), which allows us to observe the variation of the event times in the regions and compare the relative times of the groups of events that happened in the different regions.

Apart from the time moments or intervals of the existence, events may have various other attributes, for example, as shown in the table in Fig. 8.3. These attributes can be analysed as multivariate data considered in Chapter 6.

However, data having temporal components are not limited to events only. The most ubiquitous type of temporal data are *time series*, which consist of attribute values specified for a sequence of time steps (moments or intervals), usually equally spaced. We can easily obtain time series from our set of photo taking events: we just divide the whole time span of the data into equal time intervals, such as days or weeks, and compute for each interval how many events happened and, possibly, other statistics, such as the number of distinct users or the average time of the day when the photos were taken. Such aggregation can be done for all events taken together or separately for each region. In the latter case, we shall obtain multiple time series; thus, Figure 8.4 shows time series of weekly counts of the photo taking

	Number of positions	UserID, N different values	Start date	End date	Duration (day)	Start date+time: day of year	DateTaken, median: day of year	End date+time: day of year
2014 Northeast	411	233	06/01/2014	17/06/2014	161	6	103	168
2012 California	119	70	14/01/2012	25/05/2012	133	14	105	146
2015 Vancouver-Seattle	151	77	29/01/2015	25/04/2015	86	29	74	115
2010 California	100	61	30/01/2010	02/05/2010	92	30	87	122
2011 Vancouver-Seattle	217	131	30/01/2011	19/05/2011	108	30	99	139
2009 California	98	70	31/01/2009	28/04/2009	87	31	97	118
2007 California	57	43	03/02/2007	07/05/2007	92	34	86	127
2011 California	117	65	03/02/2011	15/05/2011	101	34	95	135
2010 Vancouver-Seattle	129	88	06/02/2010	26/04/2010	79	37	69	116
2016 Vancouver-Seattle	102	60	08/02/2016	20/04/2016	72	39	79	111
2011 Northeast	711	399	09/02/2011	04/06/2011	115	40	93	156
2013 California	101	65	10/02/2013	14/05/2013	93	41	103	134
2009 Northeast	497	340	11/02/2009	19/05/2009	87	42	95	139
2015 Northeast	374	246	11/02/2015	19/05/2015	97	42	103	139
2008 Northeast	461	290	12/02/2008	02/06/2008	111	43	96	154
2012 Northeast	635	367	12/02/2012	06/05/2012	83	43	94	127
2010 Northeast	646	354	13/02/2010	27/05/2010	103	44	93	147
2008 California	72	52	17/02/2008	27/04/2008	70	48	99	118
2013 Vancouver-Seattle	186	137	17/02/2013	21/05/2013	93	48	91	141
2008 Vancouver-Seattle	111	83	21/02/2008	24/05/2008	93	52	94	145
2007 Vancouver-Seattle	116	71	27/02/2007	05/07/2007	128	58	91	186
2014 Vancouver-Seattle	143	79	27/02/2014	21/05/2014	82	58	95	141
2013 Portland	87	48	02/03/2013	26/04/2013	55	61	83	116
2012 Vancouver-Seattle	150	104	02/03/2012	19/05/2012	78	62	98	140
2013 Northeast	629	387	04/03/2013	25/05/2013	82	63	101	145
2016 Northeast	283	183	05/03/2016	11/05/2016	67	65	98	132
2009 Vancouver-Seattle	161	106	08/03/2009	24/05/2009	77	67	103	144
2007 Northeast	299	226	15/03/2007	03/06/2007	80	74	93	154
2012 Toronto	68	56	30/03/2012	13/05/2012	44	90	104	134
2010 Toronto	63	46	06/04/2010	10/05/2010	34	96	113	130
2011 Toronto	76	68	10/04/2011	01/06/2011	52	100	128	152
2013 Toronto	72	55	21/04/2013	29/05/2013	38	111	125	149
2015 Toronto	59	42	30/04/2015	18/05/2015	18	120	127	138

Fig. 8.3: A table display shows various attributes of the mass photo taking events.

events by the regions. Here we apply a standard visualisation techniques known under multiple name variants: line chart or line graph, time series plot (or graph), or, simply, time plot or time graph (Section 3.2.4). The horizontal axis represents time, and values of a numeric attribute are represented by vertical positions. The positions corresponding to consecutive time steps are connected by lines. The lines in Fig. 8.4 are coloured according to the regions.

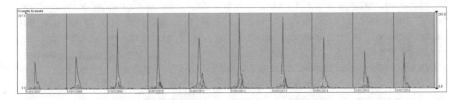

Fig. 8.4: Time series of the weekly counts of the photo taking events by the regions.

Not only time series can be generated from events, but also events can be extracted from time series. Thus, in our time graph in Fig. 8.4, we see a number of peaks, which can be treated as events, extracted from the time series, and explored separately.

Having these examples in mind, let us proceed with a more general discussion of the types and properties of temporal data.

8.2 Specifics of temporal phenomena and temporal data

As we have seen by examples of events and time series, the temporal aspect of a phenomenon may have different meanings. It may be the time interval or moment when something existed or happened. Things (entities) that happen at some times or exist during a limited time period are called *events*. It can also be a succession of time steps (moments or intervals) in the life of an evolving phenomenon whose state (properties and/or structure) changes over time. A chronological sequence of states of such a phenomenon is called a *time series*.

In analysing events and time series, it is necessary to account for the structure and properties of time. Generally, time is a continuous sequence of linearly ordered elements (moments), and it is often represented as an axis along one dimension of a visual display. However, the time in which everything (including us) exists is not only linear but also consists of recurrent cycles: daily, weekly, seasonal, etc. Many phenomena and events depend on some of these cycles; for example, vegetation phenomena (such as cherry blossoming) depend on the seasonal cycle, and activities of people depend on the daily, weekly, and seasonal cycles. There may also be other, domain-specific, temporal cycles, such as cycles of climatic processes, development cycles of biological organisms, or manufacturing cycles in industry.

Most temporal phenomena have the property of *temporal dependence*, also known as *temporal autocorrelation*: values or events that are close in time are usually more similar or more related than those that are distant in time. For example, the current air temperature is usually quite similar to what was ten minutes ago and may differ more from what was a few hours ago. The temporal dependence needs to be taken into account, but it should not be thought that the degree of similarity or relatedness is just proportional to the distance in time. It is more reasonable to expect that the dependence exists within some quite narrow time window the width of which depends on the dynamics of the phenomenon, that is, how fast it changes. The time range of the temporal dependence is often modified by temporal cycles. For example, the air temperature observed at noon may be more similar to the temperature that was at noon a day ago than to the morning temperature of the same day.

Statistics has developed specific methods for time series analysis that account for the existence of temporal dependence and cycles. Thus, *exponential smoothing*[2] is an approach to time series modelling that gives higher weights to more recent values in predicting the next value. There are methods that model periodic variation (called "seasonality" in statistics, irrespective of the length and meaning of the period) and can thus account for temporal cycles. In visualisation, temporal dependence and its effect range can be revealed if distances in time are represented by distances in the display space. The existence of periodicity can be detected from displays where times are represented by positions along an axis, like in Figs. 8.1, bottom, and 8.4. Comparisons between periods (cycles) can be easier to do after aligning the periods along a common relative time scale, as in Fig. 8.2.

As any data, temporal data may have errors, such as wrong recorded values or wrong time references, and other problems, in particular, missing data. Visualisation can often exhibit such problems quite prominently. We recommend you to read the paper expressively called "Know Your Enemy: Identifying Quality Problems of Time Series Data" [61]. It describes a range of automatic and visual methods for problem detection.

In looking for possibly wrong values, you need to account for the property of temporal dependence: not only a value that is dissimilar to all others may be wrong but also a value that appears quite realistic by itself but differs much from its nearest neighbours in time. It may be reasonable to remove suspicious values in order to avoid wrong inferences and the risk of missing important pieces of true information, which may be less prominent than the erroneous data.

Since most of the existing analysis methods cannot work when some data are missing, it is quite usual to "improve" data by inserting ("imputing") some estimated values. Many methods exist for this purpose; see, for example, [74]. Making use of the property of temporal dependence, they estimate the missing values based on available values in the temporal neighbourhood. However, different methods may produce quite different estimations. The plausibility of the estimations depends on how well the character of the variation is understood and accounted for in the chosen method. If you really need to apply value imputation, we highly recommend you first to explore the data to understand the variation pattern and to be able to choose the right method. Having applied the method, it is wise to compare the previous and new versions of the data using visualisations: the visible patterns should be similar, and no additional patterns should appear in the new version.

---

[2] https://en.wikipedia.org/wiki/Exponential_smoothing

8.3 Transformations of temporal data

In our example (Section 8.1), we applied two transformations to data describing events: *integration* of multiple elementary events (individual photo taking actions) into larger events (periods of mass photo taking, Fig. 8.1) and *aggregation* of a set of events by time intervals into time series of event counts (Fig. 8.4). Integration is usually applied to temporal clusters of events, that is, many events happening during a time interval with a much higher temporal frequency than before and after that. Such a cluster may be treated as a higher-level event. Additionally to the temporal density, other criteria may be applied for defining event clusters. Thus, in our example, we grouped the events by the regions in which they occurred. If our dataset included photos of different blooming tree species (magnolias, apple trees, lindens, ...), we could group the events by the tree species. After event groups are defined, each group is transformed into a single event whose existence time spans from the time of the appearance of the earliest event of the group until the time of the disappearance of the latest event. The new events may be characterised by multiple attributes, such as the group size and various statistics that can be derived from attributes of the member events, as the number of distinct user IDs in our example (Fig. 8.3).

For aggregating events into time series, as in Fig.8.4, the time range of the data is divided into intervals, usually of equal length, for example, 7 days (one week), as in Fig.8.4. The events are grouped by the time intervals. For these groups, event counts and other possible statistics are obtained, as in integration. The sequence of the event counts or values of another statistics makes a time series. The aggregation by time intervals can be done not only for the whole set of events but also for subsets defined based on various criteria, such as event types or, as in our example, the regions in which the events happened. In such a case, the aggregation result consists of multiple time series, one time series per each subset. In our example, we obtained one time series for each region.

Besides aggregation by intervals along the time line, as in Fig.8.4, it is also possible to aggregate events by intervals within time cycles, for example, by months of a year, by days of a week, by hours of a day, or by both days of the week and hours of the day. In such aggregation, events from time intervals that are disjoint on the time line but have the same position in a cycle are put together. For example, events that happened from 9 till 10 o'clock on any Saturday form a single aggregate. As an example, we have aggregated the photo taking events by the regions and the combination of the day of the week and hour of the day. As a result, we obtain time series with the length of 168 time steps (= 7 days × 24 hours); the previous time series in Fig.8.4 had the length of 521 time steps (≈ 3,635 days : 7 days). In Fig. 8.5, the new time series are represented in two ways: by lines in a time graph and by heights of bars in 2D histograms with the rows corresponding to the days of the week from 1 (Monday) to 7 (Sunday) and columns to the hours of the day from 0 to 23. Both the lines in the time graph and the bars in the histograms are coloured

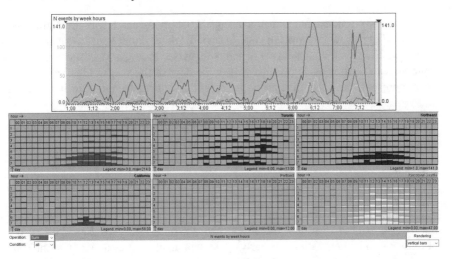

Fig. 8.5: Time series of event counts by the days of the week and the hours of the day are represented by lines in a time graph (top) and by bar heights in 2D histograms with the rows corresponding to the days and the columns corresponding to the hours (bottom).

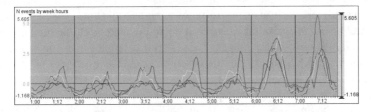

Fig. 8.6: The time series shown in Fig. 8.5, top, have been smoothed and normalised.

according to the regions the time series refer to. The histogram with the dark grey bars summarises the time series for all regions. Not surprisingly, both the time graph and the histograms show that people took more photos of cherry blossoms at the daytime than in the night and, generally, more events happened on the weekend than on the week days. However, there are differences between the regions. For example, in California (red), very few photos were taken in the weekdays, and notably more events happened on Sundays than on Saturdays. The photos were mostly taken from hour 10 till hour 15. On the Northeast (blue), the weekend activities also dominate, but many events happened in the weekdays as well, especially in the afternoons and evenings of Fridays. The weekend events were more spread along the day time: the counts increase starting from hour 7 and remain high until hour 19 on Saturday and hour 18 on Sunday. The weekend patterns in the Vancouver-Seattle region (yellow) are similar to those on the Northeast but the activities during the weekdays were also relatively high (compared to the highest counts for this region).

Generally, two-dimensional views as in Figs. 2.6, bottom, and 8.5, bottom, are good for observing periodic patterns in data. In such displays, either both dimensions represent two time cycles (as the weekly and daily cycles in Fig. 8.5), or one dimension represents some time cycle (as the cycle of the days of the year in Fig. 2.6) and the other represents the linear time progression (as the sequence of the years in Fig. 2.6).

As already mentioned in Section 8.1, it is possible not only to construct time series using events but also extract events from time series for a separate analysis. Potentially interesting are such events as peaks, drops, and abrupt increases or decreases.

Figure 8.2 demonstrates another transformation that is possible for temporal data: replacement of the absolute time references, which refer to positions along the time line, by references to positions in a temporal cycle. In this way, we aligned the events of mass photo taking within the yearly cycle, which allowed us to compare the start and end times and the durations of the events in the different years. This can be seen as transforming the event times from absolute to relative. Relative times are also involved in time series obtained by aggregating events by time cycles.

Various transformations can be applied to values in time series. *Smoothing* is used to diminish fluctuations and make the overall patterns more clear. Smoothing methods are well described in statistics textbooks and in educational materials available on the web. *Normalisation* may be useful when there are multiple time series with substantially differing value ranges. Thus, in our examples, the time series of the Northwest has much higher values than the others, and the time series of Toronto and Portland have much smaller values. If we want to compare the temporal variation patterns in these regions, we can replace the absolute counts by the differences from the means of the time series (for each time series, its own mean is used). The ranges of these absolute differences also differ much between the time series; therefore, we divide them by the standard deviations of the respective time series. We have applied such normalisation (after a little bit of smoothing) to the time series shown in Fig. 8.5; the result can be seen in Fig. 8.6. Now we can see the similarities and differences between the variation patterns. Thus, we see quite high similarity between the patterns of Vancouver-Seattle (yellow) and Portland (green). We prominently see the different pattern of Toronto (purple) with its high peaks in the evenings of Thursday and Friday (days 4 and 5) and the difference of California (red) with its particularly high peak at the Sunday noon. There may be other approaches to value normalisation (some have been discussed in Section 4.2.9), for example, relative to the minimal and maximal values of the time series. Besides, other transformations can be applied to the values. Thus, the values can be replaced by the differences with respect to the previous values, or with respect to the values for a selected time step, or with respect to the values in a chosen time series, and so on.

Figure 8.7 schematically summarises the possible transformations that can be applied to two major types of temporal data, events and time series.

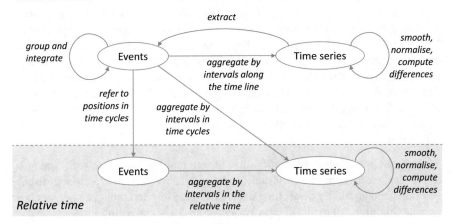

Fig. 8.7: A summary of the possible transformations of temporal data.

8.4 Temporal filtering

As discussed in Section 3.3.4, data filtering (or querying) is a very important operation in data analysis. A specific kind of querying/filtering that is used in analysing
temporal data is, obviously, selection of data based on their temporal references, i.e.,
temporal filtering. There are two common approaches to temporal filtering. First, it
can be done by selecting a continuous time interval within the time range of the
data. This kind of filter adheres to the linear view of time, in which time is treated
as a continuous linearly ordered sequence of time instants. Another approach is to
filter data according to the positions of their time references within one or more
time cycles. These two types of filtering can be called linear and cyclic, respectively. Please note that temporal filtering, in essence, involves two operations: first,
selection of one or more time moments or intervals and, second, selection of data
whose temporal references correspond to the selected time(s). The linear and cyclic
methods of filtering differ in the criteria used for selecting times. A less common
method of temporal filtering is selecting times based on fulfilment of some query
conditions defined in terms of time-related components of data, such as values attained by time-variant attributes, or existence of particular entities, or occurrence of
certain events [11, 9]. Similarly to cyclic temporal filtering, such conditions-based
temporal filtering can select multiple times.

Irrespective of the method of selecting times, the correspondence between time references in data and the times selected can be defined in terms of different temporal
relationships, depending on what you wish to select. It can be exact coincidence,
containing or being contained, overlapping, or even being before or after the selected times within a certain temporal distance.

When there are several temporal datasets, temporal filtering can be applied to all of them simultaneously or to some of them, according to the decision of an analyst. Simultaneous temporal filtering of two or more datasets is useful for exploring relationships between the respective phenomena and for having a coherent view of everything that happened in the selected times. Individual application of temporal filtering to some of the datasets may be useful for seeing particular parts of these datasets in the overall context of the remaining data. It is also possible to apply different filters to different datasets, which may be helpful for finding temporally lagged relationships between phenomena or asynchronous similarities in their developments.

When you apply a filtering method that selects multiple time intervals, you need to see the selection results in visual displays. When you have a temporal display showing the data for the whole time under study, the selected times can be somehow marked (highlighted) in this display. When you use a visualisation showing the data in aggregated way, the aggregation needs to be applied to the selected data, and the display needs to be updated. Aggregation of the selected data pieces is quite often a useful operation that helps to obtain an overview of what has been selected.

In the following section, we shall introduce a selection of visual analytics approaches to analysing temporal data of various kinds.

8.5 Analysing temporal data with visual analytics

8.5.1 Events

In our example, we were dealing with a large number of events of the same kind. We were interested to know *how many* events happened when and where but did not care about the specifics of the individual events. In such cases, helpful operations are integration or aggregation of events based on their temporal neighbourhood (possibly, in combination with some other criteria). However, these operations are not applicable when it is important to know *what* events appeared when and/or in what sequence. Here we shall discuss three kinds of analysis tasks referring to event sequences:

- In a single event sequence, observe patterns of event repetition.

- Having a large number of event sequences, find common patterns of event arrangement.

- Having a set of standard event sequences with the same relative ordering, detect temporal anomalies.

Repeated sub-sequences in a single sequence. The first task can refer not only to a sequence of events but, generally, to any sequence whose elements can be treated as symbols from some "alphabet" in a broad sense, that is, some finite set of items. Thus, it may be a sequence of nucleotides in a DNA molecule, notes in a melody, or words in a verse. On the one hand, each occurrence of a particular symbol can be seen as an event. Such an event is usually not related to the absolute time (in terms of the calendar and clock), but it has a specific position in the "internal time" of the sequence. The "internal time" may be a simple chain of discrete indivisible and dimensionless units, each containing one event, or, as in a melody, the time can be continuous, and events (including pauses) may differ in duration. On the other hand, to represent a sequence of happenings which we intuitively regard as events, it is often possible to create a suitable "alphabet". For example, it is not difficult to imagine an alphabet suitable for representing events in a football (soccer) game, such as passes, shots, tackles, faults, etc. It may be quite a large alphabet if we want to encode event details, such as who passed the ball to whom, but even in such a case we can expect that particular symbols and sub-sequences of symbols may occur repeatedly. If only the frequencies of the re-occurrences are of interest, methods from statistics and data mining suffice. A need in visualisation arises when we want to see where in the whole sequence the repetitions occur.

Fig. 8.8: Using arc diagrams for revealing the structure of a musical piece (left) and a text (right). Source: [147]

Figure 8.8 demonstrates a technique called Arc Diagram [147], which addresses this need. A computational analysis algorithm is applied to find matching pairs of longest non-overlapping sub-sequences of symbols. In the visualisation, one of the display dimensions (horizontal in Fig. 8.8) represents the internal time of the sequence. Matching sub-sequences are connected by semi-transparent arcs whose widths are proportional to the lengths of the sub-sequences and heights to the distances between them. The amount of shown detail can be controlled through filtering by the sub-sequence length or by inclusion of a particular symbol.

It is not always necessary that matching sub-sequences are identical; they may be similar or related in some other way. Thus, in analysing a football game, all sequences of passes starting with some player A passing the ball and ending with the same player A receiving the ball back may be considered similar irrespective of the intermediate passes. Generally, to match sub-sequences, one needs to define a *distance function* (as introduced in Section 4.2), that is, a method that, for two given sequences, returns a numeric value expressing the degree of their similarity or relatedness: the more similar (related), the smaller the value. If the value (called 'distance') is below a chosen threshold, the sequences are considered as matching and, hence, are linked in the visualisation.

A suitable distance function needs to be chosen depending on the data specifics and the analysis goals. Thus, if it is sufficient to treat the data as mere sequences of symbols, it is possible to apply the sequence-oriented distance functions that have been mentioned in Section 4.2.6. More sophisticated distance functions need to be defined when it is necessary to account for the durations of the events and the time intervals between them. There may be a need in a quite specific distance function, as, for example, for assessing the similarity between sequences of ball passes based on the first passer and the last pass receiver.

Common patterns of event arrangement in multiple sequences. Let us imagine that we have a lot of event sequences, for example, electronic health records of thousands of patients describing happenings, such as a stroke or a heart attack, as well as medical tests, treatments, and prescriptions. We (or medical doctors) need such a representation of the data that commonalities and differences between the sequences could be studied. Figure 8.9 demonstrates a possible approach. It involves two major operations: alignment and aggregation. Alignment means selecting in each sequence a reference point, such as occurrence of a particular kind of event, and transforming the original times of all events to relative times with respect to the time of the reference point, which is assumed to be zero. Hence, the times of the earlier events will be negative, and the times of the later events will be positive. In this way, all sequences are put in a common temporal reference system. Thus, in Fig. 8.9, event sequences from health data records are aligned in relation to the event 'stroke'. On the right of the display, the sequences are represented one below another so that the alignment points have the same horizontal position.

For the aggregation, the sequences are grouped according to the occurrence of sim-ilar events (for example, events of the same kind) at close relative times. The occur-rences are then represented in an aggregated manner as demonstrated in the mid-dle of Fig. 8.9. Each coloured block represents a group of occurrences of similar events. The horizontal position of a block corresponds to the relative times of the occurrences, and the height is proportional to the number of the occurrences. The colour represents the kind of the events. The red colour in Fig. 8.9 corresponds to taking drug A, dark blue to drug B, and light blue to headache. Thus, we see that about 60% of the patients were taking drug A during some time prior to the stroke, but there was an interval before the stroke when the drug was not taken any more. The remaining 40% began to take drug A shortly before the stroke and continued taking it after the stroke. About two thirds of the former group of patients (40% of all) received drug B shortly after the stroke. About a half of the second group had

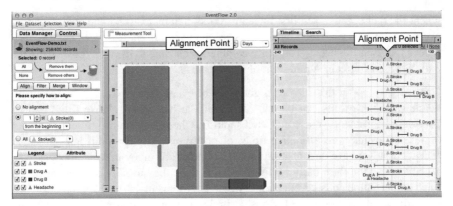

Fig. 8.9: Aggregation of event sequences in EventFlow. Source: [101]

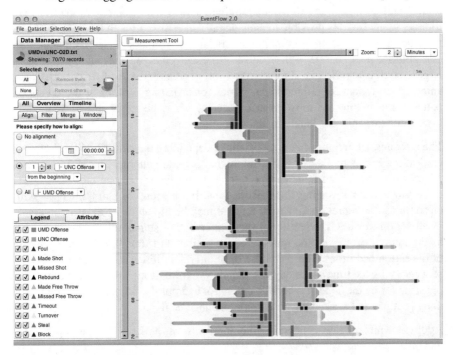

Fig. 8.10: An example of using EventFlow for analysing a basketball game. Source: [101]

headache before beginning to take drug A, and about one fourth of it started taking drug B after the stroke while continuing to take drug A.

This simple example demonstrates the main idea of the approach. In Fig. 8.10, it is applied to data describing a basketball game, which includes such events as successful and unsuccessful shots, rebounds, steals, etc. The game is divided into episodes where the ball is possessed first by team A and then by team B. The corresponding event sequences are aligned by the possession change from team A to team B. The blocks representing event groups are coloured according to the team possessing the ball: light red for team A and light blue for team B.

Generally, to apply this idea to a given set of event sequences, it is necessary to define (a) what events can be considered similar, (b) what relative times can be considered sufficiently close, (c) what should be the representative time for a group of similar events whose relative times of occurrence are close but not the same. Imagine that we have data resulting from a population survey in which people were asked to describe their daily routines: what actions they usually do at what times. Each daily routine consists of actions (which are events in this case) and times when they are usually performed. Obviously, similar actions are first of all actions of the same kind: wake up, exercise, take shower, shave, etc. However, it is also possible to treat taking shower and shaving as similar actions as they belong to the hygienic routine. Likewise, it is possible to consider all kinds of sport activities as similar, or to distinguish exercising at home, outdoors, and in a fitness centre. The definition of similarity may change in the course of the analysis: you may start with taking larger action categories for finding high-level general patterns of daily behaviours and then progressively refine the categories for revealing variants of these general patterns. For finding the most general patterns, it may be reasonable to omit some events, for example, the lunch break during the work.

The closeness of action times is also defined depending on the desired level of generalisation. For example, for extracting highly general patterns, time differences of 30 minutes or even an hour can be tolerated, but the difference threshold needs to be lowered for revealing finer distinctions. For a group of similar actions with close times, the representative time interval may be the union of the time intervals of all actions, or the intersection of these intervals, or something in between, such as the interval from the first quartile of the action start times to the third quartile of the action end times. There are also different possibilities for aligning the action sequences: by the time of the day, for example, starting from 3 AM, when most people sleep, or by the time of a certain action, for example, waking up or beginning of daily occupation (work, study, household management, etc).

Additional information concerning the approach can be found in the paper [101] and on a dedicated web site[3].

Anomalies in standard event sequences. Let us imagine a factory where some products are manufactured. To produce each item, a certain standard sequence of operations needs to be performed, and each operation takes a certain time. Ideally, the operation sequences for producing each item are identical: the same operations are done in the same order and at the same relative times with respect to the sequence start time. Moreover, the intervals between the start times of the consecutively performed sequences should also be equal. However, the real production process can

---

[3] https://hcil.umd.edu/eventflow/

deviate from this ideal. To detect deviations and understand how to improve the performance, is necessary to analyse data comprising lots of operation sequences. The data need to be represented so that anomalies (that is, deviations from the ideal procedure) are easy to spot.

To tackle this problem, Xu et al. [153] take the idea presented in Fig. 8.11. This is a widely known visualisation of the daily schedule of the trains connecting Paris and Lyon created by Étienne-Jules Marey[4] in the 1880s. The train stops are positioned along the vertical axis according to their distances from each other. The horizontal axis represents time. The train journeys are represented by diagonal lines running from top left to bottom right (Paris – Lyon) and bottom left to top right (Lyon – Paris) respectively. The slope of the line gives information about the speed of the train – the steeper the line, the faster the respective train is travelling. Horizontal sections of the trains' lines indicate if the train stops at the respective station at all and how long does a stop last. Besides, the density of the lines provides information about the frequency of the trains over time.

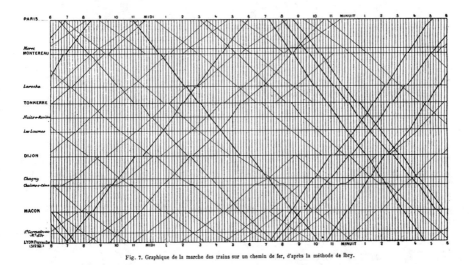

Fig. 7. Graphique de la marche des trains sur un chemin de fer, d'après la méthode de Ibry.

Fig. 8.11: Étienne-Jules Marey's graphical train schedule. Source: [96, Fig. 7], discussed by E.Tufte [137].

Operations performed in a production process can be analogised to stations along a train route, and the production of each item can be compared to the course of a train. So, the idea of the Marey's graphical train schedule can be applied to multiple operation sequences, as demonstrated in Fig. 8.12. On the top left, we see how the resulting graph should look in the case of perfect performance: all "train lines" are parallel and equally spaced. Anomalies can manifest as shown in the top centre

---

[4] https://en.wikipedia.org/wiki/%C3%89tienne-Jules_Marey

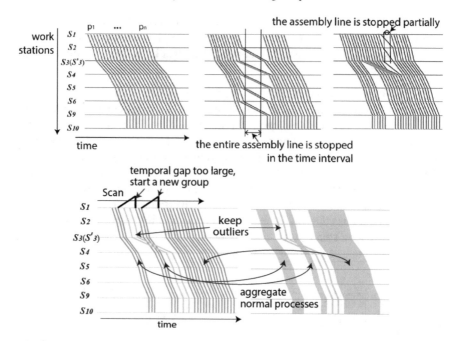

Fig. 8.12: The idea of the Marey's graph is applied for visualising the performance of a production process. Source: [153]

and right, and they are very easy to spot. To simplify the display and make the anomalies even more conspicuous, groups of nicely parallel equally spaced lines can be aggregated and represented by bands, as shown in the lower part of Fig. 8.12.

8.5.2 Univariate time series

The most common representation for a time series of numeric values is the line chart (Fig. 8.13A); however, the heatmap technique, in which data values are represented by colours (Fig. 8.13B), is now also quite popular; see the descriptions of both techniques in Section 3.2.4. The heatmap technique may be better suitable for long time series and for multiple time series that need to be compared. However, since the resource of display pixels is limited, so is the maximal length of a time series that can be represented without reduction. Besides, it is hard to argue that values and their changes can be represented more accurately in a line chart than in a heatmap.

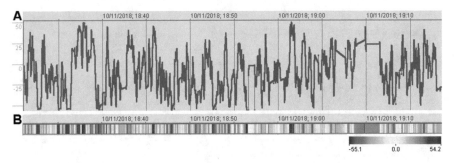

Fig. 8.13: Representation of a numeric time series (A) in a line chart, or time graph and (B) as a heatmap.

Long time series may result from long-term observations or measurements, but they may also be long due to a high frequency of obtaining data. For example, the duration of the time series in Fig. 8.13 is only 47 minutes while the time space between the measurements is 40 milliseconds; hence, the length of the time series is 70,500 time steps (the data are positions of the ball between the two opposite goals in a football game). As you can guess, both representations of this time series involve substantial losses because the same horizontal positions have to be used for representing multiple values. To see information without losses and more distinctly, temporal zooming is needed, which means selection of a small time interval for viewing in more detail. However, when you zoom in to a chosen interval, you do not see the other data beyond this interval; hence, you cannot compare this part of the time series to another one. A viable approach to dealing with this problem is to create separate displays for selected time intervals and juxtapose these displays for comparison (Fig 8.14). It is also possible to represent selected parts of the time series by lines superposed in the same display area, as shown on the bottom right of Fig 8.14. For this purpose, the selected intervals need to be temporally aligned, that is, the absolute times need to be transformed to relative with respect to the starting times of the intervals.

It can be noticed in Fig. 8.14 that different parts of a long time series may have quite similar patterns of temporal variation, which manifest as similar line shapes. Such repeated patterns are called "motifs". Purely visual discovery of motifs in long time series may be a daunting task. It is more sensible to apply one of many existing computational techniques for motif discovery [136] and then mark the detected motifs, for example, as shown in Fig. 8.15. Apart from detecting repetitions, motif discovery techniques may also be helpful for finding anomalies in time series that are expected to be highly regular, for example, in electrocardiograms. An anomaly is a segment whose shape differs from the main motif.

An interesting property of temporal phenomena is that different variation patterns may exist at different *temporal scales*. As an obvious example, let us consider the weather phenomena. You can focus on the changes at the daily scale and see that

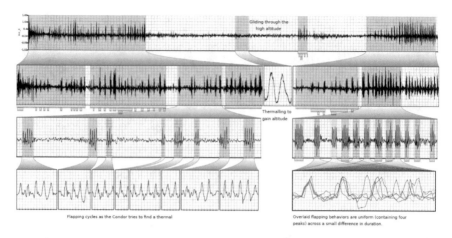

Fig. 8.14: Selecting parts of a long time series and representing the selected parts in separate displays. On the bottom left, graphs of several selected parts are juxtaposed for comparison; on the bottom right, other selected parts are represented by lines superposed in the same plot area. Source: [145].

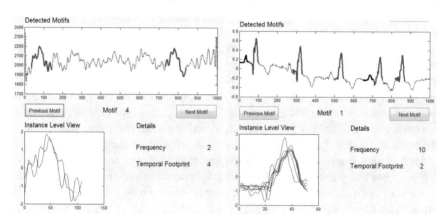

Fig. 8.15: Examples of algorithmically discovered occurrences of similar patterns of temporal variation (motifs) in time series. Source: http://www.utdallas.edu/~arvind/support/exploreMotif.html.

nights are typically cooler than days. Raising to a higher scale, you will see seasonal changes. To observe the manifestation of the El Niño–Southern Oscillation[5], which occurs every few years, you need to go further up. Yet a higher scale is required for seeing long-term trends, such as global climate changes. When you apply computational analysis methods, such as motif discovery, to detailed, low-level time series

---

[5]   https://en.wikipedia.org/wiki/El_Ni%C3%B1o%E2%80%93Southern_Oscillation

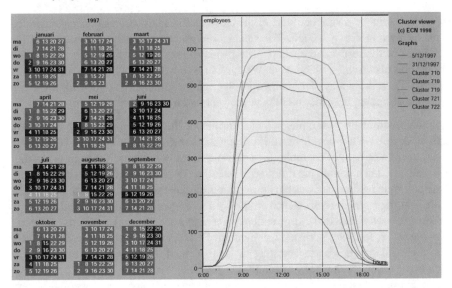

Fig. 8.16: Right: patterns of temporal variation in groups of similar daily time series. Left: the distribution of the distinct daily patterns over a year. Source: [151].

(such as hourly temperature values), you will only be able to find patterns of the lowest possible scale. When time series are represented visually, there is a chance that your vision (which includes perception by the eyes and processing by the brain) will perform temporal abstraction allowing you to observe higher-scale patterns; however, chances are quite low when the time series are long. Besides, the skill of seeing the forest for the trees, that is, abstracting from details, may require training. Therefore, it is reasonable to use data transformation methods that reduce details and produce higher level constructs.

Let us consider an example of an approach applicable when it is known or expected in advance at what temporal scale interesting patterns may exist. The approach is demonstrated in Fig. 8.16 [151]. Here, the data describe the variation of the power demand of a facility during a year with a hourly temporal resolution. It is expectable that the variation may be related to the temporal cycles: daily, weekly, and seasonal. The analysts first divided the whole time series into 365 daily time series. Then they applied a clustering method to these time series and obtained groups (clusters) of similar time series. For each cluster, they computed the average time series. On the right of Fig. 8.16, the average time series of the clusters are plotted together for comparison, each cluster in a distinct colour. We see that all but one clusters have similar variation patterns, with low values in the nigh hours, steep increase by hour 9, and more gradual decrease in the afternoon. The patterns differ in the magnitude of the increase. The remaining cluster has a pattern of constant low value throughout the day (blue line).

Having assigned distinct colours to the clusters, the analysts created a calendar display shown in Fig. 8.16, left. The squares representing the days are painted in the colours of the respective clusters. The display thus shows which patterns of daily variation occurred in which days. We can observe weekly patterns (particularly, all Saturdays and Sundays are painted in blue) and a seasonal pattern, with the green colour occurring in colder months of the year and magenta in the summer.

Contemplating this example, we can formulate the following general approach. First, decide what time unit can be appropriate to the time scale at which you wish to discover patterns. For example, for observing weekly patterns, the appropriate unit is one day. Then, divide the time series into chunks of the length of the chosen unit. Apply some method, such as clustering, for grouping the chunks by similarity. In the original time series, replace each chunk by a reference to the corresponding group. You can now visualise the time series representing the groups by colours, similarly to Fig. 8.16, left, but, perhaps, using another arrangement if your data do not refer to a calendar year. Please note that the transformed time series can be treated as an event sequence; hence, you can try the Arc Diagram method (Fig. 8.8). You can also apply suitable methods for computational analysis, such as sequential pattern mining.

However, the scales at which interesting patterns may exist are not always known in advance, or you may want to see patterns at several scales simultaneously. There is a technique that may help you [125]. As shown in Fig. 8.17, you create a two-dimensional view where the horizontal axis represents the time of your time series and the vertical dimension will be used for the time scale. At the bottom of the graph, you represent your time series at the original level of detail using the heatmap technique; each value is represented by a coloured pixel. This is the time scale of level 1. Then you go up along the time scale axis. For each time scale of level w, starting from level 2, you take all value sub-sequences consisting of w consecutive values from the previous level and derive a summary value from each sub-sequence using a suitable aggregation operator, such as the mean, variance, or entropy. This summary value is represented visually by a coloured pixel. Its vertical position is w and the horizontal position is the centre of the sub-sequence represented by this pixel. Obviously, the number of pixels decreases as the time scale increases; therefore, you obtain at the end a coloured triangle as in Fig. 8.17, right. The colour variation patterns at different heights reflect the value variation patterns existing at different temporal scales.

To give an example of interpreting such a visualisation, let us describe what can be seen in Fig. 8.17, right. Here, the values for higher temporal scales are obtained by computing the means from the underlying original values. A positive global trend can be spotted in the upper part of the triangle, where the colours progress from white to light red. It is notable that the time scales in regions $T1a$ and $T1b$ do not reveal a global trend while regions $T2a$ and $T2b$ clearly show it. The small time scales in region S depict the seasonal cycle in the data. Periods of rather low and rather high values alternate. The fluctuations between these two states appear to be

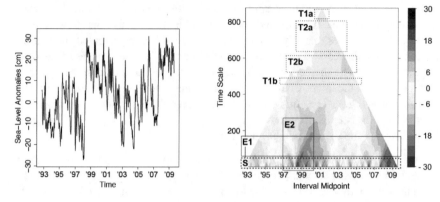

Fig. 8.17: Multiscale aggregation of a time series of sea-level anomalies from October 1992 through July 2009 exposes patterns at different time scales. Source: [125].

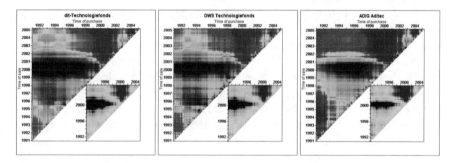

Fig. 8.18: The triangular displays represent the changes of each value with respect to all previous values. Source: [80].

of varying intensity. This is caused by a strong influence of the El Niño/La Niña events, which induce high or low sea-levels, respectively. The El Niño and La Niña patterns can be detected at larger time scales (see region $E1$). Extreme sea-level heights from 1997 to 2000 can also be identified. A short but very intense period of relatively low sea level (1997/98 El Niño) is followed by a long period of rather high sea level (La Niña condition). We can also observe the high influence of these extreme events on the larger time scales.

The visualisation in Fig. 8.18 [80] looks similar to that in Fig. 8.17, but is has been built in a different way and shows a different kind of information. Both axes in each triangular display represent the sequence of time steps of a time series (therefore, this kind of representation is sometimes called time-time plot, or TT-plot). For each pair of steps i and j, where $j > i$, the display shows the change of the value at the step j with respect to the value at the step i. The value expressing the change may be the difference of the values or the ratio, if all values are positive. The changes are represented using the heatmap technique, with two distinct colour hues corresponding to the value increase and decrease. In Fig. 8.18, where the shown data are financial time series, namely, asset prices, the growth is represented in shades of green and the decrease in shades of red. The figure shows the changes of the prices of three assets. The smaller triangles shows the changes computed for each asset individually, and the larger triangles show the changes normalised against the market median growth rates. For this specific kind of data, the visualisation represents the potential gains (in green) or losses (in red) from selling the assets at different times depending on the time of purchasing them.

In the latter example, several time series were compared by representing each one in an individual display and juxtaposing the displays. This approach can work when you have a few time series, but what to do when they are numerous? The earlier discussed example represented in Fig. 8.16 demonstrates a viable approach: clustering by similarity. The original data in this example consisted of a single but very long time series. The analysts divided it into many shorter time series. By applying a clustering algorithm to these multiple time series, groups of similar time series were obtained. For each group, a single representative time series was derived by averaging. Instead of a very large number of individual time series, a small number of representative time series could be visually explored and compared. This approach can be applied to sets of *comparable* time series. This means that the time series must have the same length (number of time steps) and temporal scale (years, days, minutes, or something else but common for all). Even more importantly, it should be meaningful to compare the values of the attributes by judging which of them is larger and which is smaller. Thus, it would be senseless to compare the use of electric power to the number of people at work (while it may be interesting to look for a correlation), but it may be quite reasonable to compare the power usage in different departments of a company.

When you have time series of diverse incomparable attributes, you may wish to know whether these attributes *change* in similar ways. In this case, you can transform the original attribute values into relative values expressing the variation, for example, into the standard score, also known as z-score, which is the difference from the attribute's own mean value divided by the standard deviation (we applied this transformation in Fig. 8.6), or use another form of normalisation, as discussed in Section 4.2.9. A different case is when you wish not to compare the ways in which the individual attributes vary but to understand how they all vary together, that is, how the combinations of the attribute values change over time. This case is discussed in the following.

8.5.3 *Time series of complex states*

Time series reflecting changes of complex phenomena or processes usually consist of complex elements, which describe different states of these phenomena or processes. In many cases, the states can be represented by combinations of values of multiple attributes. For example, weather conditions are represented by values of temperature, precipitation, wind speed and direction, cloud coverage, etc. In other cases, for example, evolving texts, changing relationships in a social network, video recordings, different representations may be required. Are there common approaches to dealing with complex time series differing in the nature and structure of their elements? Yes, there are. The key idea is assessment and numeric expression of the (dis)similarities between the states. The method that can be used for assessing the (dis)similarities depends on the kind of the elements a time series consists of, but as soon as you have found a suitable method (i.e., a *distance function*; see Section 4.2), the following approaches become applicable, as was explained in Chapter 4:

- *clustering* of similar states and representation of the temporal distribution of the clusters using temporal displays;

- *embedding* (or *projection*): positioning of the states on a plane according to their similarities and representation of the state progression as a trajectory.

We shall use our cherry blossom data to illustrate these approaches, and then give further examples from the visual analytics literature.

Our time series of the weekly photo counts in several regions (Fig. 8.4) can be seen as a single time series the elements of which are the combinations of the counts of the photos in the five regions and on the remaining territory, 6 numeric values in total. Suppose that we want to use this time series to study how the phenomenon of cherry blossoming and the corresponding photo taking activities evolved during the ten years. In our simple example, the elements of the time series are the combinations of numeric values. The similarity between them is easy to measure; the common measures (distance functions) representing the clusters suitable for such data have been introduced in Section 4.2.1. We shall use the Euclidean distance.

Let us first try the clustering approach. We take one of the most popular clustering algorithms k-means and apply it to our sequence of 521 states. The main parameter of k-means is k, which is the cluster number. We try different values and compare the results regarding the cluster distinctiveness but also looking at the cluster sizes, as we would not like to deal with many clusters most of which have one or two members. For assessing the cluster distinctiveness, we use a 2D projection of the cluster centres, which are the cluster means in our case (Fig. 8.19, top left). Close positions of two or more points in the projection means that the profiles of the respective clusters are similar; hence, it is better that all points representing the clusters are quite far from each other. Following the approach introduced in Section 4.5.2, we put the

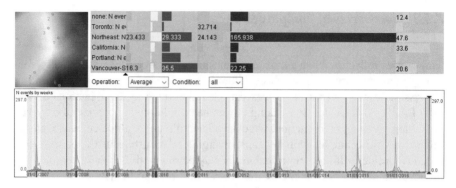

Fig. 8.19: Clustering of states by similarity.

points on a continuously coloured background and use the colours corresponding to the positions of the points for representing the respective clusters in other displays. Thus, the colours are used for painting the bars that represent the average values of the counts for each cluster in Fig. 8.19, top left. We use this display for comparing the cluster profiles. We also use the colours of the clusters for painting the background of the time graph showing the time series (Fig. 8.19, bottom). This allows us to judge whether the clusters are meaningful from the perspective of the temporal distribution.

The largest cluster (labelled 2), which has got an orange colour, consists of 395 states. The cluster profile is not visible in the bar chart display because the mean values are very close to zero. In the time graph, we see that the states from this cluster take place throughout the whole time except for the periods of high photo taking activities. In these periods, the states from the other clusters take place. The states from cluster 3 (yellow), which is close to cluster 2 in the projection and thus has a similar colour, correspond to increased but not very high activities. These states happen before and after the major peaks. The clusters that have got the green, magenta, cyan, and blue colours are the most distinct from cluster 2. They correspond to different distributions of the photo taking activities among the regions.

As we know, periodic temporal distributions and variations can be studied more conveniently using 2D displays where one or two dimensions represent time cycles. In our case, the variation is certainly periodic. It would be good to align the temporal patterns of the different years, for example, in a stack, but the time steps correspond to 7-day intervals. As we know, 365 or 366 days of a year cannot be divided into 7-day intervals without a remainder. In Fig. 8.20, we have created an arrangement of squares that are put in rows with 52 squares per row. Each row roughly represents one year. The squares represent the states; they have the colours of the clusters, and their sizes show how close the states are to the centres of their clusters: the closer, the bigger. With this representation, we can compare the relative times of the occurrence of the states from the different clusters. Thus, we see that the states from the green cluster, which is characterised by high activities in Toronto, usually happened later

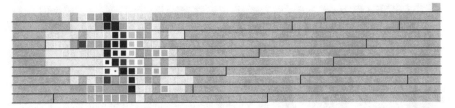

Fig. 8.20: The temporal distribution of the clusters of the states is represented by a 2D arrangement of squares painted in the cluster colours. Each row consists of 52 squares corresponding to 7-days (one week) intervals; hence, the rows roughly correspond to the years. The sizes of the squares show how close the states are to the centres of their clusters; the closer, the bigger.

in the year than the states from the other clusters characterised by high activities in the other regions. High activities in California (cyan cluster) happened later than on the Northeast (blue cluster), and high activities in Vancouver-Seattle happened sometimes later and sometimes earlier than the major peaks of the activities on the Northeast.

By this example, we see how clustering of complex states by similarity and visu-alisation of the temporal distribution of the clusters can be helpful in investigating the development of a phenomenon or process. Let us now try the other way of using the similarity measure: data embedding, or projection (this general approach to data analysis has been described in Section 4.4). We use the same similarity measure as for the clustering. The projection we have obtained is shown in Fig. 8.21, left. The states are represented by dots on a 2D projection plot. The majority of the dots are densely packed in a small area in the lower left quadrant of the plot, which means high similarity of the corresponding states. In the remaining three quadrants, the dots are highly dispersed, which means low similarity between the states. For com-parison with the previously obtained results of the state clustering, we have painted the dots in the colours of the clusters; we shall refer to them later on.

In the middle of Fig. 8.21, the dots are connected by straight lines in the chronologi-cal order. Theoretically, this should allow us to trace the state succession; practically, we cannot do this due to the display clutter. On the right, the dots are connected by curved line segments instead of the straight ones. This have improved the display appearance, and some traces have become slightly better separated. Furthermore, we have coloured the line segments according to the years when the states hap-pened, from dark green for 2007 through white for 2011 to dark purple for 2016. This improves the traceability of the trajectory.

Bach et al. [28] call a trajectory representing the succession of states of a phe-nomenon in a projection plot a *time curve*. They discuss the geometric character-istics of time curves (Fig. 8.22, top) and how these characteristics can yield mean-ingful visual patterns (Fig. 8.22, bottom). Clusters appear if a curve segment has

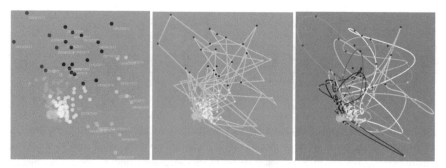

Fig. 8.21: Left: a 2D projection of the states by similarity. Centre: the projected points are connected by straight lines in the chronological order. Right: the projected points are connected by curved lines.

a significantly denser set of points than its neighbourhood, or when it has a significantly higher degree of stagnation. Related to clusters are transitions, i.e., curve segments between clusters with a high degree of progression. The presence of multiple clusters and transitions evoke a dynamic process with different continuing states. A cycle refers to the situation where a time curve comes back to a previous point after a long progression. U-turns indicate reversal in the process, and outliers indicate anomalies. There are nice examples of time curves in the literature [28, 141], some of which are reproduced in Fig. 2.12 and Fig. 7.17 in our book.

Since a time curve shows how a phenomenon evolves, its appearance, naturally, depends on the character of this evolution. A curve may have a clear and easily interpretable shape if the phenomenon evolves gradually, so that consecutive states are similar, or it has several periods of stability and transitions between them. Although our time curve (Fig. 8.21, right) does not look as nice as the examples from the papers, we still can see interpretable geometric shapes and patterns. The curve shows us there were periods when no or very small changes happened; the corresponding trajectory segments are very short and confined in a small area. Apart from these periods, there were times of big changes, when consecutive states where quite dissimilar from each other. These changes are represented by long jumps. Most of the jumps out of the dense area are followed by jumps to other positions in the sparse parts of the displays rather than returns back; hence, periods of stability alternated with periods of big changes.

Even a time curve with a simple and clear shape may not be good enough for exhibiting periodic variation patterns. As we have shown in Fig. 8.20, periodic patterns can be studied using appropriate temporal displays for representing clustering results. The same can be done for state embedding. The idea is demonstrated in Fig. 8.23: we apply continuously varying colouring to the background of the projection display, as we did previously for clusters (Fig. 8.19) and associate each state with the background colour of the corresponding projection point. Then we can use the state colours so obtained in temporal displays, analogously to the colours of the clusters

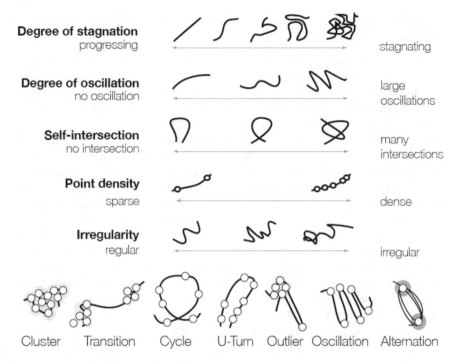

Fig. 8.22: Top: Geometric characteristics of time curves. Bottom: Examples of visual patterns in time curves. Source: [28].

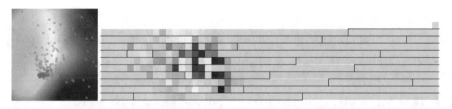

Fig. 8.23: Left: continuous colouring has been applied to the background of the projection plot. Right: the colours from the projection background are used in a display similar to that in Fig. 8.20.

in Fig. 8.19, bottom, and Fig. 8.20. This approach can be used in addition to the trajectory construction.

Let us now compare the results of the clustering and projection techniques. We see in Fig. 8.21 that the clusters tend to have their areas in the projection. The largest cluster 2 (orange) is the most compact, the areas of clusters 3 (yellow) and 1 (pale pinkish) are a bit larger, and the areas of the remaining clusters are quite large, which means high internal variance of these clusters. We also see that the clusters are not fully separated in the projection. Let us make the display more informative by representing characteristics of the states.

In Fig. 8.24, the states are represented by pie charts whose sizes encode the total event counts and the segments show the proportions by the regions. By comparing the appearances of the pie charts, we can make our own judgements concerning the similarities and differences between the states and compare our judgements with the results of the clustering and the projection. We find that in most cases neighbouring pies look similarly, but there are exceptions, for example, a relatively big and almost fully blue pie in the left part of Fig. 8.24, top left. We have a closer look into the areas where close neighbours belong to different clusters, or isolated members of some clusters are surrounded by members of another cluster (Fig. 8.24, bottom). We see that both methods do not work ideally, and their results should always be treated as approximations. In clustering, a member of a cluster is closer (i.e., more similar) to the centre of this cluster than to the centre of another cluster, but it can be quite close to some members of another cluster, as mentioned in Section 4.5.3 in discussing the assessment of cluster quality. In projection, similarities in terms of multiple characteristics usually cannot be represented by distances in a 2D or even 3D space without distortions, as discussed in Section 4.4.2. Therefore, it is useful to combine both approaches, and it is necessary to investigate the results with regard to the data that have been used for the clustering or projection.

8.6 Questions

- What are the main specific features of temporal phenomena and temporal data?

- What are the two major types of temporal data and what is the conceptual differ-ence between them?

- How is it possible to transform one type of data into the other?

- Give examples of analysis tasks relevant to the different types of temporal data.

- What computational methods can help when temporal data are too large for purely visual exploration?

- Give examples of the kinds of patterns that can exist in a temporal distribution and in a temporal variation.

- What is a common approach to exploring periodic temporal distributions and variations?

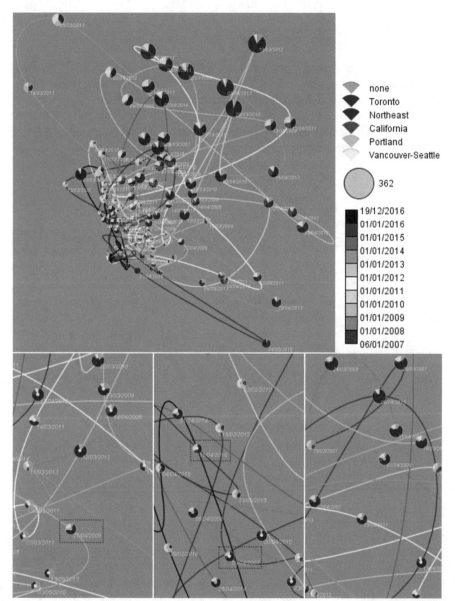

Fig. 8.24: In the projection plot, the states are represented by pie charts with the segments showing the event counts by the regions. The curve connecting the states is coloured according to the time of the state occurrence. Three fragments of the display are shown in more detail, with coloured dots on top of the pies indicating the cluster membership of the states. The frames mark cases of inconsistency between the clustering and the projection, when objects are positioned closer to members of other clusters than to members of their own clusters.

8.7 Exercises

Study temporal patterns of your email (or SMS or social media) activity. For this purpose, extract time stamps and explore their distribution over time. It can be expected that the temporal distribution reflects the natural human activity cycles, with daily and weekly periodic patterns, seasonal trends, outlier activities related to public holidays etc. Identify dates and times of unusual activities (too may or too few messages). Compare patterns for outgoing and incoming messages.

Chapter 9
Visual Analytics for Understanding Spatial Distributions and Spatial Variation

Abstract We begin with a simple motivating example that shows how putting spatial data on a map and seeing spatial relationships can help an analyst to make important discoveries. We consider possible contents and forms of spatial data, the ways of specifying spatial locations, and how to use spatial references for joining different datasets. We discuss the specifics of the (geographic) space and spatial phenomena, where spatial relationships within and between the phenomena play a crucial role. The First Law of Geography states: "Everything is related to everything else, but near things are more related than distant things", emphasising the importance of distance relationships. However, the spatial context, which includes the properties of the underlying space and the things and phenomena existing around, often modifies these relationships; hence, everything related to space needs to be considered in its spatial context. We describe techniques for transforming and analysing spatial data and give an example of an analytical workflow where some of these techniques are used but, as usual, the main instruments of the analysis are human background knowledge, capability to see and interpret patterns, and reasoning.

9.1 Motivating example

In late 18th Century London, there was sinister killer on the loose – sinister because no one could see it and one on understood how it acted. We now know that this killer was cholera, and the primary way in which it spread was through people drinking contaminated water.

John Snow[1] became famous for his investigation of the epidemic. He collected geographically-referenced data of the homes in which people died, mapped the data on a street map, analysed the distribution of the deaths, and reasoned about the pos-

---

[1] https://en.wikipedia.org/wiki/John_Snow

© Springer Nature Switzerland AG 2020
N. Andrienko et al., *Visual Analytics for Data Scientists*,
https://doi.org/10.1007/978-3-030-56146-8_9

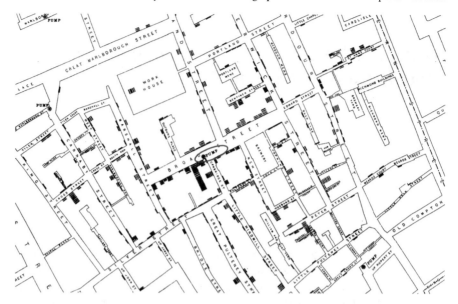

Fig. 9.1: A fragment of the John Snow's famous map which enabled him to identify a cluster of cholera cases and the potential source (we have marked it with a red ellipse).

sible source of the disease. This approach, which was novel at that time, is now the foundation of the visually-focused quantitative analysis. Snow noticed that the deaths were clustered around one of the public water pumps, which was located in the Broad Street. We have marked this location with a red ellipse on the fragment of the Snow's map shown in Fig. 9.1. To test whether this correlation was a causal effect, he removed the pump's handle, rendering it unusable. The people – no doubt grumbling as they went – had to use an alternative pump, and the number of the cholera deaths dropped. A further investigation showed that the water to this pump was contaminated from sewage from a nearby cesspit. This observation helped pave the way for modern epidemiology,

This simple but hugely influential study illustrates two important characteristics of geographical data. The first is that **non-uniform geographical patterns may indicate some underlying process**. In this case, the visually apparent spatial clustering indicated that there might be something special in the area. The second is that **different geographical datasets have a common reference frame**, as the deaths cases and the water pumps in this example, enabling data that might otherwise be unrelated, to be related. These are the reasons why studying geographical patterns in data is worthwhile. This example also demonstrates that geographic data and phenomena are studied by representing them on maps, which usually include multiple information layers. When several layers are put together (as the streets, houses, and water pumps in the Snow's map), it becomes possible to notice relationships between the things and phenomena represented by them.

9.2 How spatial phenomena are represented by data

Any data that can be meaningfully related to spatial locations (sometimes with a spatial extent) can be considered as spatial data. Data records may correspond to one or more geographical positions, linear geographical features, or areas. Examples are:

- The locations of water pumps (Fig. 9.1).

- The location at which a photograph was taken or a tweet was tweeted (Fig. 1.7).

- A building footprint (Fig. 9.1).

- The boundary of a country or an administrative district, such as a London ward in Fig. 4.18.

- The origin and destination locations of a migrant.

- Stopping positions on a journey.

9.2.1 Forms of spatial data

Spatial position can be quantified within a spatial reference frame using 2D coordinates, which will be discussed in more detail in Section 9.2.4. This may be a latitude/longitude pair or map coordinates. A single 2D coordinate describes a **point** that may conceptually represent a single location (e.g., a GPS reading) or a larger area (e.g., the location of a city on a map of a country). Spatial extents can be explicitly represented as sequences of positions that form **lines** (e.g., representing roads or rivers) or **polygons** (e.g., representing countries or districts), or implicitly, using numeric attributes. The type of spatial representation where discrete spatial objects are described by records specifying their spatial locations is known as 'vector data'. Additionally to the spatial locations, the records may also include values of various attributes, which are usually called **thematic attributes**.

An alternative representation is 'raster data', which usually represents how a continuous phenomenon, such as elevation or temperature, varies over space. It is a regular, typically fine-grained grid with cells containing attribute values. The grid itself is georeferenced using coordinates as described above. Raster data are represented visually on a map by continuous variation of colour, as demonstrated in Fig. 9.2.

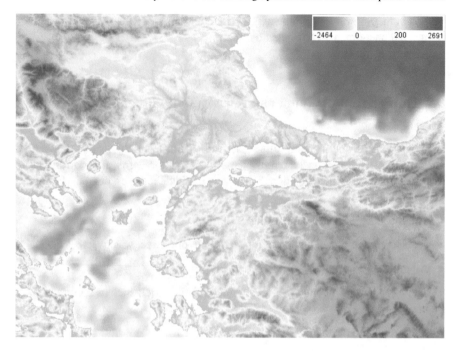

Fig. 9.2: Elevation, which is a continuous spatial phenomenon, is represented using the raster data format and depicted by continuous variation of colour, which encodes attribute values contained in the raster.

The grid in raster data is characterised by its resolution. Thus, the example raster in Fig. 9.2 has 1126 columns and 626 rows, and the cell size is 1 sq.km.

Spatial data can be *object-based* or *place-based*. Object-based data refer to entities located in space and describe properties of these entities. Place-based data refer to locations in space and describe the objects and/or phenomena located there. For an example, let's consider crime data. If data include the location and characteristics of each crime, then these would be *object-based* data. If crime data are supplied in an aggregate form by place (e.g. administrative unit) with crimes quantified as counts per place, this would be a place-based representation. It is possible to transform from object-based to place-based representations but not the other way around.

9.2.2 Georeferencing

Many spatial datasets are *explicitly georeferenced*, i.e., spatial positions are specified explicitly using coordinates. Other datasets may be *implicitly georeferenced* using *geocodes*, that is, text values for which the geographical coordinates of the corresponding points, lines or polygons can be retrieved from appropriate geographical lookup datasets. Common geocodes are country names, country codes, and postal codes. Spatial administrative and census data are often implicitly georeferenced by including codes of administrative units whose coordinates are specified elsewhere. These are usually place-based spatial data.

Ultimately, for spatial analysis, it is necessary to have explicitly georeferenced data. Obtaining explicit georeferences for data containing geocodes is achieved by joining these data with data specifying the coordinates for these geocodes. For example, customer data containing the customers' addresses and/or postcodes can be georeferenced by looking up the spatial positions of the addresses or postcodes. City-level data can be georeferenced using the city names and widely-available lookup tables. Demographic data from a census can be georeferenced by obtaining spatial positions of the census districts from administrative boundary lookups, as we did for the London wards in the example shown in Fig. 4.18.

9.2.3 Spatial joining

When two or more spatially referenced datasets are put on the same map, relationships between the data from these datasets can be detected visually, as happened when John Snow analysed the epidemic data. Such "visual join" may not always work perfectly, in particular, due to occlusions of data from one dataset by data from another dataset. However, it is possible to perform computational join of spatially coincident datasets using the spatial positions of the data items. Thus, in the example of the cholera epidemic, the dataset containing the locations of the deaths could be joined with the dataset containing the positions of the water pumps. One possible way is to determine for each death record the nearest water pump and attach an identifier of this pump to the record. Another possible way is to count for each water pump the number of the deaths within a certain maximal distance, for example, 500 metres.

Another example of spatial joining is complementing customer data with demographic data characterising the areas where the customers live. Having the customers' coordinates (which might first be obtained based on their addresses), the areas containing these coordinates are determined, and then area-specific data are retrieved from a demographic dataset and attached to the customers' records. This procedure is often used in geomarketing for guessing the likely characteristics of the customers (e.g., their income or number of children) in order to plan targeted marketing campaigns.

9.2.4 Coordinate systems and cartographic projections

A coordinate system defines a spatial reference frame that provides a possibility to specify positions in space and relate them to each other. Geographical coordinates represent 2D positions on the Earth's surface using a reference system defined mathematically as an ellipsoid surface that approximates the Earth's surface (it is an approximation because the Earth surface is not as regular as a mathematically defined ellipsoid). The most common way to represent a geographical location is using the latitude and longitude, which are coordinates in the coordinate system known as "WGS84". The latitude specifies the position with respect to the equator and the longitude with respect to a prime meridian. The latter is chosen arbitrarily. Now, the most widely used is the International Reference Meridian (IRM), also called the IERS Reference Meridian [2], which slightly deviates from the Greenwich Meridian passing through the Greenwich Royal Observatory in London.

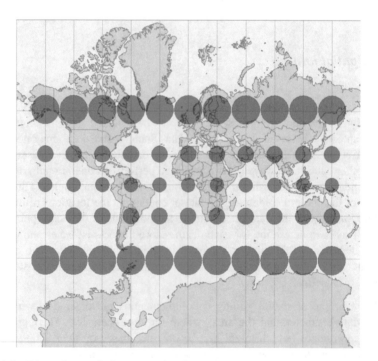

Fig. 9.3: Distortions of the areas in the Mercator projection. All red ellipses have equal sizes and shapes on the Earth surface but greatly differ when projected on a plane. Source: https://en.wikipedia.org/wiki/Tissot's_indicatrix.

---

[2] https://en.wikipedia.org/wiki/IERS_Reference_Meridian

There are also many coordinate systems that project parts of the ellipsoid onto a planar surface. This transformation causes unavoidable distortions, and the type of distortion depends on the projection techniques. Many countries whose spatial extents are relatively small have their specific Cartesian spatial reference systems in which both areas and angles are preserved within acceptable tolerances. For larger areas, it is not possible to project onto a plane whilst preserving both angles and areas. The Mercator projection is good for global coverage and where directions need to be preserved, but it maintains angle/shape at the expense of scale which continually increases towards the poles, as illustrated in Fig. 9.3. Other projections may preserve areas but distort angles and, consequently, shapes. There are also projections that try to minimise the distortions of both areas and angles, preserving neither.

As mentioned, geographical coordinates specify positions on the Earth's surface. Altitude/height and depth (in a sea) are usually measured in metres from this surface.

Implication: spatial data may not be immediately compatible. An implication of the existence of so many different spatial reference systems is that two spatial datasets may not immediately be compatible. It is relatively easy to transform between different systems into a common system, but it is necessary to know which spatial reference systems are used for different datasets. Most spatial data have this described in the metadata, and there is a standard set of codes that describe spatial reference systems which software understands. Latitudes and longitudes are usually based on the 'WGS84'[3] spatial reference system. Data provided by countries tend to use standard local mapping projections with corresponding spatial reference codes.

Implication: angular and distance distortions. The distortions introduced by projections have implications for spatial analysis because of the effects on the calculation of angles, distances, and areas. Specialised software and libraries for spatial analysis that take the spatial reference system into account will correct for this, but some will only work if the coordinates are in a particular system. The distortions also have implications for aggregation. Grid-based aggregation of geographical data in projections that are not equal-area will result in different amounts of geographical space being represented in each cell.

There are also implications for visualisation, particularly because planar projection is usually necessary for representing spatial data on a map. There are two ways in which the the effects of this can be reduced. The first is to choose an appropriate projection, which depends in the latitude, the latitude range, and the extents to which one wishes to preserve angle or area. The second of these is to choose an appropriate visualisation technique. If angle is not preserved, showing angular data may be problematic. If area is not preserved, then symbol density will not corre-

---

[3] https://en.wikipedia.org/wiki/World_Geodetic_System

spond to spatial density. These problems are reduced when the geographic extent of the territory that is visualised is small. In viewing and interpreting a map of a large territory, one should be very cautious in making judgements concerning distances, areas, shapes, and densities.

9.3 Specifics of this kind of phenomena

Geographical and, more generally, spatial phenomena have some unique characteristics that need to be understood and taken into account in analysing spatial data.

9.3.1 Spatial dependence and interdependence

Spatial dependence, also known as *spatial autocorrelation*, is the property of spatial variation when values or objects at neighbouring locations tend to be similar. This property holds for many spatial phenomena, both natural (e.g., weather attributes) and people-related (e.g., market prices). In Chapter 8, we mentioned temporal dependence, or autocorrelation, as a property of temporal phenomena. Generally, the concept of dependence, or autocorrelation, can be described in simple words as "neighbours affect neighbours". This definition can refer to distributions over any kind of base where relationships of neighbourhood exist. The concept of *spatial interdependence* refers to relationships between different phenomena distributed over a common spatial base.

The spatial distributions of phenomena tend to be both non-uniform and highly interdependent. For example, at the global scale, population is distributed on land within a particular latitudinal range. At a finer scale, population tends to cluster in cities and otherwise follow coastlines, roads, and rivers. Any phenomenon that is related to the life and activities of people will be influenced by the population distribution properties.

In the John Snow example (Fig. 9.1), the geographical distribution of the households with deaths is (unsurprisingly) along roads, because this is where the people live. Without this geographic context, these linear structures may be incorrectly interpreted as offering clues as to the cause of cholera. In fact, it was the proximity to the water pump that turned out to be significant. The skill of the analyst is to know which spatial patterns are significant and which ones are not.

The ability of shops to attract customers is another example. It depends on other geographical phenomena including the shop's accessibility, visibility, amenities (e.g.

car parking and public transport), and the position and nature of the competition. In retail modelling, 'gravity modelling' quantifies the shops' competition as a function of the sum of the inverse distances and sizes of the competing shops. This illustrates that retail modelling needs to consider other geographical data including the competition, population, catchment areas, and infrastructure.

Large cities contain more crime than smaller cities. This is because crime involves people, and there are more people in cities. Crime analysis needs to consider other geographical contextual data to account for these effects, such as where people live and how space in cities is used.

The properties of spatial dependence and interdependence are captured in the "Tobler's First Law of Geography"[4], which states: "Everything is related to everything else, but near things are more related than distant things" [134, p. 236].

Implication: put your data in spatial context. The use of maps is crucial in analysing spatial data. Maps can show not only spatial distributions of entities and attributes and the distance/neighbourhood relationships between them but, very importantly, the geographical context i.e., other things and phenomena that spatially co-occur with the things or phenomena under our study. In particular, the spatial context can interfere with the influence of the spatial dependence. For example, entities or attribute values located at a small distance from each other may not be similar or strongly related due to the presence of a spatial barrier between them, such as a river, a mountain ridge, or a country border. Seeing things in context is very important for correct interpretation and making valid inferences. Thus, as we noted concerning the John Snow example, it was important to see the streets in the map for correct interpretation of the alignments formed by the dots representing the deaths. In the example maps showing the distribution of the cherry blossom photos in Fig. 9.4, the cartographic background allows us to understand the clustering and alignment patterns observed at the larger and smaller scales.

Implication: consider possible confounding factors. When you observe a particular spatial distribution of a phenomenon on a map, you should think carefully whether this distribution can be related to another phenomenon that is not depicted on the map. Thus, the spatial distributions of crimes, diseases, sales of products, and locations of tweets may be similar to each other, which does not mean that any of them affects the others. All these are phenomena related to people; therefore, all of them will have similar spatial distributions to the distribution of the population density. Hence, it is not meaningful to analyse such distributions by themselves. To make reasonable judgements, e.g., how high or low the levels of crime or tweeting activity are in different places, it is necessary to normalise the data concerning the phenomenon under analysis in relation to the confounding phenomenon, such as the population density in this example.

---

[4] https://en.wikipedia.org/wiki/Tobler%27s_first_law_of_geography

9.3.2 Spatial precision and accuracy

Sometimes the *precision* at which data are supplied (the number of decimal places) helps indicate the accuracy at which the data were collected. However, it often does not, particularly for derived data or transformed data, where the precision of stored values may be much higher than the precision of the original data. This is particularly true for spatial coordinates, which are often recorded at a much higher precision than the equipment is able to measure.

Besides, coordinates that are present in data may have been obtained from lookup tables in which whole cities or other geographic objects are represented as points. The coordinates may look highly precise, but this impression of precision is deceptive. It is often possible to detect visually that data may refer to some set of "standard" positions: when you put the data on the map, many data items will have exactly the same position, and you will see notably fewer dots visible on a map than there are records in the dataset.

Implication: it is not always clear what spatial scale of analysis the data support. Without knowing how accurate and precise the spatial data are, it is difficult to know at what scale the spatial analysis can be done, even if the coordinates are reported at a high spatial precision. Sometimes column headings and other metadata can give clues. If not, then counting unique values of coordinates and drawing maps of these locations will help establish whether there is a finite set of locations. This can help in choosing appropriate scales of analysis.

9.3.3 Spatial scale of analysis

In a spatial distribution of a phenomenon, different patterns can be observed at different spatial scales. These patterns will evoke different interpretations and, eventually, enable derivation of different pieces of knowledge. As an example, let us take the data describing the cherry blossom photos that we used in Chapter 8. When we consider the spatial distribution of the photos at the country scale (Fig. 9.4, top), we see the major regions where cherry blossoming occurs and attracts people who take photos. When we zoom in to the northwestern part of the country (not shown in the figure), we see that the photos are clustered in and around the big cities: New York, Philadelphia, Washington, Baltimore, and Boston. At a smaller scale, as in Fig. 9.4, bottom, we see dense clusters occurring in green areas and alignments along waterfronts.

It is good when we have detailed data that can be considered at different spatial scales, as the photo data in our example. However, we often have to use aggregated data, such as counts and statistical summaries associated with some areas. Such

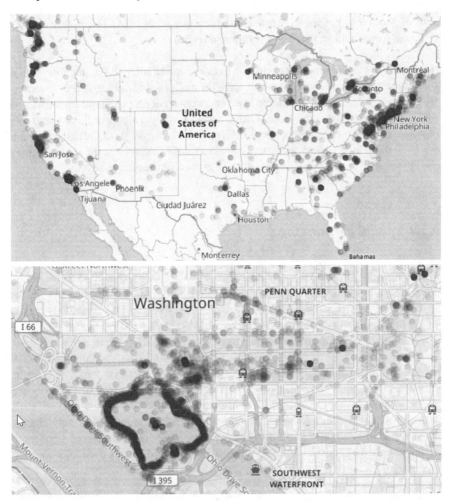

Fig. 9.4: Dependence of the spatial distribution patterns on the spatial scale. Top: the distribution of the cherry blossom photos at the country scale reveals the regions where the phenomenon mostly occurs. Bottom: at a city scale, the distribution is related to parks and waterfronts.

data have been derived from detailed data by someone else, and we have no access to the original data. By considering the example with the photos, it is easy to guess that, depending on the choice of the areas for aggregating the data, different spatial patterns may be captured while others may be destroyed. Thus, if the photo data were aggregated by the states, we would never know that the photos are clustered in particular big cities and, certainly, could not refer the distribution to parks and waterfronts. Considering economic activity, using area units of the size of about 100 sq.km would reveal geographical trends as a national scale, whereas using area units of about 1 sq.km size could disclose particular spots of high economic activity and local trends within smaller geographical regions.

Implication: spatial analysis results are scale dependent. The implication is that analysts must choose the right spatial scale for analysis depending on the data characteristics (particularly, aggregation level and/or precision), on the one hand, and the analysis goals, on the other hand. Often, a visual analytics approach can facilitate finding a scale that produces patterns helpful to the analysis. It may also happen that available data do not allow performing analysis at the required scale. Analysts may have to search for more appropriate data or to modify the analysis goals.

9.3.4 Spatial partitioning

Spatial partitioning divides space into discrete spatial units. Obviously, this can be done differently. If we then summarise data within these spatial units, the results are likely to be dependent on the way in which the space has been divided. This is called the Modifiable Areal Unit Problem (MAUP) [107]. The reason for this effect is that spatial distributions are non-uniform, as mentioned in Section 9.3.1. Spatial patterns, such as high densities, clusters, and alignments, may be concealed when covered by big units or split among several units. The fact that a pattern may be destroyed by splitting into parts indicates that not only the sizes of the units are important but also the delineation. Thus, even when a regular grid with equal cells is used for partitioning, distribution patterns of grid-aggregated data may change when the grid is slightly shifted or rotated. The notorious practice of gerrymandering[5] defines boundaries of electoral districts so as to favour particular election outcomes.

In spatial analysis, we regularly partition space. Most of the transformations (Section 9.4) involve partitioning space. Having detailed data allows us to find suitable partitioning for the intended analysis, possibly, by trying different approaches. One possible approach is **data-driven space tessellation** attempting to create spatial units that enclose spatial clusters of data items, as illustrated in Fig. 9.5.

The approach involves two major operations. First, points are organised in groups based on their spatial proximity so that the spatial extent of a group does not exceed a chosen maximal radius. The grouping (clustering) algorithm is described in paper [21]. Second, the territory is partitioned into Voronoi polygons[6] using the centroids of the point groups as the generating seeds. When there are big empty areas, as in our example in Fig. 9.5, additional generating seeds can be created in these areas, for example, at predefined equal distances from each other. The use of this

---

[5] https://en.wikipedia.org/wiki/Gerrymandering
[6] https://en.wikipedia.org/wiki/Voronoi_diagram

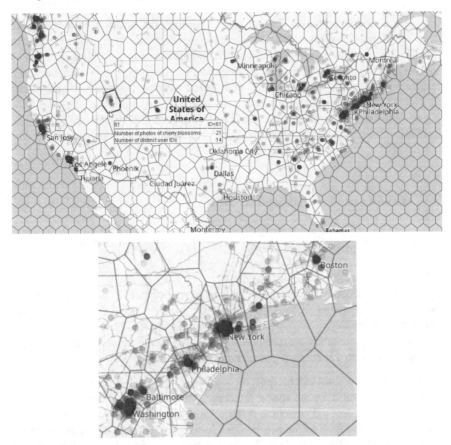

Fig. 9.5: Data-driven tessellation attempts to create spatial compartments enclosing spatial clusters of points. Top: division of the USA territory according to the spatial distribution of the cherry blossom photos. Bottom: an enlarged map fragment.

method does not guarantee a perfect result. The result depends on the choice of the maximal radius of a point group, and finding a suitable value may require several trials. A good feature of this approach is that, by choosing a larger or smaller parameter value, we can adjust the partitioning to the desired spatial scale of analysis while preserving the patterns that exist at this scale.

However, not always we can decide how to partition the territory. We often get data that have been already aggregated by some units, and we also often want to aggregate available detailed data by predetermined areas for comparability to other datasets.

Implication: data may contain data processing artefacts. The implication of this is potentially quite serious: our analysis may be dependent on how the data have been processed and prepared. Data-driven approaches that partition space taking the data distribution into account can help reduce these effects. A smoothing filter can also be used, so that data items located near boundaries can partially contribute to several neighbouring units. By using visual analytics techniques for comparing the results of summarised data partitioned in different ways, we can determine the extent to which MAUP operates. This will help us determine appropriate spatial partitioning and how much confidence we can have in the results. However, we sometimes have little choice, because the data we use may already be highly aggregated.

9.4 Transformations of spatial data

9.4.1 Coordinate transformations

There are a number of reasons why we may need or wish to do a coordinate transformation. If not all spatial datasets are in the same spatial reference frame, we would transform coordinate so that they are in a common spatial reference frame. If we need to do spatial analysis by means of a particular software tool, we should ensure that our data are in a spatial reference frame that the tool accepts.

If we are doing our own calculations on the coordinates, we need to ensure that the calculations are valid for the spatial reference frame. If the we are using a country-specific Cartesian spatial reference system, we can assume that these will preserve both angle and distance within acceptable tolerances; hence, the calculations can be based on the Euclidean geometry. If the coordinates in our data are latitudes and longitudes, which are angular spheroid coordinates, we cannot use the Euclidean geometry and instead need to use Great Circle-based calculations designed for such coordinates.

If our spatial analysis concerns angles, we need to ensure that our spatial reference frame preserves angles (Mercator or a country-level Cartesian mapping system). If our analysis is about area or density, we need to ensure that our spatial reference frame preserves distances and areas.

Apart from transformations of geographical coordinates from one projection to another, it may be interesting and useful to transform positions in a real, physical space to positions in an imaginary, artificial space. Figure 9.6 demonstrates the idea of a transformation that can be applied to positions of football (soccer) players on the pitch. To understand how the players of a team position themselves in relation to their teammates, we create an artificial "relative space" of the team [11]. The team space is a Cartesian coordinate system with the origin in the team centre, which is determined as the mean position of the team players, excluding the goalkeeper and, possibly, "spatial outliers" – players that are distant from the others. The vertical axis corresponds to the direction of the team's attacks, i.e., towards the goal of the opponent team. The horizontal axis is directed from left to right with respect to the attack direction. The players' coordinates on the pitch are then transformed into the coordinates in the new coordinate system with the axes left-right and back-front.

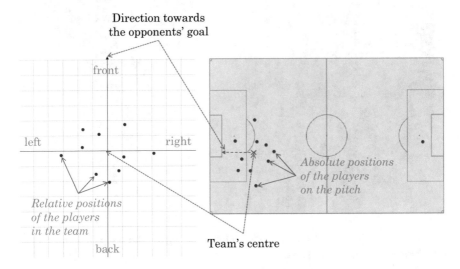

Fig. 9.6: Transformation of "absolute" spatial positions of team players on the pitch (right) into their relative positions with respect to the team centre (left).

Since the players are constantly moving in the course of the game, the position of the team centre changes every moment. Hence, the team centre's position needs to be determined individually for each time step that the players' position data refer to. The players' relative positions in each time step are calculated with respect to the position of the team centre in this step. While the team centre's position on the pitch is constantly changing, its position in the team space remains the same – it is always the point (0,0) – unlike the positions of the players. Figure 9.7 demonstrates how trajectories (tracks) that players of a team made on the pitch during a game may look like after a transformation into the team space. While on the pitch map (left) we see an incomprehensible mess made by overlapping tangled lines, the map of the team space (right) looks much tidier. We see that the lines corresponding to the players (represented by different colours) form compact rolls within certain parts of the team space. This indicates that the players have their "areas of responsibility" in the team and tend to keep within these areas. You may note that the track of the goalkeeper, which is at the bottom of the team space (in green), is extended vertically much more than the tracks of the other players. It is not because the goalkeeper moved more to the front and back than the others, but because the team centre moved to the front and back with respect to the goalkeeper's position.

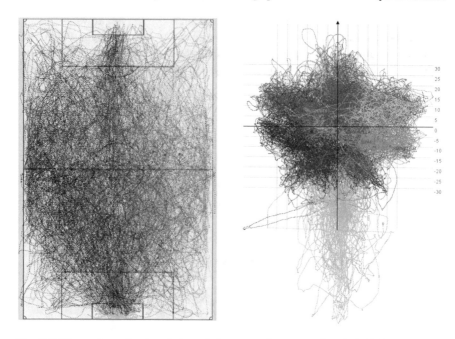

Fig. 9.7: The trajectories (tracks) made by team players on the pitch (left) have been transformed into the "team space".

Don't think that such coordinate transformations to artificial "relative" spaces are only possible for a group of simultaneously moving objects. It can also be useful for comparing properties of two or more spatial distributions that are not spatially coincident. For example, imagine that you have spatial data describing personally significant places of different individuals: home, work place, school or kindergarten attended by children, usual places for shopping, sports, entertainment, and meetings with friends. For each individual, there is a set of personal places with their categories and coordinates. How can you compare the spatial distributions of the personal places of different individuals when you are not interested where on the Earth these places are located, but you want to know how far from home each person works and performs other activities and whether their places are spatially dispersed or clustered? You can, for example, take the position of each person's home to be the origin of a new coordinate system, take the direction from the home to the work, or to the most frequently visited place, as the vertical axis, and transform the positions of all places into this coordinate system. Now you can superpose the distributions of the places of two or more individuals within this common coordinate space and compare these distributions visually and, if necessary, computationally.

9.4.2 Aggregation

You may have detailed spatial data referring to locations or objects that can be treated as points in space (i.e., the spatial extents, if any, are negligible). An example is our dataset with the data concerning the cherry blossom photos. To see the spatial distribution of your data, you may represent each data item on a map by a dot, as we did in Fig. 9.4. There is a problem here: multiple dots may overlap and conceal each other. By varying the transparency of the dot symbols, you may achieve the effect that dense spatial clusters become well visible, as it has been done in Fig. 9.4. This is useful, but the dot symbols beyond the densely filled areas are hardly visible. Another problem is that you cannot guess how many dots are in the different densely packed regions and thus cannot compare the regions. To see a spatial distribution more clearly, to quantify it, and to compare different parts of it, you may need to aggregate the data by areal spatial units.

The general idea of spatial aggregation is simple: you partition the space into some areal units (see Section 9.3.4), or you take some predefined partitioning, and calculate for each unit the count of the original data items contained in it and, when needed, summary statistics of the values of thematic attributes present in the data. In the result, you obtain a set of data records referring to the spatial units, i.e., place-based data (Section 9.2.1). You can visualise these data, for example, on a choropleth map, as in Fig. 2.8. While the procedure of counting and summarising is always the same, the partitioning of the space can be done in different ways, as will be discussed in the following. To illustrate how space partitioning affects the result of aggregation, we shall use an example dataset containing the positions of the London pubs. A fragment of a map portraying the detailed data is shown in Fig. 9.8. We do not use the photo dataset because it refers to a very large territory, which is much distorted when shown on a map due to the projection effects, as discussed in Section 9.2.4.

9.4.2.1 Gridding

Perhaps, the easiest approach is to divide a territory into uniform compartments by means of a regular grid. The cells of such a grid may be rectangular (most often, squares) or hexagonal. This approach is easy to implement on a small territory represented in a Cartesian coordinate system (i.e., in some projection) without substantial distortions of the angles, distances, and areas. When you have a large territory, you need to account for the curvature of the Earth surface; therefore, the division into equal compartments becomes quite problematic. That's why we use the example with the London pubs for our illustrations rather than deal with the large territory of the USA.

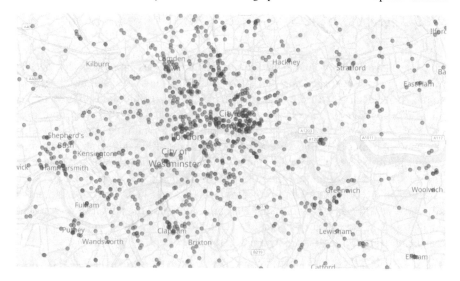

Fig. 9.8: The positions of the London pubs.

There are two variants of grid-based aggregation. One is to use a grid with quite large cells, as in Fig. 9.9, where the cell dimensions are 1 × 1 km. In this example maps, the counts of the pubs contained in the grid cells are represented by proportional sizes (areas) of the circles drawn inside the cells.

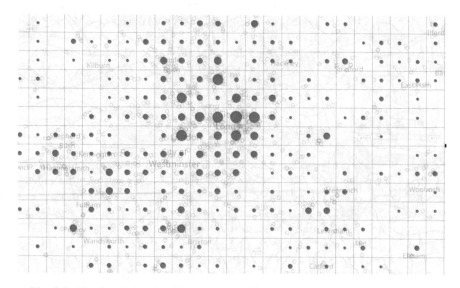

Fig. 9.9: The London pubs data aggregated by a regular grid with square cells.

Another, a very different approach is to use a grid with tiny cells and transform our data into a raster; this form of spatial data was mentioned in Section 9.2.1. A raster represents data as continuous and smooth variation of some attribute over the territory. The example in Fig. 9.10 represents the variation of the *density* of the London pub. Here, the cell dimensions are 100×100 m. As can be seen from the example, a raster is visualised in such a way that the cells are not discernible. The image looks smooth thanks to the use of interpolation for determining the attribute values corresponding to neighbouring screen pixels. Moreover, **smoothing** is involved not merely in the process of drawing a raster but, more essentially, in the aggregation procedure as such. The idea is that each data item contributes into calculating the value not only in the cell containing this item but also in the neighbouring cells. A suitable radius of the cell neighbourhood, i.e., the maximal distance at which an item may have influence on a cell value, is specified by the analyst. The neighbourhood radius is a parameter determining the degree of smoothing: the larger, the smoother.

In particular, generation of a raster representing the densities of data items is done using the statistical technique known as *kernel density estimation (KDE)*[7]. The term *kernel* refers to the function that is used for the aggregation, and the smoothing parameter (i.e., the neighbourhood radius) is called the *bandwidth*. The two images in Fig. 9.10 represent variants of the density raster generated by means of KDE with the bandwidth of 1 km (upper image) and 500 m (lower image). It is easy to notice that the larger bandwidth can be good for revealing larger-scale distribution patterns, whereas the smaller bandwidth promotes more local patterns to show up.

Since the fine-grained raster format and its visual representation aim at mimicking continuous variation of a phenomenon, the variant of aggregation that produces raster data can be called *continuous spatial aggregation*. The other variants, which do not blur the spatial compartments but keep them explicit, can be called *discrete spatial aggregation*.

If we compare the results of the continuous aggregation in Fig. 9.10 and the discrete aggregation by a regular grid in Fig. 9.9, we may note that the distribution in the discrete variant looks unnatural and, to some extent, misleading, because the areas of especially high density of the pubs are "dissolved" and/or distorted. The continuous variant is much more faithful in showing the density variation pattern. However, its disadvantage is that it is hard to judge how many pubs are in different places. We can consult the map legend to see how the raster values are represented by the colours, but the problem lies in interpreting the values, not the colours, because the values were calculated with applying smoothing.

Another disadvantage of the raster form is that you cannot see more than one raster presented on a map, which complicates comparisons between different attributes or different phenomena. Thus, if we apply discrete spatial aggregation to the distributions of the pubs, bars, and restaurants, we can obtain data where each spatial

---

[7] https://en.wikipedia.org/wiki/Kernel_density_estimation

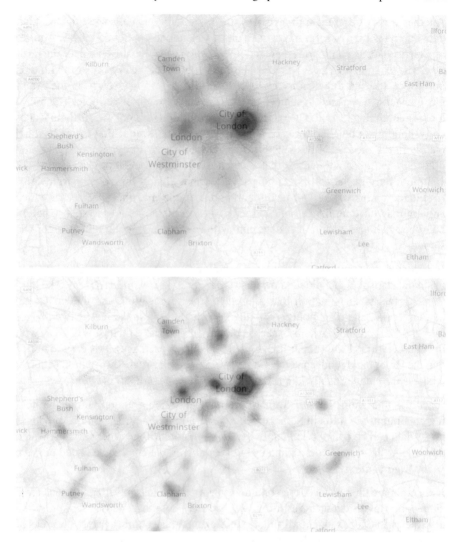

Fig. 9.10: The London pubs data aggregated into a raster (fine-grained grid) by means of KDE with the bandwidth of 1000 metres (upper image) and 500 metres (lower image).

compartment is characterised by statistics referring to all three kinds of facilities. They can be visualised together by means of the chart map technique. In Fig. 9.11, we used pie charts to represent the counts of the different facilities and, simultaneously, the total numbers and the proportions of the different facilities in the totals. Another possible representation of the same data could be a map with bar charts, as we had in Fig. 1.8. Please note that the possibility of obtaining multiple aggregate

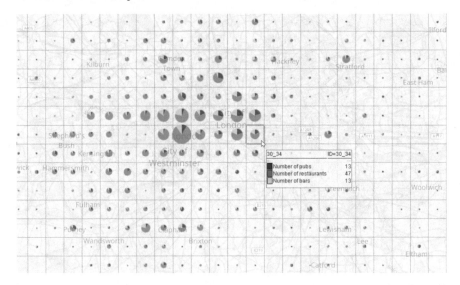

Fig. 9.11: The data about London pubs, restaurants, and bars have been aggregated by the same discrete compartments and can be visualised together. The counts of the different facilities in the compartments are represented by sectors of pier charts.

attributes for the same compartments and visualising them together does not depend on the shapes of the compartments that are used for discrete aggregation.

9.4.2.2 Data-driven partitioning

As we have noted previously, discrete aggregation by a regular grid distorts spatial distribution patterns. This can be easily noticed in Fig. 9.9 by comparing the positions of the red circles representing the aggregates with the positions of the clusters of pubs; the pubs are represented by small hollow circles in grey. In Section 9.3.4, we introduced the data-driven space tessellation as an approach allowing us to respect the distribution of data items in partitioning the space. Figure 9.12 demonstrates a result of applying this approach to the London pubs data. The visual representation of the aggregates is the same as in Fig. 9.9: the circle sizes are proportional to the counts of the pubs in the compartments.

The advantage of this approach is that the distribution patterns can be captured much better than with a regular grid. In Fig. 9.12, the positions of the big red circles or groups of such circles correspond sufficiently well to the positions of the clusters of pubs. In fact, the distribution patterns that can be perceived from the map in Fig. 9.12 is similar to what we see in the continuous density maps in Fig. 9.10.

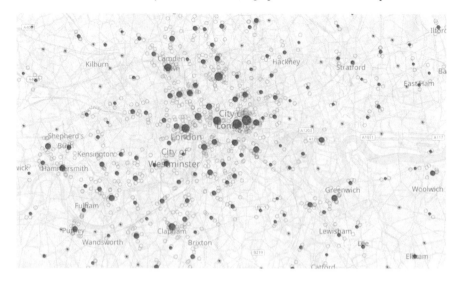

Fig. 9.12: The London pubs data aggregated by an irregular grid with polygonal cells obtained by means of data-driven tessellation.

A disadvantage of the data-driven partitioning as compared to regular gridding is that the resulting compartments are not equal in their sizes. It can be noticed in Fig. 9.10 that the compartments are smaller where the density is higher. The reason is that centres of groups of points are used as generating seeds for Voronoi polygons, and the distances between group centres may be quite big in areas where the points are sparse. The differences between the cell sizes in dense and sparse regions can even become larger, as in Fig. 9.5, if we allow the clustering algorithm to subdivide groups of points in which the maximal density is not achieved near the group centre [21]. The subdivision is controlled by a parameter specifying the minimal group radius.

The variation of the sizes of the spatial compartments complicates numeric comparisons between the compartment-associated aggregates. It is understandable that a value in a smaller cell indicates a higher density than the same or, perhaps, even higher value in a larger cell, but it is hard to relate the values to the cell sizes by purely visual inspection and to estimate the differences more accurately. Therefore, it is advisable to calculate the densities by dividing the counts by the cell areas and visualise the densities explicitly.

9.4.2.3 Predefined regions

Spatial data can also be aggregated into existing geographical units, such as units of administrative division, census areas, electoral districts, etc. The use of such units is often appropriate, as many kinds of them are specially designed for summarising population-related data, so that they include approximately equal numbers of people to whom the data refer (inhabitants, voters, workers, etc.). These units are usually smaller where population density is high. You may have to use predefined regions when you want to relate your data to other data that are only available in the form aggregated by these regions. However, it is sensible to avoid the use of predefined regions when you have no serious reasons for that. Unlike a data-driven division, predefined regions may not represent the distribution of your data well enough, and differences in the sizes of the regions will complicate comparisons and quantitative judgements, as we discussed previously regarding irregular cells. Hence, unjustified use of predefined compartments that are irrelevant to your data and analysis goals combines the disadvantages of the regular and irregular tessellations and does not offer any advantages.

Let us see in Fig.9.13, top, how the distribution of the pubs will look like after aggregation by administrative districts, namely, London wards. In the area of the City of London, where there is a dense concentration of pubs, we now see many tiny districts with small pub counts, while the largest value is attained in the City of Westminster, where the pubs are much sparser. These distortions of the distribution patterns emerge due to the differences in the district sizes. The normalisation of the values by the areas of the districts (i.e., transforming the counts into densities) decreases the distortions (Fig.9.13, bottom, but the overall pattern of clustering of the pubs is not conveyed.

Hence, there are several conflicting criteria for choosing appropriate space partitioning for aggregating your data, including

- preservation of the distribution patterns,

- possibility to calculate and compare multiple attributes,

- possibility to aggregate several datasets for joint analysis,

- possibility of comparisons between places and quantitative judgements,

- possibility to relate your data to other data available in an aggregated form.

Each time when you consider applying spatial aggregation to your data, you need to weigh the importance of these different criteria for your data and analysis goals.

9.5 Analysis tasks and visual analytics techniques

The most common tasks in analysing spatial data are

- Characterise how a phenomenon varies over space and scale.

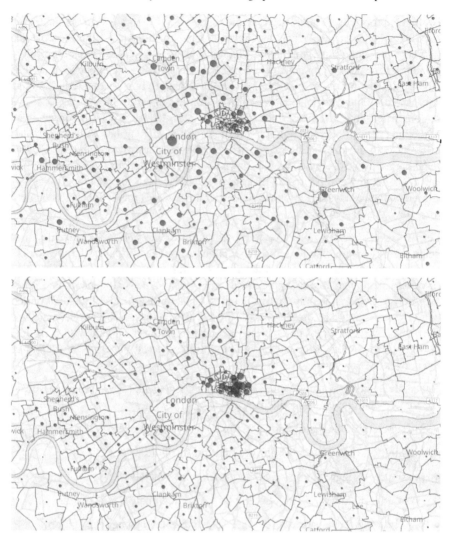

Fig. 9.13: The London pubs data aggregated by predefined districts (wards). Upper image: the circle sizes are proportional to the pub counts. Lower image: the circle sizes are proportional to the pub density, i.e., count divided by the land area.

- Discover and characterise relationships between two or more spatial phenomena.

The most important technique in analysing spatial phenomena is representing the data that reflect the phenomena visually on maps. There are many examples of maps in this chapter as well as in Chapter 1, where we analysed the spatial distribution of the epidemic in Vastopolis, in Chapter 2, where we illustrated different types of patterns that can exist in spatial distributions and variations, in Chapter 3, where we showed examples of choropleth maps with different colour scales, and in Chapter 4, where we used spatial data to describe the process of cluster analysis and demonstrate the impact of the parameter setting on the result. We have earlier mentioned (in Section 9.3.1) and would like to emphasise again the importance of representing spatial data in the relevant spatial context, that is, to put your data on a background map showing general geographic information (land and waters, countries and populated sites, roads and parks, etc.) that is appropriate to the spatial extent of your data and the scale of the intended analysis. While the contextual information should be visible on a map, it should be depicted so that your data can be easily seen on top of the background, and you can focus on your data without being distracted by the background. Hence, the colours in the background should not be too bright or too dark, and the lines and labels should not be too prominent.

Currently, there are numerous online servers providing georeferenced map tiles, i.e., images containing pieces of large maps, which can be stuck together for creating a map of the territory relevant to your analysis[8]. Many of the map tile servers use the open data from OpenStreetMap[9] (OSM), which is a database of crowdsourced worldwide geographic information and a set of data-based services. You can choose between different map styles; not all of them are suitable for background maps. Sometimes it may also be useful to put your data on satellite images, which are also accessible in the form of georeferenced tiles.

There is a trick how to reduce the prominence of a background map that interferes with the visibility of your data: put a semi-transparent grey rectangle on top of the background map and underneath your data. By varying the shade of grey and the degree of transparency, you can create a suitable map display for your analysis. We have applied this trick in generating the maps in the figures from 9.8 up to 9.13.

The OSM server can be queried for obtaining not only map tiles (i.e., images) but also the geographical data in the vector form. It is possible to get the coordinates and attributes of various geographical objects represented by points, lines, or polygons. Thus, our example dataset with the locations of the London pubs have been created by retrieving the data from OSM.

As we discussed in the previous section, before putting your data on a map, you may apply transformations. In particular, aggregation and smoothing are very often used in geographic data analysis. Visual analytics workflows may involve specialised computational methods for spatial analysis[10], such as spatial statistics[11]. Geographic Information Systems[12] (GIS) include many tools for spatial data transformation, visualisation, and analysis. Professional analysis of spatial information

---

[8] https://wiki.openstreetmap.org/wiki/Tiles

[9] https://en.wikipedia.org/wiki/OpenStreetMap

[10] https://en.wikipedia.org/wiki/Spatial_analysis

[11] https://en.wikipedia.org/wiki/Geostatistics

[12] https://en.wikipedia.org/wiki/Geographic_information_system

is mostly done using this kind of software. Implementations of some methods for spatial analysis can also be found in open packages and libraries.

In analysing spatial data, it may also be necessary to perform specific spatial calculations, including computation of numeric measures (distances, lengths, areas, angles), generation of geometric objects, such as spatial buffers, convex hulls, and Voronoi polygons, checking various geometric conditions (point on line, point in polygon, crossing, intersection, etc.).

A very important analytical operation is data selection and filtering (Section 3.3.4) based on the spatial positions and footprints of the data items. There are multiple possibilities for specifying conditions for spatial queries and spatial filtering:

- containment in a given bounding box;

- containment in a specific area or one of several areas;

- being within a given distance from some location or object;

- lying on a given line;

- intersecting or overlapping a given spatial object or one of multiple objects;

- containment of a given location or object, or one of multiple locations or objects.

In the following, we shall describe an example of an analytical workflow that includes visualisation, spatial filtering, data transformations, and joint consideration of two spatial datasets.

9.6 An example of a spatial analysis workflow

We shall analyse the spatial distribution of the London pubs using the dataset that we used previously for illustrating the transformations of spatial data. From all example maps showing these data, the best representation of the distribution is provided by the raster maps in Fig. 9.10. We observe that the highest spatial density of the pubs (a spot in the darkest shade of red) is in the area of the City of London. We also see, especially in the map created with the smaller smoothing bandwidth (the lower image in Fig. 9.10), two smaller dense clusters west of the largest and densest City cluster. Our knowledge of the geography of London tells us that all these three "dark spots" on the map are in the areas where not many people live but there are many visitors, in particular, tourists. It can be conjectured that the pubs are here primarily for serving visitors rather than locals from the neighbourhood.

We want to explore the other areas of high density of pubs and try to understand the reasons for their existence. However, a raster map is not the most convenient tool for such an exploration, because we cannot see well the map background covered by the raster. When we increase the transparency of the raster image, we see the background, but the dark spots of high pub density become harder to see. To see both the pub-dense areas and the spatial context well enough, we shall generate explicit outlines of these areas using the following approach. First, we apply density-based clustering (Section 4.5.1) to the locations of the pubs based on the spatial distances between them. Since the spatial positions in our data are specified by geographic coordinates according to the WGS84 reference system (Section 9.2.4), the spatial distances are calculated using the great-circle distance function (Section 4.2.8). We try several combinations of the parameters 'neighbourhood radius' and 'number of neighbours' and conclude that the result obtained with 300 metres and 5 neighbours is sufficiently good: there are 23 clusters (not too many) which are quite dense and spatially compact (Fig 9.14). The sizes of the clusters (counts of the members) vary from 6 to 79. In total, the clusters include 327 pubs (14.45% of all), and 1,936 pubs are treated as "noise".

We filter out the "noise" and generate polygons enclosing the clusters. We first build convex hulls around the clusters and then extend the hulls by adding 150 m wide spatial buffers around them. The buffering operation outlines areas *around* the clusters, whereas the convex hulls are too "tight", so that many pubs have borderline positions.

Now we switch off the visibility of the map layer containing the individual pubs a points and focus our exploration on the cluster areas whose outlines we have obtained. We use zooming and panning to see the areas in detail, and we consider background maps generated from different variants of map tiles to find relevant spatial context information. We find that the background showing the transportation lines and hubs is definitely relevant to our investigation. In Fig. 9.15, the "pubby" areas are shown as yellowish semi-transparent shapes on top of the background map with the transit information. The sizes of the red dots represent the counts of the pubs in the areas. We have reduced the range of the values represented by the dot sizes to the maximum of 26 pubs. In the three most "pubby" areas that we have identified earlier, the corresponding counts (79, 37, and 28) are not depicted by proportional symbols.

The map tells us that many of the pub clustering areas are situated at transit hubs. It seems quite likely that a large part of the customers of these pubs are commuters. However, there are also clusters that are not spatially associated with transportation facilities. A careful investigation of different kinds of spatial context of these clusters (e.g., the cluster located north of the metro station Angel) did not reveal anything special about these areas, except that not only pubs but also restaurants and small shops are densely clustered there. We can conjecture that London seems to have particular areas where people come to shop, dine, and socialise.

Now we want to investigate whether there is any association between the spatial distribution of the pubs and the variation of the characteristics of the resident population. We have downloaded open population statistics from the London Datas-

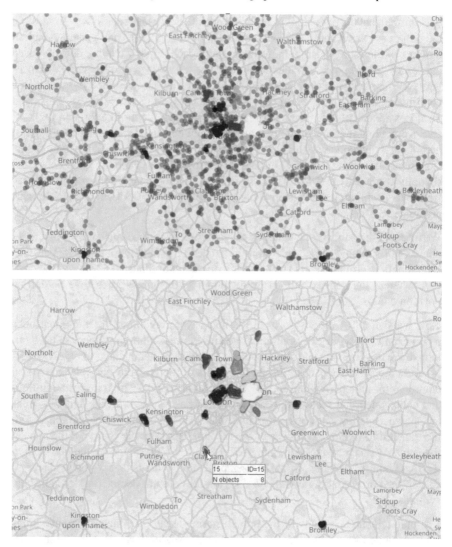

Fig. 9.14: A result of density-based clustering of the London pubs based on the spatial distances between them. Top: the coloured dots represent the members of the clusters and the grey dots represent the "noise". Bottom: the "noise" has been filtered out, and the clusters have been enclosed in polygonal outlines.

tore [13]. In particular, there are data describing the deprivation of the population. The data consist of numeric scores associated with small districts, called Lower Layer Super Output Areas (LSOA)[14]. The deprivation is characterised in terms of several

---

[13] https://data.london.gov.uk/

[14] https://en.wikipedia.org/wiki/Lower_Layer_Super_Output_Area

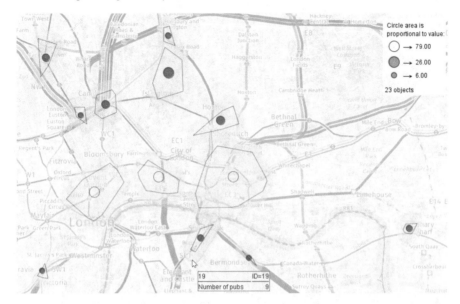

Fig. 9.15: A fragment of a map with the outlines of the pub clusters and the background showing the transportation lines.

criteria, which are combined into an integrated score called Index of Multiple Deprivation (IMD)[15]. The component criteria, which have different weights in the IMD, are deprivation in terms of income, employment, education, health, crime, barriers to housing and services, and living environment. The choropleth maps in Fig. 9.16 show the spatial distributions of the IMD scores (top left) and some of its component scores, namely, education, living environment, and crime. Higher values of the scores indicate higher deprivation.

We use choropleth maps in which the value ranges of the attributes are divided into five equal-frequency class intervals (Section 3.3.1), i.e., each of the corresponding classes (groups of areas) contains approximately one fifth of the total number of the areas. Such groups are called quintiles. In all maps, we apply a colour scale with the shades of blue corresponding to lower deprivation scores and shades of red to higher scores.

As we know that the largest clusters of pubs in the central London, apparently, serve mostly visitors rather than local population, we want to exclude the central areas from the further exploration, as they can skew our observations. We exclude these 65 areas (1.3% of the 4,835 areas that we have) by means of spatial filtering (Fig. 9.17). After that, we re-divide the value ranges of the deprivation attributes into five equal-frequency intervals based on the remaining 4,770 areas, i.e., each interval contains approximately 954 areas (the group sizes may slightly vary, because

---

[15] https://en.wikipedia.org/wiki/Multiple_deprivation_index

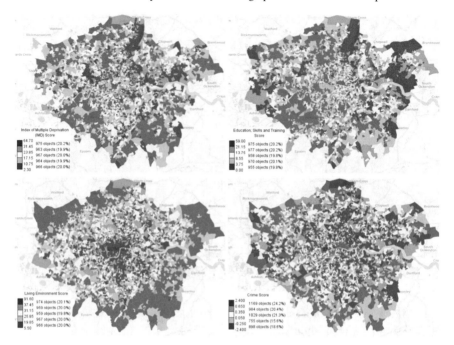

Fig. 9.16: Spatial distributions of the compound deprivation score – IMD (top left) and its components: education, skills, and training (top right), living environment (bottom left), and crime (bottom right).

multiple areas may have the same attribute value and thus need to be in the same group, even when the group becomes bigger than others). In the following analysis, we shall investigate how the pubs are distributed among the area groups defined according to the different deprivation scores. To be able to do that, we apply the spatial calculation operation that counts the number of pubs in each area.

The existence of associations between the distribution of the pubs and the variation of the area statistics cannot be detected straightforwardly by computing the statistical correlation coefficients or looking at a scatterplot. It is because the statistical distribution of the per-area pub counts is extremely skewed: 77.4% of the areas do not contain pubs at all, 15.5% of the areas contain 1 pub each, 4.3% have 2 pubs per area, and only 2.8% have 3 or more pubs per area. In total, there are 1,611 pubs in 4,770 areas. In such a situation, a viable approach is to consider the distribution of the pubs by groups of areas rather than over the individual areas. Our approach is to define groups of areas based on values of area attributes, such as the deprivation scores, and see how many pubs there are in each group. For comparability, the groups should be equal in size (number of areas) or population. In the particular case of the output areas designed for reporting the population statistics, equal-size groups of areas are expected to have approximately equal population counts. It would still

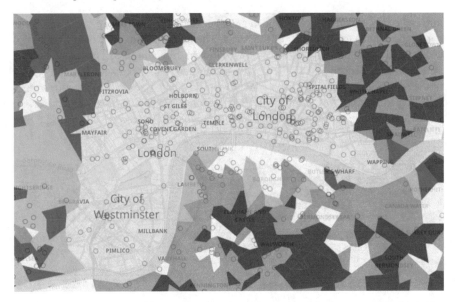

Fig. 9.17: The areas in the centre of London have been excluded from the further analysis by means of spatial filtering.

be appropriate to check whether this expectation holds and adjust the division if necessary; however, 193 records in the data that we use lack the population counts. Therefore, we shall divide the areas into equal-size groups each containing approximately 20% of the areas, and we shall look at the population counts in the groups keeping in mind that these counts may be incomplete.

Our approach to the exploration is based on the following idea. We divide the areas into equal groups according to values of some attribute characterising the groups, and we compare the total counts of the pubs in the groups. As there are 1,611 pubs distributed over 5 groups of areas, the average number of pubs per group should be around 322. If the counts do not differ much from the average and from each other, it means that there is no association between the distribution of the pubs over the set of the areas and the characteristic of the areas that was used for the grouping. Significant differences among the pub counts indicate the existence of an association. The association is positive when the groups of areas with higher values of the attribute have more pubs. If there are fewer pubs in the areas with higher attribute values, it means that there is a negative association.

We first consider the deprivation attributes, including the composite score IMD and its components. The distributions of the pubs over the quintiles of the areas defined according to different deprivation attributes are visually represented in Fig. 9.18, top. The lower image shows the distribution of the population over the same groups of the areas. Each distribution is represented by a segmented bar; the whole bar length corresponds to the total number of the pubs in the upper image and to the total number of the population in the lower image. Each bar is divided into 5 segments corresponding to the quintiles of the areas. The colours from dark blue to dark red correspond to the attribute values from the lowest to the highest. The lengths of the segments are proportional to the counts of the pubs (upper image) and the counts of the population (lower image) in the groups. The bars in both images correspond, from top to bottom, to the following deprivation attributes: IMD score; education, skills, and training score; health deprivation and disability score; crime score; living environment score.

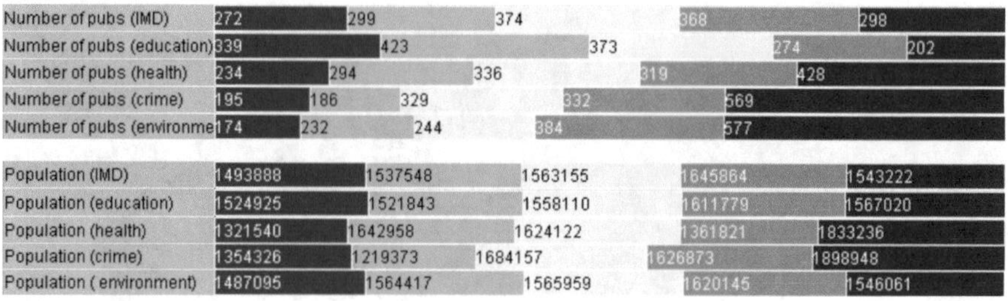

Number of pubs (IMD)	272	299	374	368	298
Number of pubs (education)	339	423	373	274	202
Number of pubs (health)	234	294	336	319	428
Number of pubs (crime)	195	186	329	332	569
Number of pubs (environme...	174	232	244	384	577
Population (IMD)	1493888	1537548	1563155	1645864	1543222
Population (education)	1524925	1521843	1558110	1611779	1567020
Population (health)	1321540	1642958	1624122	1361821	1833236
Population (crime)	1354326	1219373	1684157	1626873	1898948
Population (environment)	1487095	1564417	1565959	1620145	1546061

Fig. 9.18: Distribution of the London pubs (top) and population (bottom) over quintiles of areas defined according to different deprivation attributes.

Number of pubs (% white)	186	284	351	347	355
Number of pubs (% BAME)	355	346	351	280	191
Number of pubs (% 0-15)	487	340	264	224	208
Number of pubs (% 16-29)	201	207	240	349	526
Number of pubs (% 30-44)	156	225	270	356	516
Number of pubs (% 45-64)	367	328	306	297	225
Number of pubs (% 65+)	366	295	290	272	300
Population (% white)	1647017	1605065	1576275	1498505	1456815
Population (% BAME)	1453890	1498615	1570669	1607760	1652743
Population (% 0-15)	1417162	1471267	1519848	1585898	1789502
Population (% 16-29)	1366430	1458318	1506868	1639062	1812999
Population (% 30-44)	1352074	1461451	1554847	1611256	1804049
Population (% 45-64)	1483045	1512391	1563950	1578960	1645331
Population (% 65+)	1543086	1572099	1569042	1541569	1557881

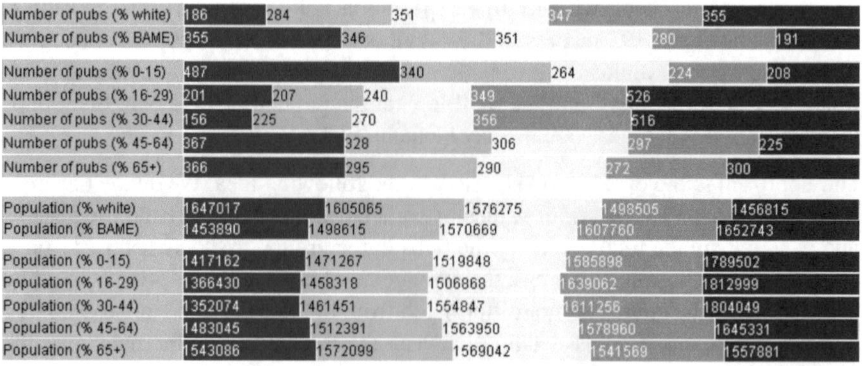

Fig. 9.19: Distribution of the London pubs (top) and population (bottom) over quintiles of areas defined according to different attributes describing the population structure.

Concerning the distributions of the population, which are shown in the lower image, we can see that the groups defined based on IMD, education, and living environment have quite similar population counts whereas the groups based on health and crime are less equal, but the differences are not as large as we see in some of the bars in the upper image. The latter tells us that there is no association between the distribution of the pubs and the composite deprivation index IMD. The same refers to the income and employment deprivation (not shown in Fig. 9.18), which are the IMD components having the highest weights. However, there are notable differences between the per-group pub counts for the groups defined based on the education, health, crime, and living environment deprivation. These differences indicate the existence of associations. There is a strong negative association between the pub distribution and the education deprivation, i.e., there are notably fewer pubs in the areas with higher education deprivation. For the remaining three deprivation attributes shown in Fig. 9.18, the associations with the pub distribution are positive. The association of the living environment deprivation is especially strong, also taking into account the approximately uniform distribution of the population over the groups of the area. Hence, higher numbers of pubs are associated with worse living environment and with higher levels of crime and health deprivation.

We apply the same approach to investigate how the spatial distribution of the pubs is associated with the population structure of the output areas. Similarly to Fig. 9.18, Figure 9.19 represents the distributions of the pubs over the area groups defined according to the percentages of the white population and the national minorities (abbreviated as BAME – Black, Asian and Minority Ethnic) and to the percentages of the age groups 0-15 years old, 16-29, 30-44, 45-64, and 65+. We clearly see that the distribution of the pubs is negatively associated with the percentages of the national minorities, children (0-15 years), and people aged 45-64 years, and positively associated with the percentages of the white population and the age groups of 16-29 and 30-44 years old. There is no association with the percentages of the elderly people (65+ years).

This completes our investigation. Now we can report the following findings:

- The spatial distribution of the London pubs is strongly clustered. The largest and densest clusters are located in the centre, where there are many tourists and other visitors of the city. Many clusters are related to the transportation facilities.

- The spatial distribution of the pubs is associated with some aspects of the deprivation of the resident population, namely, education (negative association), health, crime, and living environment (positive association).

- The spatial distribution of the pubs is also associated with the ethnic and age structure of the resident population: there are more pubs in the areas with higher percentages of white people and age groups 16-29 and 30-34 and fewer pubs in the areas with higher proportions of ethnic minorities, children, and 45-64 years old people.

In this investigation, we used multiple analytical techniques and operations: data transformations, density-based clustering, basic spatial computations, filtering, and, obviously, visual representations, including different types of maps. As it is usual for visual analytics workflows, our major activity was to observe and interpret distribution patterns, derive knowledge by means of analytical reasoning, and plan further steps in the analysis based on what we have observed and learned so far.

9.7 Conclusion

In analysing spatial distributions and variations, it should always be kept in mind that spatial phenomena are never independent of the properties of the underlying space (e.g., whether it is land or water, a flat surface or a rocky mountain ridge), objects existing in it, and other spatial phenomena. The spatial context modifies the fundamental spatial relationship of distance: the distance between two points measured mathematically may not be the same as the physical distance indicating how reachable one point is from the other. Moreover, spatial relationships may be anisotropic: the physical distance from A to B may differ from the distance from B to A. Due to the non-trivial effects of the spatial context, spatial distributions and variations are challenging to analyse. Potentially relevant spatial information cannot be completely and properly encoded for machine processing. Hence, the involvement of human knowledge and understanding of space and spatial relationships is crucial for analysis. To utilise their spatial knowledge and capabilities for understanding and analysis, humans need maps that show the data in the spatial context.

The fact that many things and phenomena share the same underlying space means not only that you can put them together on the same map for inspecting their relationships visually, but you can also join different datasets based on the spatial references of the data items. Such spatial joins allow you to analyse the data also by means of various computational techniques. These can be very helpful, but, irrespective of how advanced they are, they can only be a complement to the power of human understanding and reasoning.

The space and spatial relationships are so important for human beings that humans can intuitively perceive and understand spatial relationships and spatial patterns. To involve these capabilities also in analysis of non-spatial data, *spatialisation* (Chapter 2) is widely used. The properties of artificial spaces (Section 4.4.1) are not the same as the properties of the physical space, which requires human analysts to verify and amend their intuitive impressions and judgements based on their knowledge of how the spaces have been generated.

Space, spatial phenomena, and spatial analysis are challenging but very interesting. We hope that you will be able to do many discoveries by looking at spatial relationships between various things, and that you will enjoy the process.

9.8 Questions and exercises

- Which of the following distributions and variations allow or disallow the use of spatial smoothing? Why?

  - Measurements of air quality at sample locations throughout a town.

- Measurements of water quality at sample locations throughout a river basin.

- Observed locations of plants from a certain species growing in different alpine valleys.

- Locations of public transport stops in a town.

• Analyse (or imagine yourself analysing) a dataset with locations of schools (in London or elsewhere) or another kind of objects that are not so strongly clustered as the pubs in London. How the specifics of the spatial distribution that you may observe will affect your analysis workflow? What techniques you will try and which techniques may not be very useful?

• Describe the spatial relationships between different sites of taking measurements of water quality in a river basin. How would you take these relationships into account in analysing the data?

Chapter 10
Visual Analytics for Understanding Phenomena in Space and Time

Abstract There are different kinds of spatio-temporal phenomena, including events that occur at different locations, movements of discrete entities, changes of shapes and sizes of entities, changes of conditions at different places and overall situations across large areas. Spatio-temporal data may specify positions, times, and characteristics of spatial events, represent trajectories of moving entities, or consist of spatially referenced time series of attribute values. It is possible to transform spatio-temporal data from one of the forms to another and thus adapt them to analysis tasks. After presenting a motivating example of analysis, we discuss the specifics of spatio-temporal data and possible data quality issues that often appear in real data sets and influence analysis processes and results. We introduce and discuss the visual analytics techniques suitable for spatio-temporal data and present another example of an analytical workflow.

10.1 Motivating example

We already analysed spatio-temporal data in our very first example presented in Chapter 1. The data were about spatially referenced microblog posts. Each post is a **spatial event**, i.e., an event (as defined in Chapter 8) having a location in space. We analysed the distribution of these events and variation of their characteristics (specifically, the keyword occurrences) in space and time.

Here we shall use an example of data reflecting a different kind of spatio-temporal phenomenon, namely, movements (i.e., changes of spatial positions) of discrete entities. Our dataset contains about 18M positions of 5,055 vessels that moved between the bay of Brest and the outer sea [114] during 6 months from the 1st of October, 2015 till March 31, 2016. Each data record specifies a vessel identity (a so-called

MMSI[1]), a time moment, the geographic position of the vessel at that time moment, and the navigation status reporting the activity of the vessel such as 'at anchor', 'under way using engine', 'engaged in fishing', and others. Being arranged in the chronological order and connected by lines, the positions of each individual moving entity make a **trajectory** of this entity.

Our task in this analysis is to study when, where, and for how long the cargo vessels were anchoring outside of the port and understand whether the events of anchoring may indicate waiting for an opportunity to enter or exit the bay (through a narrow strait) or the port of Brest. Naturally, we hope that the attribute 'navigational status' will tell us which positions in the data correspond to anchoring. However, by visual inspection of a sample of the data (Fig. 10.1), we notice that the values of this attribute are quite unreliable. The upper left image shows a selection of points from multiple trajectories in which the navigational status equals 'at anchor'. It is well visible that the selection includes not only stationary points but also point sequences arranged in long traces, which means that the vessels were moving rather than anchoring. On the opposite, the upper right image demonstrates several trajectory fragments that look like hairballs. Such shapes are typical for stops, when the position of a moving object does not change but the tracking device reports each time a slightly differing position due to unavoidable errors in the measurements. The hairball shapes thus signify that the vessels were at anchor or moored at the shore, but the recorded navigational status was 'under way using engine'. The lower image shows trajectory fragments of several vessels. The character of their movements (back and forth repeated multiple times) indicates that they were fishing, i.e., the navigational status should be 'engaged in fishing', but the value attached to the positions is 'under way using engine'. This means that we need to identify anchoring events without relying on the reported navigational status.

Since we are interested in anchoring of cargo vessels in the vicinity of the port of Brest, we need to select only relevant data from the database. In accord with our analysis goals, we exclude the data referring to fishing vessels and, for the remaining vessels, select only those trajectories that passed through the strait at least once. From these trajectories, we select only the points located inside the bay of Brest, in the strait, and in the area extending to about 20 km west of the strait. As a vessel might not be present in the study area all the time, there may be long temporal gaps between some of the selected positions of this vessel. To exclude such gaps from consideration, we split the trajectories into parts at all positions in the record sequences where the time interval to the next point exceeds 30 minutes or the spatial distance exceeds 2 km. Next, we further divide the trajectories by excluding the stops (segments with near-zero speed) within the Brest port area. From the resulting trajectories, we select only those that passed through the strait and had duration at least 15 minutes. As a result of these selections and transformations, we obtain 1718 trajectories for the further analysis. Of these trajectories, 945 came into the bay from

---

[1] https://en.wikipedia.org/wiki/Maritime_Mobile_Service_Identity

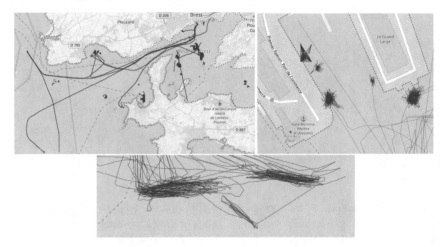

Fig. 10.1: Fragments of trajectories with wrongly reported navigation status: under way using engine (upper left and bottom), at anchor (upper right).

the outer area, 914 moved from the bay out, and 141 trajectories include incoming and outgoing parts.

The analysis goal requires us to identify the *anchoring events*. As we cannot rely completely on the navigation status in position records (see Fig. 10.1), we apply the following heuristics. We know that, in the vicinity of ports or major traffic lanes, vessels do not anchor at arbitrarily positions. There are special areas, called anchoring zones, where vessels are allowed to anchor. Hence, we need to identify these anchoring zones. For this purpose, we find the areas containing spatial clusters of vessel positions with the corresponding navigational status being 'anchoring' (Fig. 10.2) and ignore occasional solitary occurrences of records reporting anchoring in other places. Next, we assume that any sufficiently (at least 5 minutes) long stop in an anchoring zone corresponds to anchoring. We extract the stops made in the anchoring zones, irrespective of the value of the attribute 'navigational status', into a separate dataset for the further analysis. In this way, we get a set of 212 anchoring events (we shall further call them shortly "stops") that happened in 126 trajectories. Fig. 10.3 shows these trajectories in bright blue and the positions of the stops in red.

Since we want to understand how the stops are related to passing the strait between the bay and the outer sea, we find the part corresponding to strait passing in each trajectory. For this purpose, we interactively outline the area of the strait as a whole and, separately, two areas stretching across the strait at the inner and outer ends of it, as shown in Fig. 10.4. The segments of the trajectories located inside the whole strait area (painted in yellow in Fig. 10.4) are treated as *strait passing events*. For these events, we determine the times of vessel appearances in the areas at the inner and outer ends of the strait (painted in red in Fig. 10.4). Based on the chronological order of the appearances, we determine the direction of the strait passing events:

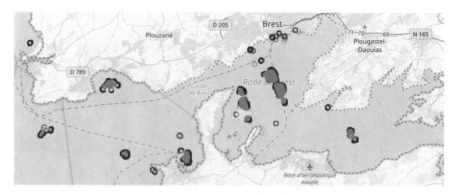

Fig. 10.2: Delineation of anchoring zones: The violet dots show all positions reported as anchoring. The yellow-filled polygons outline anchoring zones containing dense concentrations of the anchoring points beyond the port and major traffic lanes.

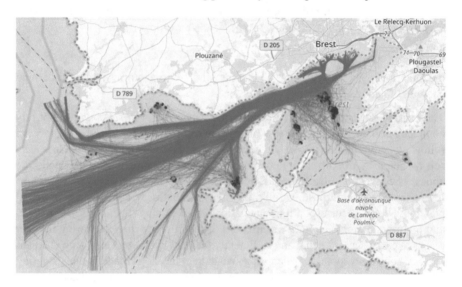

Fig. 10.3: The trajectories selected for analysis with the anchoring events (stops) marked in red.

inward or outward with respect to the bay. Then we categorise the stops based on the directions of the preceding and following events of strait passing.

The pie charts on the map in Fig. 10.5 represent the counts of the different categories of the stops that occurred in the anchoring areas. The most numerous category 'inward;none' (105 stops) includes the stops of the vessels that entered the bay, anchored inside the bay, and, afterwards, entered the port. The category 'outward;inward' (36 stops) contains the stops of the vessels that exited the bay, anchored in the outer area, then returned to the bay and came in the port. 34 stops took

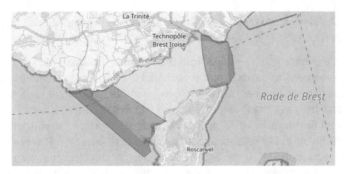

Fig. 10.4: Interactively specified areas used for identifying fragments of trajectories corresponding to travelling through the strait (yellow) and determining the travelling direction (red).

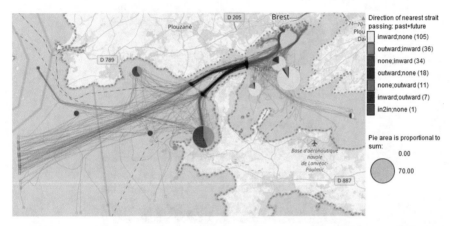

Fig. 10.5: The pie charts represent the counts of the stops in the anchoring zones categorised with regard to the directions of the preceding and following strait passing by the vessels.

place before entering the bay ('none;inward'), 18 happened after exiting the bay ('outward;none') and 11 before exiting the bay ('none;outward'). In 7 cases, vessels entered the bay from the outside, anchored, and then returned back without visiting the port ('inward;outward'), and there was one stop that happened after entering the strait at the inner side and returning back ('in2in;none').

We see that the majority of the stop events (yellow pie segments) happened after entering the bay and, moreover, a large part of the stops that took place in the outer area happened after exiting the bay and before re-entering it (orange pie segments). It appears probable that the vessels stopped because they had to wait for being served in the port. Most of them were waiting inside the bay but some had or preferred to wait outside. Hence, the majority of the anchoring events can be related to waiting for port services rather than to a difficult traffic situation in the strait.

Additional evidence can be gained from the 2D time histogram in Fig. 10.6, where the rows correspond to the days of the week and the columns to the hours of the day. It shows us that the number of the anchoring vessels reaches the highest levels on the weekend (two top rows) and on Monday (the bottom row). It tends to decrease starting from the morning of Wednesday (the third row from the bottom of the histogram) till the morning of Thursday (the fourth row), and then it starts increasing again. The accumulation of the anchoring vessels by the weekend and gradual decrease of their number during the weekdays supports our hypothesis that the stops may be related to the port operation.

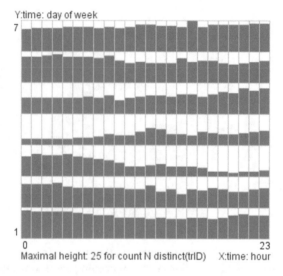

Fig. 10.6: A 2D time histogram represents the counts of the anchoring events by the hours of the day (horizontal axis) and days of the week (vertical axis) by the heights of the corresponding bars.

Now we want to look at the movements of the vessels that made stops on their way. We apply dynamic aggregation of the vessel trajectories by a set of interactively define areas, which include the anchoring zones, the port area, the areas at the outer and inner ends of the strait, and a few additional regions in the outer sea. The aggregation connects the areas by vectors and computes for each vector the number of moves that happened between its origin and destination areas. The result is shown on a flow map, where the vectors are represented by curved lines with the widths proportional to the move counts (Fig. 10.7). The curvatures of the lines are lower at the vector origins and higher at the destinations.

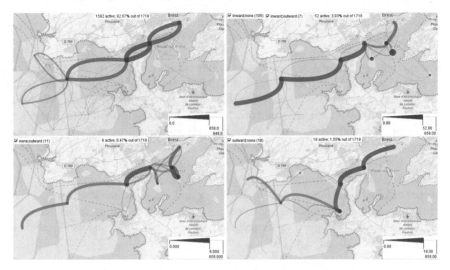

Fig. 10.7: The trajectories under study are represented in an aggregated form on flow maps. Top left: trajectories without stops. The remaining images represent subsets of the trajectories having stops after entering the bay (top right), before (bottom left) and after exiting the bay (bottom right).

The aggregation we have applied to the trajectories is dynamic in the sense that it re-acts to changes of the filters that are applied to the trajectories. As soon as the subset of the trajectories selected by one or more filters changes, the counts of the moves between the areas are automatically re-calculated, and the flow map representing them is immediately updated. The four images in Fig. 10.7 represent different states of the same map display corresponding to different query conditions. The upper left image represents the 1592 trajectories (92.67% of all initially selected trajectories) that did not include stops. This flow map can be considered as showing normal, uninterrupted traffic to and from the port of Brest. The flows between any two ar-eas look symmetric, i.e., the lines have equal widths, which means approximately the same numbers of moves in the two opposite directions. The remaining images in Fig. 10.7 show aggregates of different selections of the trajectories depending on the relative times (with respect to the strait passing) when they had anchoring events. These flow maps demonstrate which parts of the routes are common in a given selection, and which parts vary and proportions of the varying parts.

Concluding the exploration of the stops, we can summarise that most of them are likely to have happened because the vessels had to wait for being served in the port of Brest. The vessels coming from the outer sea were waiting mostly inside the bay (Fig. 10.7, top right), and the vessels that had been unloaded in the port and had to wait for the next load or another service were waiting mostly outside of the bay. The vessels that had to wait for port services tended to accumulate over the weekend, and their number reduced during the weekend (Fig. 10.6, right).

10.2 Specifics of this kind of phenomena/data

The motivating example in Section 10.1 demonstrates a variety of forms of spatio-temporal data. We have been dealing with trajectories of moving entities, with spatial events (stops), and with flows representing aggregated movements. We have applied data transformations, some of which converted one form of data to another. Thus, we extracted events from trajectories and aggregated trajectories into flows between places. Another transformation that we applied was division of trajectories into smaller trajectories representing different trips or travelling between particular places. We have also identified spatially condensed groups of events and used spatial boundaries of these groups to define relevant places, which were used for aggregation of the events and trajectories and for filtering of the trajectories. Now we shall consider the possible transformations systematically, following the framework introduced in the monograph [10].

10.2.1 Data structures and transformations

Basically, spatio-temporal data are temporal data, as discussed in Chapter 8, which include references to spatial locations, as in spatial data (Chapter 9). We consider the following types of spatio-temporal data:

- *Spatial time series* consist of attribute values specified at multiple different spatial locations and different times, i.e., these are attribute time series, as considered in Chapter 8, referring to spatial locations. Typically, the locations do not change over time.

- *Spatial event data*: data characterising spatial events. Events in general were also considered in Chapter 8, Spatial events are events occurring at some locations in space, i.e., they can also be treated as *spatial objects*.

- *Movement data (trajectories)*: data characterising changes of spatial positions of moving entities. In fact, these data are *time series* consisting of spatial positions and, possibly, values of other attributes. Each data item describes a *spatial event* of an entity being at a specific location at some time; hence, trajectories are composed of spatial events. Furthermore, a trajectory itself is a spatial object: it can be considered as a directed line in space representing the travel route. It is also a temporal object, i.e., an event, as it has its specific time of existence, which is the time interval from the beginning to the end of the travel.

As can be seen from these notes, there are intrinsic relationships between different types of spatio-temporal data, and it is not surprising that there are multiple ways to transform data from one type to another. Figure 10.8 (adapted from [10]) presents the general scheme of possible transformations.

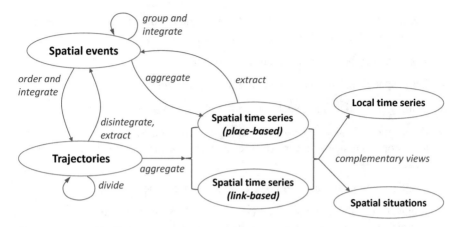

Fig. 10.8: Transformations between different types of spatio-temporal data.

If you compare this transformation scheme with the transformation scheme of temporal data presented in Fig. 8.7, you will certainly notice commonalities. It is not surprising, because spatio-temporal data are a special class of temporal data. Thus, like events in general, spatial events can be grouped and integrated into larger events. These larger events are also spatial events because they have spatial locations, which can be defined as unions of the locations of the component events or as spatial outlines of the event groups. All transformations that are applicable to time series (Section 8.3) can, obviously, be applied to spatial time series. Events extracted from spatial time series are spatial events: the spatial locations of these events are the places the original time series refer to. Generally, the whole system of possible transformations of temporal data is applicable, in particular, to spatio-temporal data, with the specifics that the derived data will usually have spatial references. Since spatio-temporal data are also a special subclass of spatial data, all transformations applicable to spatial data (Section 9.4) can be applied, in particular, to spatio-temporal data.

Special notes are required concerning movement data. Such data are usually available originally as collections of records specifying spatial positions of moving objects (e.g. vessels) at different times. Such records describe *spatial events* of the presence (or appearance) of the moving objects at certain locations and specify the times when these events occurred. When all records referring to the same moving object are put in a chronological order, they together describe a trajectory of this object. Hence, trajectories are obtained by integrating spatial events of object appearance at specific locations. The trajectories can be again disintegrated to the component events. Particular events of interest, such as stops or zigzagged movement, can be detected in trajectories and extracted from them. A trajectory describing movements of an object during a long time period can be divided into shorter trajectories, for example, representing different trips of the object. Trajectories may be divided according to different criteria:

- Temporal gap between consecutive positions;

- Large spatial distance between consecutive positions;

- Visiting particular areas, such as ports or fuel stations;

- Occurrence of particular events or changes in the movement, e.g., a sufficiently long stop, a change of a car driver, or a change of the ball possession in a sport game;

- Beginning of a new time period, e.g., a new day or a new season.

The relevance of the division criteria depends on the nature of the moving objects, the character of their movements, and the analysis goals. For example, if you want to compare daily mobility behaviours of individuals, their trajectories need to be divided into daily trips. If you wish to find repeated trips, you need to divide the trajectories by stop events. For studying temporal patterns of stops, the stops need to be extracted from complete trajectories without any division: this is necessary for correct estimation of the stop durations.

Aggregation of spatio-temporal data combines temporal and spatial aggregation. Section 8.3 discusses possible ways for aggregating *events* into *time series* using either absolute or relative (cyclic) time references. Section 9.4.2 considers aggregation of spatial data by areas, which can be defined using different ways of space partitioning discussed in Section 9.3.4. Having divided the space into *compartments* (shortly, places) and time into *intervals*, it is possible to aggregate either spatial events or trajectories by places and time intervals. *Place-based* aggregation involves counting for each pair of place and time interval (1) the events that occurred in this place during this interval, or (2) the number of visits of this place by moving objects and the number of distinct objects that visited this place or stayed in it during the interval. Additionally, various summary statistics of the events or visits can be calculated, for example, the average or total duration of the events or visits. The result of this operation is time series of the aggregated counts (e.g. counts of stops or counts of distinct visitors) and statistical summaries associated with the places.

Link-based aggregation summarises movements (transitions) between places and, thus, can be applied to trajectories. For each combination of two places and a time interval, the number of times when any object moved from the first to the second place during this interval and the number of the objects that moved are counted. Additionally, summary statistics of the transitions can be computed, such as the average speed or the duration of the transitions. The result of this operation is time series of the counts and statistical summaries associated with the pairs of places. The time series characterise the links between the places; therefore, they can be called *link-based*. The term "link between place A and place B" refers to the existence of at least one transition from A to B.

Spatial time series have an interesting specific property as compared to time series of attribute values, which were considered in Chapter 8. Both place-based and link-based time series can be viewed in two complementary ways: as spatially distributed *local time series* (i.e., each time series refers to one place or one link) and as temporal sequence of *spatial situations*, where each situation is a particular spatial distribution of the counts and summaries over the set of places or the set of links. These perspectives require different methods of visualisation and analysis. Thus, the first perspective focuses on the places or links, and the analyst compares the respective temporal variations of the attribute values such as counts of distinct vessels in ports over days. The local time series can be analysed using methods suitable for attribute time series, such as those considered in Chapter 8. The second perspective focuses on the time intervals, and the analyst compares the respective spatial distributions of the values associated with the places or links. The spatial distributions can be analysed using methods suitable for spatial data analysis, such as those mentioned in Chapter 9.

Technically, the two complementary views of spatial time series can be supported by transposing a table containing the time series. The table format in which the rows correspond to places and the columns to time steps is suitable for considering the data as local time series referring to places. Each table row describes one time series. The format in which the rows correspond to time steps and the columns to places is suitable for analysing data as spatial situations referring to time steps. Each table row describes one spatial situation. Methods of time series analysis (see Section 8.5.2) are applicable to local time series, and methods for analysis of time series of complex states (Section 8.5.3) can be applied to spatial situations.

The types of transformations applicable to temporal data include transforming time references from absolute to relative (Section 8.3). Similarly, spatial references in spatial data can also be transformed from absolute to relative (Section 9.4). Obviously, both kinds of transformations can be applied to spatio-temporal data. Let us show a few examples of how this can be done and for what purpose.

In Fig. 10.9, we have trajectories of seasonal migration of several white storks. The STC on the left shows the trajectories according to the original time references, and the STC on the right shows the result of transforming the time references to relative times with respect to the beginning of each migration season. After this transformation, we can better see the similarities and differences between the seasonal movement patterns of the same and different birds across different seasons. The main purpose of cycle-based time transformations is to support exploration of routine behaviours and repetitive processes. Other potentially useful transformations of time references in trajectories are transformations to relative times with respect to trip starts, or trip ends, or particular reference time moments, such as the moments when aircraft reach their planned cruise altitudes. Such transformations are useful for comparing movement dynamics, e.g., speed patterns, among trips.

In Section 9.4, we had an example of a spatial coordinates transformation that created a so-called "team space" (or "group space") and represented relative move-

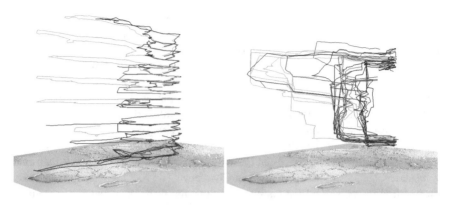

Fig. 10.9: Trajectories of seasonal migration of 17 white storks (depicted in distinct colours) over 8 seasons are shown in space-time cubes according to the calendar dates (left) and relative times with respect to the beginnings of the migration seasons (right).

ments of multiple individuals with respect to others as trajectories in this space (Figs. 9.6 and 9.7). This kind of transformation is useful for analysing collective movements, such as movements of team members in a sports game[11] or joint movements of collective animals [20].

In Chapter 7, we demonstrated an example of transforming spatial positions of moving entities into nodes of a *state transition graph* (Figs. 7.7 and 7.10 to 7.12). The general approach is to assign meaningful categories to places, represent the categories by graph nodes (i.e., states), aggregate the individual movements between the places into collective flows between the place categories, and represent the flows (i.e., transitions) by edges of the graph [19]. This kind of transformation can support semantic analysis of movement behaviours. Importantly, it allows us to compare movement behaviours that took place in different geographical regions.

If we create a meaningful arrangement of the nodes of a state transition graph on a plane, i.e., so that the relative positions represent some semantic relationships, we can treat this arrangement as a special kind of space – a semantic space. The conceptual difference of a space from a mere arrangement is that distances and other spatial relationships in a space are supposed to be meaningful. Having constructed such a semantic space, we can translate movement data or spatial events from physical space to the semantic space, and we can use cartographic visualisation techniques for representing the data in the semantic space. However, we should be careful in interpreting such visualisations, keeping in mind that a semantic space consists of discrete locations ("semantic places"), and there is nothing between them, differently from the continuous physical space.

An example of the use of a semantic space is shown in Fig. 10.10. We have constructed a semantic space for representing human daily mobility. The semantic

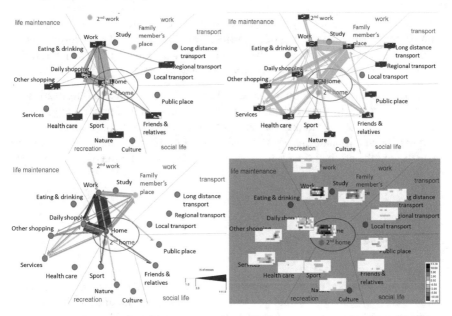

Fig. 10.10: Mobility data of two persons living in different geographical regions have been translated from the geographical space into a semantic space of human activities and aggregated into flows between "semantic places" and time series of presence in these places by the days of the week and hours of the day. The upper two maps show the behaviours of the two persons, and the lower two maps represent the differences between their flows (left) and the patterns of the presence in the "semantic places".

places in this space are the home, work, shopping, health care, sports, and other concepts corresponding to activities of people. We have arranged these "places" into "regions": work, transport, life maintenance, recreation, and social life. We got data from two persons living in different geographical regions who tracked their movements over long time periods. By analysing these data, it was possible to detect repeatedly visited places and identify the meanings of these places based on the typical times, frequencies, and durations of the visits. We substituted the place coordinates in the data by the semantic labels of the places and in this way translated the data from the geographic space to the semantic space. We applied spatio-temporal aggregation to the transformed data and represented the mobility behaviours of the two individuals by aggregate flows between semantic places and by weekly time series of the presence in these places. The time series are represented on the maps in Fig. 10.10 by 2D time charts, or "mosaic plots", as introduced in Fig. 3.12. Although these two persons live very far from each other, we can now easily compare their behaviours. In the upper part of Fig. 10.10, two maps representing the behaviours are juxtaposed for comparison. In the lower left image, the aggregated inter-place movements of the two persons are superposed on the same map. In the lower right image, the daily and weekly patterns of the presence in the "semantic places" are compared by explicit encoding of the differences. The shades of blue correspond to higher values of the person whose behaviour is shown on the top left, and the shades of orange correspond to higher values of the other person.

Apart from the transformation of the spatial coordinates into positions in an artificial "semantic" space, this example also demonstrates that spatio-temporal aggregation can be applied not only to original data describing movements or spatial events but also to data with transformed spatial positions and/or temporal references (here, we transformed the original time references into positions within the daily and weekly cycles).

10.2.2 General properties

Since spatio-temporal phenomena and data are both spatial and temporal, they have all properties pertinent to spatial and temporal phenomena and data, as discussed in Chapters 9 and 8, respectively. Specifically, spatial events combine the properties of general events and discrete spatial objects, and spatial time series combine the properties of general time series and spatial phenomena described by spatial variation of attribute values. Spatio-temporal data may also have the same kinds of quality problems as spatial and temporal data.

Unlike spatial events and spatial time series, movement data have specific properties that are not pertinent to either purely spatial (static) or purely temporal (space-irrelevant) data taken alone. Therefore, we consider the properties of movement data separately.

The first group of properties relates to the data structure. The essential components of the movement phenomenon are moving entities, shortly, *movers*, the space in which they move, and the time when they move, particularly, the times when the movers are present at different spatial locations. These components of the movement phenomenon are represented by corresponding components of movement data. A set of movement data can represent movements of a single entity; in this case, there is no need in a special data component referring to movers. When data describe movements of several objects, each mover needs to be denoted by some identifier. Since, in general, there are three essential components in movement data, we put the properties into subgroups referring to these components.

- Properties of the mover set:
  - number of movers: a single mover, a small number of movers, a large number of movers;
  - population coverage: whether there are data about all movers of interest for a given territory and time period or only for a sample of the movers;

- representativeness: whether the sample of movers is representative of the population, i.e., has the same distribution of the properties as in the whole population or is biased towards individuals with particular properties.

- Spatial properties:

  - spatial resolution: what is the minimal change of position of an object that can be reflected in the data?

  - spatial precision: are the positions defined as points or as locations having spatial extents (e.g. areas)? For example, the position of a mobile phone may be represented by a reference to a cell in a mobile phone network;

  - position exactness: how exactly could the positions be determined? Thus, a movement sensor may detect the presence of a mover within its detection area but may not be able to determine the exact coordinates of the mover in this area. The position of the mover may be recorded as a point, but, in fact, it is the position of the sensor and not the mover's true position;

  - positioning accuracy: how much error may be in the measurements?

  - spatial coverage: are positions recorded everywhere or, if not, how are the locations where positions are recorded distributed over the studied territory (in terms of the spatial extent, uniformity, and density)?

- Temporal properties:

  - temporal resolution: the lengths of the time intervals between the position measurements;

  - temporal regularity: whether the length of the time intervals between the measurements is constant or variable for particular movers and for the whole data set;

  - temporal coverage: whether the measurements were made during the whole time span of the data or in a sample of time units, or there were intentional or unintentional breaks in the measurements;

  - time cycles coverage: whether all positions of relevant time cycles (daily, weekly, seasonal, etc.) are sufficiently represented in the data, or the data refer only to subsets of positions (e.g., only work days or only daytime), or there is a bias towards some positions;

  - missing positions: whether there are unusually long time intervals between consecutive position records;

  - meanings of the position absence: whether absence of positions corresponds to absence of movement (i.e., stops), or to conditions when measurements were impossible, or to device failure, or to private information that has been removed.

The properties in the subgroups "spatial" and "temporal" can refer, respectively, to spatial and temporal data in general. The properties in the subgroup "mover set" are, basically , the properties of a statistical sample in relation to a statistical population[2].

The properties of movement data are strongly related to the data collection methods, which may be

- Time-based: positions of movers are recorded at regularly spaced time moments.

- Change-based: a record is made when mover's position, speed, or movement direction differs from the previous one.

- Location-based: a record is made when a mover enters or comes close to a specific place, e.g., where a sensor is installed.

- Event-based: positions and times are recorded when certain events occur, for example, when movers perform certain activities, such as cellphone calls or posting georeferenced contents to social media.

Only time-based measurements can produce temporally regular data. The temporal resolution may depend on the capacities and/or settings of the measuring device. GPS tracking, which may be time-based or change-based, typically yields very high spatial precision and quite high accuracy while the temporal and spatial resolution depends on the device settings. The spatial coverage of GPS tracking is very high (almost complete) in open areas. Location-based and event-based recordings usually produce temporally irregular data with low temporal and spatial resolution and low spatial coverage. The spatial precision of location-based recordings may be low (positions specified as areas) or high (positions specified as points), but even in the latter case the position exactness is typically low. The spatial precision of event-based recording may be high while the accuracy may vary (cf. positions of photos taken by a GPS-enabled camera or phone with positions specified manually by the photographer).

Before analysing spatio-temporal data, it is necessary to find out what data collection method was used. Due to the complexity of this kind of data, they are often provided together with a description and/or metadata, which cab be used to check if the described data collection procedure and data properties match the actual properties of the data.

Irrespective of the collection method and device settings, there is also indispensable uncertainty in any time-related data caused by their discreteness. Since time is continuous, the data cannot refer to every possible instant. For any two successive instants t_1 and t_2 referred to in the data, there are moments in between for which there are no data. Therefore, one cannot know definitely what happened between t_1 and t_2. Movement data with fine temporal and spatial resolution give a possibility of interpolation, i.e., estimation of the positions of the moving entities between the

---

[2] https://en.wikipedia.org/wiki/Sampling_(statistics)

measured positions. In this way, the continuous path of a mover can be approximately reconstructed.

Movement data that do not allow valid interpolation between subsequent positions are called *episodic* [24]. Episodic data are usually produced by location-based and event-based collection methods but may also be produced by time-based methods when the position measurements cannot be done sufficiently frequently, for example, due to the limited battery lives of the recording devices. Thus, when tracking movements of wild animals, ecologists have to reduce the frequency of measurements to be able to track the animals over longer time periods.

Whatever the measurement frequency is, there may be time gaps between recorded positions that are longer than usual or expected according to the device settings, which means that some positions may be missing. In data analysis, it is important to know the meaning of the position absence: whether it corresponds to absence of movement, or to conditions when measurements were impossible (e.g., GPS measurements in a tunnel), or to device failure, or to private information that has been intentionally removed.

10.2.3 Possible data quality issues

As all other kinds of data, spatio-temporal data may have quality problems. The most common types of data quality problems have been mentioned in Sections 5.1 and 5.2. These may be measurement errors, insufficient precision, missing values or missing records, gaps in spatial or temporal coverage, etc. Some examples of data quality problems given in Sections 5.1 and 5.2 have been made using spatio-temporal data: georeferenced social media posts in Fig. 5.1, mobile phone calls in Fig. 5.5, Fig. 5.6, and Fig. 5.7, trajectories of moving entities in Fig. 5.9 and Fig. 5.10.

The most frequent problems that we encountered dealing with spatio-temporal data were insufficient temporal resolution, variable temporal resolution (fine in some parts of data and coarse in other parts), temporal gaps, and erroneous coordinates. Thus, the dataset with the vessel trajectories that we considered in the introductory example (Section 10.1) had the latter two kinds of quality issues. Figure 10.11, top, demonstrates how vessel trajectories with temporal gaps between consecutive records appear on a map. The lower image shows how the problem can be dealt with by splitting the trajectories into parts. The map in Fig. 10.12 demonstrates position errors, i.e., wrong coordinates. Generally, wrong coordinates exhibit themselves as outliers. When you have a set of spatial events, solitary events located far from all others may have wrongly specified positions. In movement data, positioning errors lead to appearance of local outliers, i.e., positions that are spatially distant from other positions in their temporal neighbourhood. Many such outliers existed in the

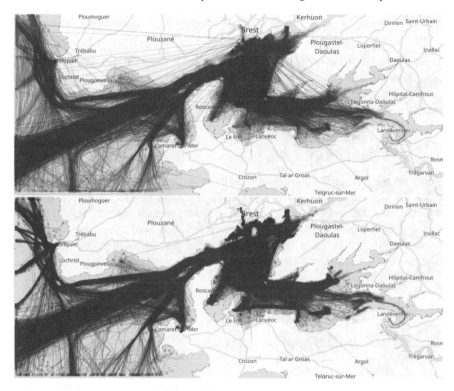

Fig. 10.11: Top: long straight line segments in trajectories of vessels correspond to *temporal gaps*, i.e., long time intervals in which position records for the vessels are missing. Bottom: the result of dividing the trajectories by the spatio-temporal gaps in which the spatial distance exceeded 2 km and the time interval length exceeded 30 minutes.

vessel dataset, and Figure 10.12 shows that some of the erroneous positions where even on land, where vessels could never come in reality. While it may be easy to notice obviously wrong positions in places where moving entities (or spatial events) could never be, many other positioning errors may not be so obvious. They can be detected by calculating the speed of the movement from each recorded position to the next position as the ratio of the spatial distance to the length of the time interval between them. Unrealistically high speed values signify position errors, which need to be cleaned.

Since spatio-temporal data may include thematic attributes (i.e., attributes whose values are neither spatial locations nor time moments or intervals), errors may also occur in values of these attributes, or some values may be missing. In our introductory example in Section 10.1, we had issues with the attribute 'navigational status'. Another group of problems, which is pertinent to movement data in particular, relates to identifiers of entities. It may happen that two or more entities are denoted

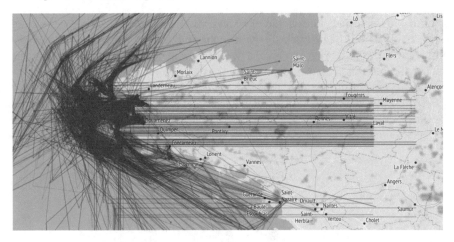

Fig. 10.12: Vessel trajectories affected by positioning errors. Some trajectories even include positions located on land far from the sea.

by the same identifier, as in Fig. 5.10, but it may also happen that the same entity is denoted by multiple different identifiers. The latter kind of circumstance is not always a result of errors in data, but it can be a specific feature of a dataset resulting from the data collection method or, sometimes, even intentionally designed. For example, when data about people's mobility are collected by tracking mobile devices worn by people, the data will contain identifiers of the devices rather than identifiers of the individuals wearing them. Hence, individuals wearing two or more devices will be represented in the data multiple times with different identifiers. Another example, which we also encountered in our practice, is intentional periodic change of identifiers denoting people, which is done for protecting people's privacy by excluding the possibility to identify individuals based on the locations they visit. Thus, this technique was applied to the dataset with the mobile phone calls considered in Section5.2 before the data were provided for analysis. Obviously, data with such a property are not suitable for studying long-term individual mobility behaviours. As usual, you need to be aware of the methods used for collecting and preparing data for understanding if the data are appropriate to your analysis goals.

Paper [12] systematically considers the kinds of errors that may occur in movement data and mentions the methods that can be used for detecting and, whenever possible, fixing the errors. Similar principles and approaches are applicable to other types of spatio-temporal data. Generally, it is necessary to consider all major components of the data structure, namely identities, spatial locations, times, and thematic attributes, and their combinations. Any unexpected regularity or irregularity of distributions requires attention and explanation. Calculation of derived attributes (e.g. speed, direction) and aggregates over space, time, and categories of objects provides further distributions to be assessed.

We shall demonstrate an example of a workflow for data quality assessment using trajectories of 4,521 cars tracked within an area including Greater London and surrounding territory during two regular weeks in winter 2017. The dataset contains 4,284,493 position records in total. From the description of the data collection procedure received from the data provider, we learn that the cars where tracked only when the engines were working, which means that long time intervals containing no records about a car normally corresponds to stops of the car. This information needs to be taken into account in inspecting and analysing the data.

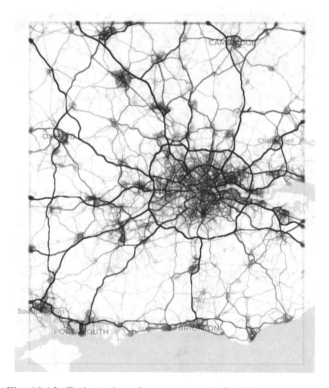

Fig. 10.13: Trajectories of cars represented as lines on a map.

To see the spatial distribution of the available data, we create a map with the trajectories represented by lines drawn with low opacity (Fig. 10.13). The map shows that the dataset covers the whole area with no obvious omissions, and the trajectories mostly stick to the road network.

To see the temporal distribution of the data, we generate a time histogram of the times specified in the position records (Fig. 10.14). The histogram shows us that the temporal coverage is also good; there are no obvious gaps without recordings, except for the night times, when the intensity of car movement is normally low. The periodic pattern of the variation of the number of records nicely adheres to

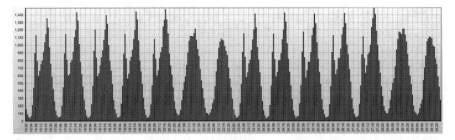

Fig. 10.14: A time histogram shows the temporal distribution of the time stamps in the car dataset.

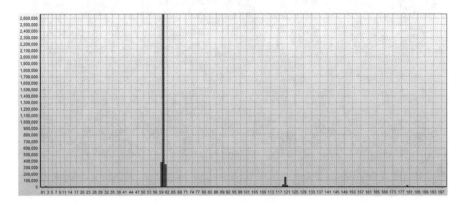

Fig. 10.15: A frequency histogram of the distribution of the sampling rates in the position recordings of the cars.

the daily and weekly cycles of human activities. Both the spatial and the temporal distributions correspond well to our common-sense expectations.

To understand the temporal resolution of the position recordings, it is necessary to compute the lengths of the time intervals between consecutive records in each trajectory and inspect the overall frequency distribution of the interval lengths. As we know that long time intervals in this dataset correspond to stops, such intervals need to be excluded from the investigation. Figure 10.15 demonstrates that the most frequent sampling rate (i.e., the time between recordings) is around 1 minute. There is a much smaller subset of points where the time interval to the next point is 2 minutes, and only a few points have 3 minutes intervals to the next points. All other interval lengths occur in the data infrequently. Next, it is useful to check if the sampling rates are common for all cars. For this purpose, we calculate the median sampling rate for each trajectory. The results demonstrate that more than 98% of the cars in this data set have the median sampling rate of one minute ± one second. However, we have identified a few outliers, which include about 100 cars that had only a few positions recorded and, accordingly, quite irregular sampling rates; 9 cars with many recorded

positions but the median sampling rates of 3 to 5 minutes; and 2 cars with very high
sampling rates (13 seconds). Such outliers need to be separated from the others in
the further analysis. We have also identified several thousands of duplicate pairs of
a car identifier and a time stamp, which were indicated by zero time intervals between records with the same car identifiers. We removed the duplicates from the
dataset.

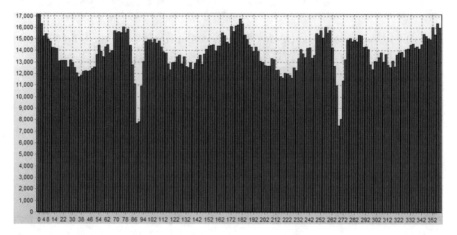

Fig. 10.16: A frequency histogram of the heading values recorded together with the
positions of the cars.

When position records contain thematic attributes that will be involved in the analysis, their properties need to be inspected. Our example dataset contains measurements of the vehicle headings, specified as compass degrees; their frequency distribution is represented by a frequency histogram in Fig. 10.16. There are two strange
pits around the values 90 and 270 degrees. It is quite unlikely that these directions
were indeed much less frequent than the others. The pits may be due to the method
that was used by the tracking devices for determining the vehicle heading. The
method might calculate the angle based on the ratio of the longitude and latitude
(x and y) differences between two consecutively measured positions, of which the
second position was not recorded. Such a method would fail in cases when the y-
difference equals zero. Whatever the reason is, the measured heading values cannot
be trusted.

In addition to pre-recorded attributes, calculation of derived attributes and studying
their distributions is a useful instrument for further checks. Unrealistic values or unexpected properties of distributions may indicate certain problems in the data under
inspection. Thus, as we already mentioned, unrealistic values of calculated speeds
can indicate errors in the recorded positions or the use of the same identifier for
denoting several moving entities. Small sizes of spatial bounding boxes containing
sub-sequences of positions from substantially long time intervals can indicate actual

absence of movement, although the recorded positions may not be exactly the same due to unavoidable measurement errors.

10.3 Visual Analytics Techniques

The major analysis goal specific to spatio-temporal data is to understand how the phenomenon is distributed in space and time and how its characteristics vary over space and time.

10.3.1 Spatio-temporal distribution of spatial events

In Chapter 1, we had an introductory example in which we analysed the spatio-temporal distribution of spatial events, namely, georeferenced microblog posts. We used a combination of visualisation techniques (time histogram and dot map) that showed us separately the temporal and spatial aspects of the distribution. We also used temporal filtering to select particular days of the epidemic and observe the spatial distributions of the messages in these days. Another possibility is to use spatial filtering for selecting areas on a map and inspect the temporal distribution of the events in these areas. A specific technique that is capable to show the spatio-temporal distribution of spatial events as a whole is the space-time cube, as in Fig. 1.9. However, it may not be very helpful when the events are very numerous or the time period is long. Filtering of the events (e.g., attribute-based) and focusing on shorter time intervals can make the view clearer, but the overall picture of the distribution needs to be mentally reconstructed from multiple views of different subsets and intervals. Spatio-temporal aggregation and analysis of the resulting spatial time series (as will be discussed in Section 10.3.2) is often the most viable approach to dealing with large amounts of spatial events.

However, this approach will not fulfil your need when you want to detect spatio-temporal concentrations (clusters) of spatial events. This kind of spatio-temporal pattern is often of special interest, because such concentrations may have specific meaning and importance. For example, a spatio-temporal cluster of "stop-and-go" events in movements of vehicles signifies a traffic congestion at particular place and time. In Chapter 8, we considered an example of events of mass photo taking, when many people took photos of blossoming cherry trees. These large-scale events, which happened at different times in different regions of the USA, were indicative of the seasonal natural phenomenon of cherry blossoming. We detected the events of mass photo taking by analysing a dataset describing elementary events of one person taking one photo, The mass events are spatio-temporal clusters of the

Fig. 10.17: Finding dense spatio-temporal clusters of photo taking events. The events are represented by dots on a map (top) and in a space-time cube (bottom). The coloured dots represent events belonging to clusters and the grey dots represent noise. The clusters are enclosed in spatio-temporal convex hulls.

elementary events. The clusters were detected by means of density-based clustering (Section 4.5.1) of the elementary events according to the spatial and temporal proximity (Section 4.2.8). Figure 10.17 shows the result of the density-based clustering with the spatial distance threshold 100 km, temporal distance threshold 14 days, and the minimal number of neighbours 3. Only the clusters with at least 50 events are taken for consideration; the smaller clusters have been joined with the "noise" (the concepts relevant to density-based clustering are explained in Section 4.5.1).

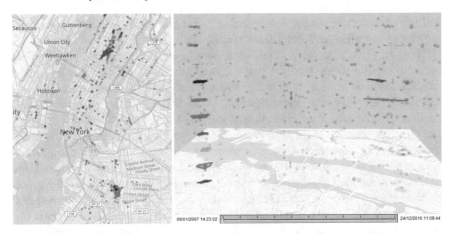

Fig. 10.18: By setting larger or smaller parameter values for density-based cluster-ing, it is possible to detect large-scale, as in Fig. 10.17, or local, as in this figure, spatio-temporal clusters of spatial events. The view direction in the STC (right) is from the east.

A nice feature of density-based clustering is that we can control the spatial and tem-poral scales of the clusters that will be detected by making appropriate parameter settings. Thus, using the same clustering algorithm and the same distance function (spatio-temporal distance) as previously by decreasing the spatial distance thresh-old to 500 m, we discover 78 local clusters of the size from 10 to 383 events that occurred in 8 cities. As an example, Figure 10.18 shows the local clusters that oc-curred in the New York City. By decreasing the temporal threshold from several days to an hour, it may be possible to detect events of multiple photos taken in the same place close in time. We have detected 73 such local clusters with the sizes from 5 to 24 elementary events and duration from 17 seconds (during which 16 photos were made, all by a single photographer) to 6 hours (19 photos made by 15 photographers).

Computationally detected spatio-temporal clusters of spatial events can be treated as higher-order events, as it was with the mass photo taking events that we consid-ered in Chapter 8. The higher-order events can be characterised by the counts of the component events, overall duration, spatial extent (e.g., the area of the polygon enclosing the cluster in space), density, and statistical summaries of the attributes of the component events. Thus, in our examples with photo taking, we were interested to know how many distinct photographers contributed to each aggregate event. For the lately mentioned set of local short-term clusters, only 5 clusters consisted of photos of single photographers and the remaining 68 clusters were indeed events of collective photo taking that involved from 3 to 24 distinct people.

Thanks to providing the possibility to derive higher-order events from elementary ones, density-based clustering can be used as an important instrument in analytical workflows. An example is a workflow aiming at identifying and characteris-ing places of traffic congestion from elementary events of slow movement of ve-hicles [14]. First, the analysts applied density-based clustering to the elementary events using a distance function that combined the distances between the events in space and time and the difference in the movement direction (in a traffic jam, af-fected vehicles move in the same direction). After filtering out the noise, the analysts applied again density-based clustering taking into account only the spatial distances and direction differences between the events but ignoring the times. In this way, the traffic jams that happened in the same places at different times where put together, and the spatial boundaries of the unions defined the places of the traffic jams. At the next step of the analysis, the places were characterised by time series of the occur-rence of the low-speed events. The use of a distance function that combined spatial and temporal distances with differences between values of a thematic attribute (di-rection) demonstrates the potential of the clustering-based approach to analysis to be adapted to various goals and applications.

However, not all kinds of spatio-temporal patterns that may be of interest in anal-ysis can be detected by means of standard density-based clustering with a suitable distance function. Thus, we may be interested to reveal dynamic patterns of cluster formation, growth and decline, as well as merging and splitting. A result of density-based clustering does not immediately provide us this information. A possible way to reconstruct the dynamic patterns of cluster changes is to cut the clusters we have received from the clustering algorithm into temporal slices and to trace the slices through the time. To avoid seeing the changes as abrupt, the slices can be made partly overlapping in time. The slices of a cluster can be treated as spatio-temporal events, which can be represented by spatio-temporal hulls and characterised by at-tributes derived from the component elementary events. These slice-events can be visualised on a map and in a space-time cube. By connecting each slice-event with the preceding and following in time events, we can build trajectories reflecting the cluster evolution.

A similar result can be obtained by performing the clustering in a specific, incremen-tal way along the time. A clustering algorithm that tracks cluster evolution and ex-plicitly represents the changes by data records is described in paper [22]. The kinds of dynamic patterns that can be revealed by either incremental spatio-temporal clus-tering of events or slicing clusters obtained by means of a standard density-based clustering algorithm can be vividly demonstrated by example of reconstructing evolution of a thunderstorm from a dataset describing elementary lightning strike events. In the space-time cube in Fig. 10.19, the states that event clusters had in different time interval (in other words, temporal slices of event clusters) are repre-sented by spatio-temporal convex hulls enclosing the component events. The hulls are stacked along the temporal (vertical) dimension. Additionally, the centres of the consecutive cluster states are connected by lines, which form trajectories showing the movements of the clusters in space over time. A complementary view of the cluster evolution is provided by an animated map display, as in Fig. 10.20.

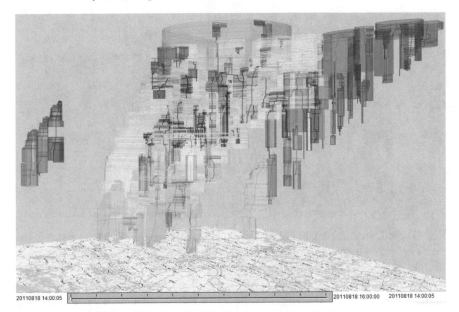

20110818 14:00:05 |————————————————————————————————| 20110818 16:00:00 20110818 14:00:05

Fig. 10.19: Dynamic patterns of cluster changes are represented in a space-time cube by stacked spatio-temporal convex hulls of the cluster states corresponding to different time intervals. The lines connecting the centres of the consecutive cluster states form trajectories representing cluster movements in space.

This example has several interesting aspects to pay attention to. Its primary purpose was to demonstrate dynamic spatio-temporal patterns of clustering of spatial events. Besides, it demonstrates that trajectories may represent movements of not only discrete entities treated as points (because their spatial dimensions are not relevant or not specified in data) but also spatially extended phenomena, such as thunderstorms. This example also shows that trajectories may not be originally described in data but reconstructed by means of sophisticated data processing and analysis.

In the following section, we shall see what kinds of patterns can be found when spatial events are aggregated into spatial time series.

10.3.2 Analysis of spatial time series

Spatial time series may be the form in which spatio-temporal data are available originally, or they may result from aggregation of spatial event data or trajectories of moving entities, as described in Section 10.2.1. Spatial time series consist of attribute values associated with spatial locations and time steps. The analysis goal is

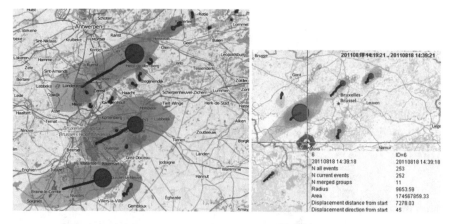

Fig. 10.20: Dynamic patterns of evolving clusters of spatial events are represented on an animated map. The violet-painted shapes show the outlines of the clusters at the end of the time window currently represented in the map, and the grey-painted shapes show the cumulative outlines including all component events that happened during the whole time window. The trajectory lines show the cluster movements, and the sizes of the red dots represent the cluster sizes. Characteristics of the cluster states can be accessed by mouse hovering on the trajectory lines (right).

to discover and understand the patterns of the spatio-temporal variation of the at-tribute values. This goal is quite difficult to achieve, as we need to see the values both in space and in time. There is no visualisation technique that could provide a comprehensive view of all values in space and time. The space-time cube, which quite often turns to be helpful in exploring spatial events or trajectories, is not suit-able for representing spatial time series. It would be necessary to represent the value for each place and time moment by some mark, and most of the marks would not be visible due to occlusions.

Hence, to visualise the variation, it is necessary to decompose it into components that could be represented by available visualisation techniques, particularly, maps showing the spatial aspect of the distribution and temporal displays showing the temporal aspect. As we said in Section 10.2.1, spatial time series can be viewed in two complementary ways: as a set of local time series associated with places or links between places and as a chronological sequence of spatial situations, in other words, spatial variations of attribute values that took place in different time steps. In order to understand the spatio-temporal variation as a whole, it is necessary to consider both perspectives. As we noted previously, these perspectives require different visualisation and analysis techniques.

When the time series are short, the sequence of the spatial situations can be repre-sented by multiple maps each showing the situation in one time step. With such a representation, the situations can be easily compared, and the temporal patterns of their changes can be seen. However, this approach becomes daunting when there are many time steps. When the spatial time series involve a small number of spatial locations, it may sometimes be possible to represent the local time series on a map by temporal charts, as, for example, the 2D time charts in Fig. 10.10. However, this technique will not work when the locations are numerous, and also when the time series are very long.

When existing visualisation techniques cannot scale to the amounts of the data that need to be analysed, the most common approach is to use clustering, as described in Section 4.5. In the case of spatial time series, clustering can be applied to each of the two complementary perspectives, that is, to the local time series and to the spatial situations. The results of this two-way clustering will provide complementary pieces of information enabling understanding of the spatio-temporal variation. Let us demonstrate the two-way clustering by the example of the data that were originally available as positions of cars in the Greater London and around (Section 10.2.3, Fig. 10.13). We partitioned the territory by means of data-driven tessellation as described in Section 9.4.2.2. Then, we aggregated the car trajectories by the resulting space compartments and time intervals of the length of one hour into place-based and link-based spatial time series, as described in Section 10.2.1. The time series include 312 time steps. The place-based time series refer to 3,535 distinct places (space compartments), and the link-based time series refer to 12,352 links between the places.

Figure 10.21 demonstrates a result of applying partition-based clustering to the place-based local time series of the hourly counts of the place visits, and, similarly, Figure 10.22 demonstrates the application of clustering to the link-based time series of the hourly magnitudes of the flows (i.e., the counts to the moves). The 2D time charts represent statistical summaries (namely, mean hourly values) of the time series belonging to each cluster. This combination of visual displays allows us to see the temporal patterns of the variation in the 2D time charts and the spatial distribution of these patterns in the map. Thus, in the map in Fig. 10.22, we can observe consistency of the cluster affiliation along chains of links following the major roads; hence, the traffic has common patterns along the major transportation corridors formed by the most important motorways. We can also notice pairs of opposite links that were put in distinct clusters, which means that the temporal patterns of the respective flows differ.

When we take the other perspective, i.e., viewing spatial time series as temporal sequences of spatial situations, we have time series of complex states, as considered in Section 8.5.3. Each spatial situation is one complex state described by multiple features; in this case, by attribute values in different places. The analysis methods that are applicable to time series of complex states are thus suitable to spatial time series. These include clustering of the states according to the similarity of their features. In Fig. 10.23, partition-based clustering has been applied to the hourly spatial situations in terms of the place visits (i.e., place-based) and in terms of the flows between the places (i.e., link-based). The results are clusters of time steps, which

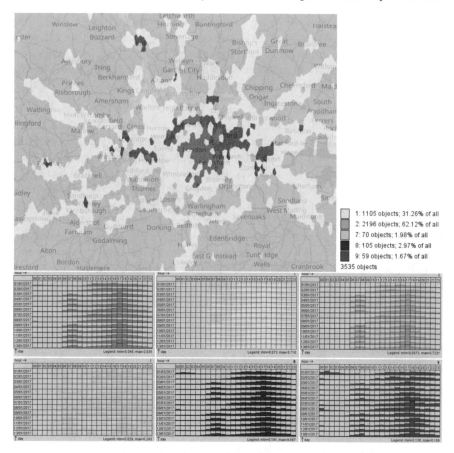

Fig. 10.21: Partition-based clustering have been applied to local time series associated with space compartments. The map fragment shows the cluster affiliations of the compartments, and the 2D time charts show the average hourly values computed from the time series of each cluster.

are represented in 2D time charts. In Fig. 10.24 and Fig. 10.25, the averaged spatial situations of these time clusters are shown on multiple maps. The temporal patterns visible in the 2D time charts clearly correspond to the human activity cycles, daily and weekly, and represent the specifics of the two different kinds of aggregates. For example, commuting flows are characterised by similar magnitudes in the mornings and evenings, but their directions are opposite. The daytime patterns of the working days differ much from the nights according to the place visits, while the movement patterns of regular commuters are similar, as the commuters don't move much at night and in the middle of their work days.

In analysing local time series, one may be interested in detecting specific kinds of temporal variation patterns, such as peaks. There are computational techniques for

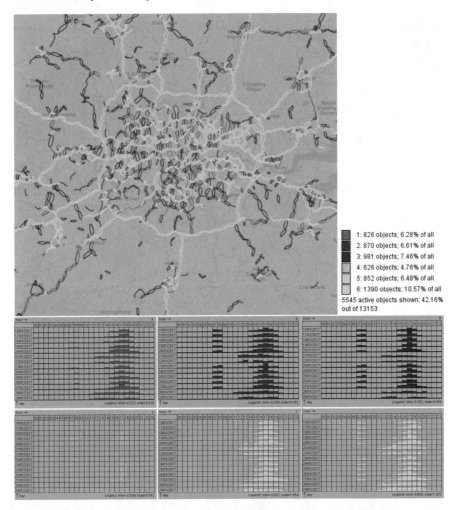

Fig. 10.22: The links have been clustered according to the similarity of the normalised time series of the flow volumes. The map fragment shows the cluster affiliations of the links, and the 2D time charts show the average hourly values computed from the normalised time series of each cluster.

pattern detection (also called pattern mining) in time series, as, for example illustrated in Fig. 8.15. In the case of spatial time series, analysts are interested to know not only when certain patterns occurred but also where in space this happened. The analysis of the pattern locations in space can be supported by the technique of *event extraction* from time series. Each occurrence of a pattern is an event, which has a specific time of existence. It can be extracted from the time series to a separate dataset for further analysis. As explained in Section 10.2.1, events extracted from spatial time series are spatial events having the same spatial references as the time

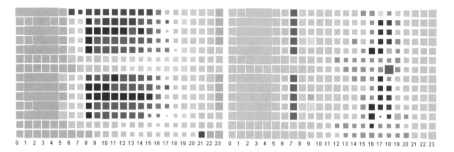

0 1 2 3 4 5 6 7 8 9 10 11 12 13 14 15 16 17 18 19 20 21 22 23　0 1 2 3 4 5 6 7 8 9 10 11 12 13 14 15 16 17 18 19 20 21 22 23

Fig. 10.23: 2D time charts represent the temporal distributions of the clusters of the place-based (left) and link-based (right) spatial situations. The colours correspond to different clusters, and the sizes of the coloured rectangles represent the closeness of the cluster members to the cluster centroids (the closer, the bigger).

series from which they have been extracted. Spatial events extracted from spatial time series can be analysed using the methods suitable for spatial event data, as considered in Section 10.3.1. An example of extracting spatial events from spatial time series is presented in Fig. 10.26. The time series were obtained by spatio-temporal aggregation of elementary photo taking events that occurred on the territory of Switzerland in the period from 2005 to 2009. A computational peak detection method was applied for detection and extraction of high peak patterns from the time series. The peaks signify abrupt increases of the photo taking activities. The map shows the spatial locations of the peak events. Most of them happen in nature areas in summer, but some happened in December in big cities. This example demonstrates an interesting chain of data transformations: spatial events – spatial time series – spatial events. Unlike the original elementary events, the latter events are abstractions obtained through data analysis.

10.3.3 Analysis of trajectories

We have explained and demonstrated previously that trajectories can be aggregated into spatial time series, which can then be analysed. However, a lot of important information concerning individual movements is lost due to the aggregation.

In our introductory example, we also demonstrated the extraction of stop events from trajectories. In that example, not the trajectories by themselves but the events were of our primary interest; we did not analyse the remaining parts of the trajectories. However, it is also possible to use event extraction operations for a more comprehensive exploration of movement data. In fact, any filter applied to positions or segments of trajectories extracts spatial events, i.e., the parts of the trajectories satisfying the query conditions. By repeatedly performing filtering with different

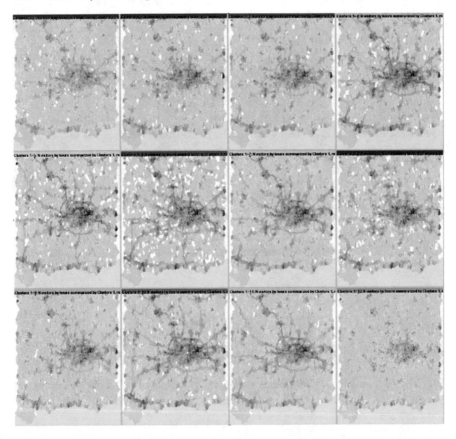

Fig. 10.24: The place-based spatial situations corresponding to the time clusters shown in Fig. 10.23, left, are represented on multiple choropleth maps, each showing the average situation for one of the clusters. The light blue colouring corresponds to zero values.

query conditions and considering the spatial and temporal distribution of the resulting events, we can gain multiple complementary pieces of knowledge about the movements.

An example using the London cars dataset is demonstrated in Fig. 10.27 and Fig. 10.28. We are looking at the spatial (Fig. 10.27) and temporal (Fig. 10.28) distributions of all positions from the trajectories of the cars and of different kinds of events: stops for 1 hour or more, low speed (less than 20 km/h), medium speed (60 to 90 km/h) and high speed (over 90 km/h). We can see that the spatial distribution of the low speed events is similar to the overall spatial distribution of all car positions but the temporal distribution differs, having the highest frequencies in hours 8 and 17-8 of the working days. The stop events, expectantly, are concentrated in the cities and towns, and they happen especially often in the morning hours

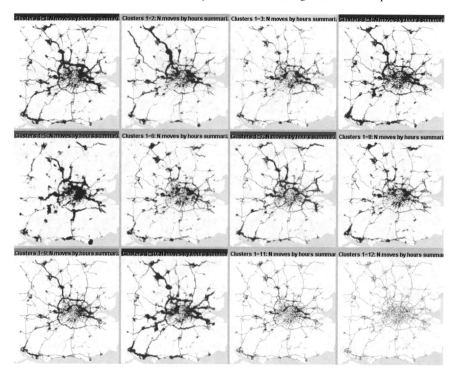

Fig. 10.25: The link-based spatial situations corresponding to the time clusters shown in Fig. 10.23, right are represented on multiple flow maps, each showing the average situation for one of the clusters.

7 to 9 of the working days. In the afternoon and evening hours of the working days, the stop frequency increases in hour 15, reaches the highest values in hours 17-18, and then gradually decreases. The evening patterns on Fridays differ from the remaining working days, and the weekend days can be easily distinguished due to the difference of their patterns. The spatial distributions of the medium-speed and fast movement events notably differ from the others. While the temporal distribution of the medium-speed events is similar to the overall temporal distribution, the temporal distribution of the high-speed events differs a lot, with the highest frequencies attained Saturdays, Sundays, and on Monday, January 2, which was a free day in UK because the New Year holiday (January 1) happened to be on Sunday.

Apart from exploring the temporal distribution of all events throughout the whole territory, it is possible to select particular areas by means of spatial filtering and see the temporal distributions of the events in these areas. And, obviously, it is possible to apply spatio-temporal aggregation to the extracted events of any kind and to analyse the resulting spatial time series.

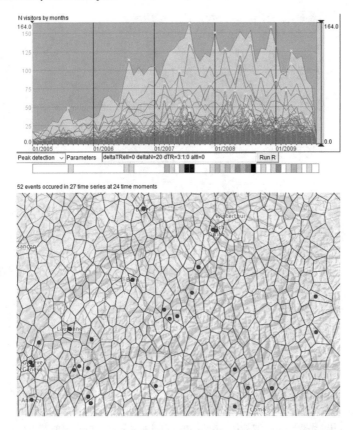

Fig. 10.26: Extraction of peak events from spatial time series of monthly counts of photo taking events in different spatial compartments in Switzerland in the time period from 2005 to 2009. Top: the time series with the detected peaks. Bottom: the spatial positions of the peaks.

While event extraction helps us to gain more refined knowledge about movements than by only aggregating the full trajectories, we still lose all information concerning the trips: their origins and destinations, routes, times when they took place, dynamics of the speed along the route, and much more. Hence, not all analysis tasks can be fulfilled by constructing spatial time series or extracting events, and there may be a need to deal with data in the form of trajectories.

Trajectories are usually visualised as lines on a map; however, this representation does not show the movements were conducted in time. For seeing both the spatial and temporal aspects of a single trajectory or a few trajectories, the space-time cube may be a helpful technique (Fig. 10.9). However, it will fail when there are many trajectories to analyse. A viable approach, as in many other cases, is to cluster the trajectories by similarity and then analyse and compare the clusters instead of

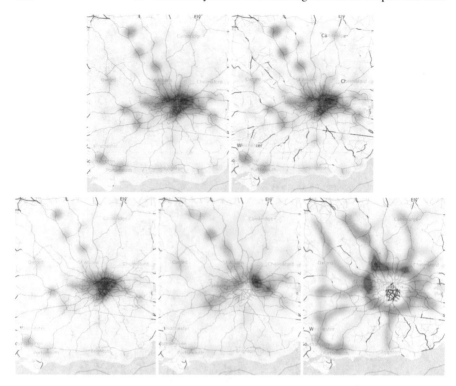

Fig. 10.27: Density fields show the spatial distributions of all positions from the trajectories of the cars (top left), the events of the stops for 1 hour or more (top right), and the events of the slow, medium, and fast movement (bottom).

dealing with the individual trajectories. The question is: what is "similarity of trajectories"? Or, in other words, how to define a distance function (i.e., a measure of the dissimilarity; see Section 4.2) for trajectories?

Trajectories are complex objects with multiple heterogeneous features: spatial (route geometry and spatial location), spatio-temporal (spatial positions at different time moments), overall temporal (time and duration of the trip), and temporal variation of thematic attributes describing the movement (speed, direction, acceleration, etc.) and, possibly, the context (road category, relief, land cover, traffic conditions, etc.). Trajectories can be similar or different in terms of any of these features. It is hard to imagine a distance function that takes all these features into account, and it is also hard to imagine the purpose for which such a function could be used. In practical analyses, analysts typically focus on one or a few features at once. When it is necessary to consider more features, the analytical work is split into several steps. The limitation for the number of the features that can be considered at once is not due to the technical difficulty of computing the dissimilarities but due to the difficulty of

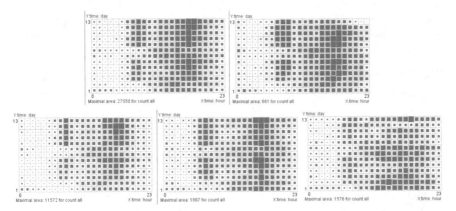

Fig. 10.28: 2D time histograms show the temporal distributions of all positions from the trajectories of the cars (left), the events of the stops for 1 hour or more, and the events of the slow, medium, and fast movement.

interpreting the results of these computations and the clusters obtained using such a distance function.

Therefore, the approach to analysing trajectories (or other complex objects) by means of clustering consists in using **multiple distance functions** accounting for different types of features. These functions can be used in different steps of the analysis, allowing the analyst to obtain complementary pieces of knowledge concerning different aspects of the movements. Furthermore, according to the idea of **progressive clustering**, results of earlier steps of clustering can be refined in following steps by applying clustering with a different distance function to selected clusters obtained earlier. The concept of progressive clustering was introduced in Section 4.5.3 as an approach to refining "bad" (insufficiently coherent) clusters, and we also mentioned that progressive clustering may involve different distance functions at different steps. Trajectories is a kind of objects that may need to be analysed using this approach. A simple example is finding in the first step clusters of trips according to the destinations and refining the major clusters in the next step by means of clustering according to the similarity of the routes taken for reaching the destinations [116].

There exists an extensive set of distance functions designed for trajectories of moving objects. Paper [109] provides a review of these functions and examples of their application for clustering. Many of the functions are space-aware adaptations of distance functions originally designed for time series (Section 4.2.3).

Obviously, results of clustering need to be visualised to enable interpretation and knowledge construction. The visualisation needs to show the features that were accounted for in the clustering. Thus, maps are needed to show clusters according to spatial features, and temporal displays are necessary for representing clusters according to temporal and dynamic features. Often, spatial and temporal displays need to be combined, similarly to the case of spatial time series considered in the previous section.

The following section presents an example of a workflow in which a set of trajectories is analysed with the use of clustering.

10.4 Analysis example: Understanding approach schemes in aviation

To forecast future trends in airspace utilisation, aviation researchers need to understand patterns of air traffic, quantify them, and identify driving forces that cause disruptions and undesired effects. This study is based on a data set consisting of flight trajectories that arrived during 4 days in 5 airports of London. One of specific questions of interest is to find the major approach routes that that precede the landing, quantify the intensity and effectiveness of their usage, and identify when which routes are used. The task of identifying the major routes requires an abstraction from individual trajectories to general movement patterns. As we know from previous chapters, clustering can be a powerful tool for data abstraction. For our goals, we need to apply clustering to the trajectories using an appropriate distance function that compares trajectories in terms of the followed routes.

After exploring the data properties and cleaning the data, as described in Section 10.2.3, a visual inspection of the trajectories (Fig. 10.29) uncovers a challenging problem: many trajectories include holding loops, which are made by aircraft as commanded by air traffic controllers while waiting for the possibility to land. The loops are not planned, and they do not belong to the proper landing approach that has been intended. Hence, they must be filtered out before doing the clustering; otherwise, they will strongly affect the clustering results: trajectories following the same route but differing in the number of loops will not be put in the same cluster.

To understand how to mark the loops in the data and filter them out, the analyst interactively explores the data and determines that one full loop takes approximately 5 minutes. So, the analyst applies a tool that computes for each trajectory position the sum of the turns in the next 5 minutes. For a position at the beginning of a loop, this sum should theoretically be about $\pm 360°$ (positive and negative values correspond to right and left turns, respectively). In practice, this value cannot be reached in discretely sampled data, where not all turning points are present due to time gaps between the records. By interactive filtering and observing the results on a map display, the analyst ascertains that loop starts can be extracted using a threshold value of $\pm 240°$ for the summary turn over a 5-minute window. These positions of the loop starts and the following positions within the 5 minutes intervals are marked in the data as loops by creating a binary attribute based on the current filter.

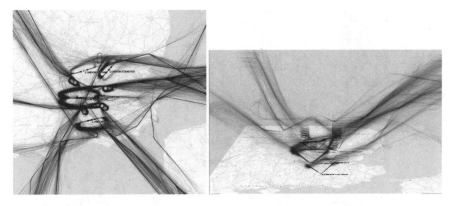

Fig. 10.29: An overview of aircraft trajectories that landed in London airports: a map (left) and a 3D view (right).

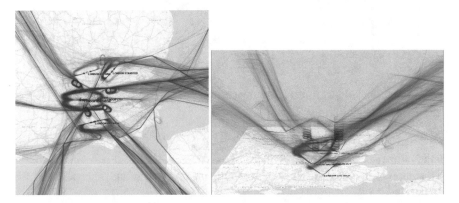

Fig. 10.30: The holding loops occurring in the flight trajectories are highlighted in red on the map and in a 3D view.

To verify the result, the analyst builds the interactive display shown in Fig. 10.30, where the segments of the trajectories are coloured according to the values of the binary attribute just created; red colour corresponds to the loops. The loops occurred in 1,484 trajectories (29.4%), including more than 50% of the flights that landed in the airport Heathrow and about 5%-10% of the flights that landed in the other airports. For reducing the impact of overplotting by numerous trajectories that share the same airspace and, respectively, screen real estate, the analyst applies a different visualisation that presents detailed holding loops (drawn as lines) in the context of the overall traffic represented by a density map (Fig. 10.31).

By filtering, the analyst hides the loops and selects the final parts of the trajectories starting from the 75 km distance to the destination. Then the analyst applies density-based clustering using the distance function "route similarity" [10], which takes into account only the selected parts of the trajectories. The "route similarity"

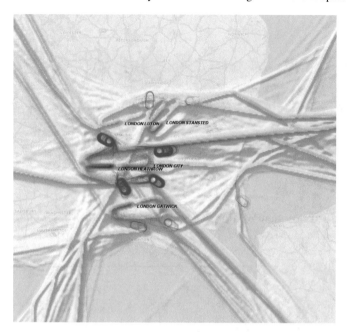

Fig. 10.31: The holding loops are shown as red lines on top of a density map representing the overall traffic.

function matches points and segments of two trajectories based on spatial proximity, computes the mean distance between the matched parts, and adds penalties for unmatched parts.

After trying several combinations of parameter values for the neighbourhood radius and number of neighbours (see Section 4.5.1), the analyst manages to obtain clusters that separate very well different approach routes to all airports except Stansted. As explained in Section 4.5.3, "bad" clusters can be improved through progressive clustering. The analyst selects the subset of trajectories ending in Stansted and applies two steps of progressive clustering with slightly different parameters separately to this subset, thus yielding good, clean clusters of different approaches into Stansted. Figure 10.32 shows the original clustering result for Stansted and the outcome of the progressive clustering. After merging the clustering results for all airports, there are in total 34 clusters including 4,628 trajectories (91.7%), and 417 trajectories (8.3%) are labelled as noise.

Each of the clusters obtained corresponds to a possible approach route. The analyst explores the temporal distribution of the use of these routes by representing the flight counts on a segmented time histogram (Fig. 10.33) where the segment colours correspond to the cluster affiliations of the flights. It is well visible that day 1 differs substantially from the following days 2–4. The analyst guesses that the changes in the landing schemes are, most likely, due to changes of wind direction. Inspection of

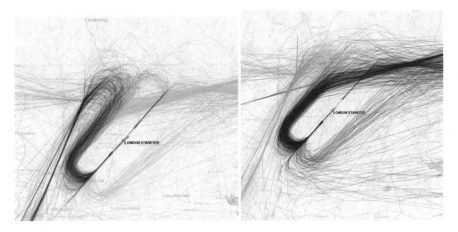

Fig. 10.32: Improving clusters of approach routes to Stansted by means of progressive clustering: before (left) and after (right).

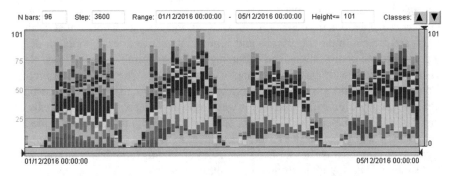

Fig. 10.33: The time histogram represents the dynamics of the use of different approach routes.

meteorological records confirms this guess. Indeed, on the second day, the wind direction has changed. Figure 10.34 demonstrates the spatial footprints of the clusters of the approaches at different days, namely day 1 (left) and 3 (right).

After identifying the major routes, the analyst proceeds with the task of quantifying the routes according to their usage and the frequency of the holding loops on each route. Counting trajectories and the corresponding holding loops is an obvious calculation. However, the results appear to be quite interesting. While the overall amounts of landings corresponded well to the expectations, it was surprising to find that only landing at Heathrow airport are highly affected by holding loops, while landings at other airports are suffering from waiting quite rarely. Specifically, on the most affected approach route 184 out of 258 trajectories (71.3%) included holding loops, with an impressive statistics of time spent in the loops: 10.2 minutes in average, 8.78 minutes median, and more than 29 minutes maximal waiting time.

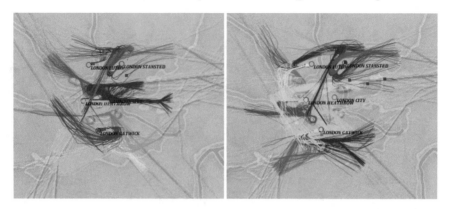

Fig. 10.34: Clusters of the approach routes taken in two different days.

Let's reconstruct the analysis workflow in this study. The analysts started with investigation of data quality. We did not discuss this step here, as it followed precisely the recommendations from Section 10.2.3. Next, the analyst identified that holding loops are present in the trajectories and applied interactive filtering tools for defining appropriate computations and excluding the loops from the trajectories. Next, visually-driven density-based clustering was applied for separating repeated routes from occasional and grouping trajectories into clusters. Global parameter settings did not work well enough, therefore it was necessary to apply progressive clustering, defining parameters specifically for selected subsets of the data. Next, temporal dynamics of the cluster lifelines was investigated, and spatial distributions have been studied for the indicative days. Finally, statistics of holding loops has been computed and studied, featuring quite surprising results.

It is necessary to stress that in this example we have applied a sophisticated workflow consisting of rather simple and easy to understand steps, combining visualisations and computations in a transparent way. Each step was quite easy to understand and produced interpretable results, which is very important when results of analysis need to be communicated to domain experts or decision makers.

10.5 Concluding remarks

In this chapter, we considered three major forms of spatio-temporal data – spatial events, spatial time series, and trajectories – and explored the opportunities provided by the possible transformations between these three forms (Fig. 10.8). Spatial events may be considered as an elementary component of spatio-temporal data. Spatial time series is a complex construct in which attribute values vary over three- to four-dimensional base composed of space and time. The complex behaviours of spatio-temporal phenomena represented by spatial time series cannot be viewed in its entirety but need to be decomposed into two complementary perspectives: a set of local time series distributed over space and a sequence of spatial situations (or spatial distributions of attribute values) varying over time. Spatio-temporal aggregation of spatial events and movement data produces spatial time series. Spatial events can be extracted from spatial time series and movement data.

The purposes of the transformations listed in Figure 10.8 can be summarised as follows:

- *Aggregation*:
  - Supports abstraction, gaining an overall view of characteristics and behaviour;
  - Reduces large data;
  - Simplifies complex data.
- *Extraction of events*:
  - Selects a portion of data relevant to a task;
  - Enables focusing of analysis;
  - Allows dealing with complex data portion-wise.
- *Integration, disintegration, division*:
  - Adapts data to analysis tasks.

Apart from the transformations, we have also demonstrated the use of clustering for analysing spatio-temporal data. It could be noticed that we applied partition-based clustering and generic distance functions, such as Euclidean or Manhattan distance, to spatial time series and density-based clustering with specific distance functions based on spatial and/or temporal distances to spatial events and trajectories. We chose density-based clustering for dealing with discrete spatio-temporal objects, such as events and trajectories, because such objects are usually distributed in space and time very unevenly, so that there are dense concentrations and sparsely populated or even empty areas. It rarely makes practical sense to put spatio-temporal objects from dense and sparse regions together in one group. Density-based clustering reveals dense concentrations of objects and under-populated regions beyond these concentrations. Spatial time series have a regular structure, which can be considered as a matrix filled with attribute values with the dimensions being the space and time. The attribute values are supposed to be everywhere in the matrix (if not too many of them are missing, which can be quite problematic for analysis); hence, the task of finding dense and sparse regions is not relevant. Clustering is done according to the attribute values, and partition-based clustering producing not too many clusters is a convenient tool for data simplification and abstraction. Density-based clustering can be used for the purpose of finding groups of very similar local time series or spatial situations and separating them from the remaining more varied data.

To conclude, visual displays, interactive filtering, data transformations, and clustering with the use of appropriate distance functions are the major instruments for analysing spatio-temporal phenomena. These instruments need to be used jointly within analytical workflows.

10.6 Questions and exercises

- Describe the expected properties of a dataset with georeferenced social media posts, such as we considered in the introductory example with the epidemic in Vastopolis in Chapter 1. Think about the accuracy of the temporal and spatial references and how the latter depend on the devices used for message posting. Can such a dataset be taken as representative of the whole population of a territory? The message posting events can be integrated into trajectories of the social media users. Describe the properties of such trajectory data. Can such data be used for characterising (a) overall movement patterns over the territory, (b) use of public or private transportation means, (c) routine mobility behaviours of individuals? What data properties are required for each of these tasks?

- Consider the georeferenced social media data from the perspective of personal privacy. How easy or difficult is it to identify personal places of individuals: home, place of work or study, places of regular shopping, sports, etc.? What analysis methods could be used for this purpose? Is it possible to reconstruct social relationships between active social media users only from the geographic positions and times of their posts? If so, what could be an approach to achieve this?

- Imagine a dataset consisting of monthly aggregates of several weather parameters (temperature, precipitation, wind direction and speed, sunlight, etc.) measured over a long time period at many weather stations distributed over a country. What type of data is this? Assume that your analysis goals are to understand regional and overall climate trends: how the climate in different regions and the country-wide spatial patterns of climate change over years. What methods are suitable for achieving these goals? What analytical workflow would you perform?

- Imagine that you have a dataset with trajectories of various vessels: cargo, ferries, liners, fishing boats, and excursion boats. Unfortunately, the data do not specify the ship types. How could you identify these types? What kinds of movement patterns would you search for? What data transformations would you use? What visualisations can be helpful to you?

Chapter 11
Visual Analytics for Understanding Texts

Abstract Texts are created for humans, who are trained to read and understand them. Texts are poorly suited for machine processing; still, humans need computer help when it is necessary to gain an overall understanding of characteristics and contents of large volumes of text or to find specific information in these volumes. Computer support in text analysis involves derivation of various kinds of structured data, such as numeric attributes and lists of significant items with associated numeric measures or weighted binary relationships between them. Computers themselves cannot give any meaning to the data they derive; therefore, the data need to be presented to humans in ways enabling semantic interpretations. While there exist a few text-specific visualisation techniques, such as Word Cloud and Word Tree, which explicitly represent words, it is often beneficial to use also general approaches suitable for multidimensional data. Collections of texts having spatial and/or temporal references are transformed to data that can be visualised and analysed using general methods devised for spatial, temporal, and spatio-temporal data. We show multiple examples of possible tasks in text data analysis and approaches to accomplishing them.

11.1 Motivating example

Optimising text readability is an important everyday problem. Authors and editors need to know whether their texts are hard to read, why, and how to improve the texts. Let us consider an example of successful application of the visual analytics approach to address this problem [106].

To build a model that estimates how easy or difficult for reading a given text is, it is necessary to obtain a set of numeric measures of text readability. Expert knowledge or even human common sense are usually good helpers in finding potentially useful features (i.e., numeric measures) for a given task. For the task of text readability assessment, there are many features that can be useful:

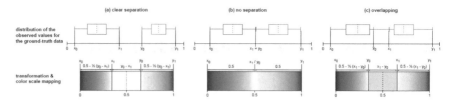

Fig. 11.1: Visual comparison of statistical distributions of feature values over two sets of texts. Source: [106].

- frequencies of different parts of speech: nouns, verbs, pronouns, adjectives, etc.;

- frequencies of common and uncommon words compared to the average usage frequencies of these words;

- features based on the sentence structure: an automatic text parser can be used to derive a sentence tree; the depth and width of this tree and the position of the verb can indicate the sentence complexity;

- use of passive voice, use of quotations, etc.

It is a typical case that the set of potentially useful features is very large. Many of them are redundant, repeatedly representing similar aspects of the text readability. Not all features are sufficiently distinctive for characterising texts with good and bad readability. Hence, the task of feature selection (Section 4.3) arises.

First, it is necessary to evaluate the distinctiveness of the candidate features and discard those that are insufficiently distinctive. An elegant and efficient strategy is to compare the statistical distributions of all potentially useful features for sets of easy and difficult sets. The authors of the approach took a set of stories for 4-6 years old children and, oppositely, a set of documents of the European Commission. After computing the value of each feature for each text, they compared the distributions of the values of each feature over the two sets. For this purpose, they used the visualisation demonstrated in Fig. 11.1, which showed them how well each feature can separate easy and difficult texts. The authors used this information to filter out the features that showed low discrimination power.

Second, from the remaining features, it is necessary to select a subset of non-redundant features, i.e., representing different aspects of text readability. As explained in Section 4.3, this is done based on the pairwise correlations between the features. In our example, the authors used the matrix display of the correlations between the attributes shown in Fig. 4.5. The rows and columns of the correlation matrix were sorted with the help of a hierarchical clustering algorithm. Dark regions in the matrix display reveal groups of correlated features. These groups were manually inspected to find out which semantic aspect they measure. For each group, the authors chose one feature as a representative, giving preference to the feature that was easiest to understand. This means that the feature must be consciously controllable

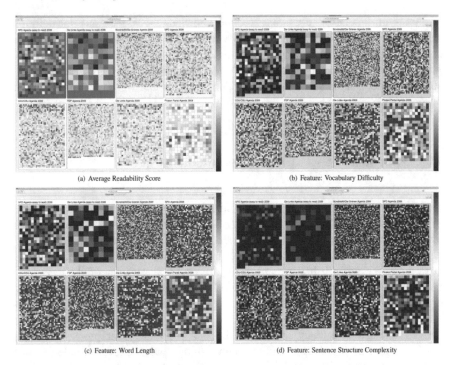

(a) Average Readability Score

(b) Feature: Vocabulary Difficulty

(c) Feature: Word Length

(d) Feature: Sentence Structure Complexity

Fig. 11.2: Visual analysis of the readability of 8 election agendas from the elections of the German parliament in 2009. Source: [106].

when writing a text, allowing the writer to improve the readability of a sentence with respect to this feature. In this way, the following list of distinctive and non-redundant features was selected: (1) word length: the average number of characters in a word; (2) vocabulary complexity: proportion of terms that are not contained in the list of most common terms; (3) ratio of the number of nouns to the number of verbs; (4) number of nominal forms; (5) sentence length (number of words); (6) complexity of the sentence structure, expressed as the number of branches in the sentence tree.

The selected features can be used for evaluation of the readability of individual sentences in a text, and it is also possible to create a "portrait" of a text document, as demonstrated in Fig. 11.2. Each sentence is represented by a pixel (a small square) coloured according to the average readability score (Fig. 11.2a) or the value of one of the selected features. Multiple documents can be compared in terms of their readability. When documents are long, the scores of the sentences can be aggregated for paragraphs or sections.

This example demonstrates that analysis of texts, on the one hand, relies on calculation of derived numeric data, on the other hand, requires involvement of human knowledge and reasoning, which needs to be supported by appropriate visualisations. In this case, human background knowledge was necessary for defining an initial comprehensive set of potentially relevant features. Then, visualisation of computation results helped the analysts to determine which features are really useful, understand the relationships and redundancies among them, and select a small subset of understandable and non-redundant features that were appropriate to the purpose.

11.2 Specifics of this kind of phenomena/data

Texts can be available in the forms of documents or collections of documents, fields in data records, as, for example, customer reviews of products, and streams that grow over time, such as news or messages in social media. Texts are meant to be read, understood, and used by humans, who are usually able to do this quite well. Problems arise, however, when it becomes necessary to deal with large amounts of text. To reduce time and effort, humans need help from computers. But... unlike humans, computers lack the ability to *understand* the contents of texts. They can just formally calculate various metrics and indicators, as in our motivating example. These derived data can nevertheless be helpful, being appropriately interpreted by human analysts and used for analysis and reasoning. In our example, even simple features derived from texts were useful for judging text readability and understanding how to improve. Currently there are computational methods capable to get much more from texts than these simple measures, but still a human is needed to give meaning to what is derived and make use of it.

11.3 Analysis tasks

As we said previously, computational techniques and visual analytics approaches need to be used for dealing with large individual text documents or collections of documents. The high-level tasks in analysing large amounts of text are:

- Efficiently gain a general understanding of text contents.

- Find relevant pieces of information and understand relationships between them.

- Compare texts and understand their similarities and differences.

- Find groups of semantically related texts.

- Observe text streams (texts appearing over time) and understand changes and trends.

- Understand the distribution of texts in time and space.

11.4 Computational processing of textual data

To analyse texts with the help of computers, you need to transform them into some kind of structured data, such as tables or graphs, and then visualise, interpret, and analyse these structured data. However, before the extraction of structured data, texts usually need to undergo several steps of preprocessing, namely:

- Tokenisation (segmentation): splits longer strings of text into smaller pieces, or tokens. Long texts are divided into paragraphs or individual sentences, and sentences may be divided into words.

- Normalisation, which includes

  - converting all text to the same case (upper or lower), removing punctuation, converting numbers to their word equivalents, and so on;

  - stemming: eliminating affixes (suffixes, prefixes, infixes, circumfixes) from words in order to obtain word stems;

  - lemmatisation: transforming words to their canonical forms; for example, the word "better" is transformed to "good";

- Removal of so-called "stop words", which do not convey significant information. These include articles, pronouns, prepositions, conjunctions, auxiliary verbs, and words with similar functions.

Software tools for performing these operations are now easy to find. Obviously, you need to choose tools that can deal with the language in which the texts are written.

Computational techniques for deriving structured data from texts can be grouped in the following classes, in the order of increasing sophistication:

- calculation of simple numeric measures, such as word length, sentence length, etc.;

- extraction of significant keywords and computation of statistical characteristics of their usage, such as frequency and specificity for a given text;

- probabilistic topic modelling (see Section 4.6);

- NLP (Natural Language Processing) techniques for

  - identification of named entities (people, places, organisations, etc.);

  - sentiment analysis (identification of emotions and attitudes and measurement of their intensity).

These techniques create the following kinds of structured data:

- numeric attributes associated with entire documents or their parts (sections, paragraphs, or sentences);

- lists of significant, semantically meaningful items, such as keywords, topics, or entities, with associated values of a numeric attribute representing the frequency, importance, or specificity of each item, or the sentiment with regard to the item;

- lists of entities and weighted relationships between them, where the weights indicating the strengths of the relationships are derived from the references to the entities occurring in the same texts.

Obviously, depending on the kind of the derived data, different visualisation and analysis techniques are applied to them.

11.5 Visualisation of structured data derived from texts

11.5.1 Numeric attributes

Text-derived numeric attributes can be visualised using any methods suitable for this kind of data. For example, in Fig. 11.2, numeric measures of text readability are represented by colour coding. When multiple numeric attributes have been derived and need to be considered together, the techniques devised for multidimensional data are applied in standard ways (see Chapter 6).

11.5.2 Significant items with numeric measures

Derived data consisting of some kind of semantically meaningful items (keywords, topics, etc.) and corresponding numeric measures can be dealt with in two ways. First, each item can be treated as a numeric attribute with the measures being the attribute values. This means that the data are treated as usual multidimensional numeric data; hence, the approaches suitable for this kind of data can be applied in standard ways. The other approach is specific to texts and exploits the human capability to understand meanings of words, in particular, textual labels denoting the items. This approaches involves showing these textual labels. The corresponding measures are most often represented by font size. A typical example is a word cloud, or text cloud, as we used in our introductory example in Chapter 1 (Figs. 1.12 and 1.13).

An example of applying the first approach is demonstrated in Fig. 11.3 (source: [105]). From a large number of customer reviews, analysts extracted significant terms that were accompanied in the texts by expressions of positive or negative opinions. These terms were treated as attributes (features), and the respective ex-

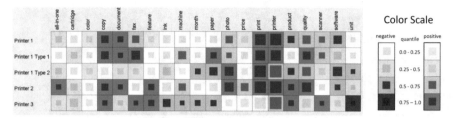

Fig. 11.3: Characteristics of printers in terms of the amounts of opinions (represented by square sizes) and sentiment values (shown by colours) in respect to their most distinguishing features. Source: [105].

pressions of opinions were represented as feature values -1 (negative) or 1 (positive). Each sentence was thus represented by a combination of the feature values for the terms; value 0 (neutral) was given to the features for which there were no opinion indicators. Then, for each product reviewed, the feature values obtained from the corresponding texts were aggregated, and the aggregates were visualised in a matrix display as shown in Fig. 11.3. The rows of the matrix correspond to different products (printers in this example) and the columns to the features, i.e., the terms extracted from the reviews. The aggregated feature values are represented in the cells by squares. The sizes of the squares represent the amounts of the expressed opinions concerning the features and the colours show how positive or negative, on the average, the opinions are.

Probabilistic topic modelling is a popular class of techniques applied for text analysis. As we explained in Chapter 4, topic modelling can be seen as a special category of dimensionality reduction methods. The original dimensions are in this case frequencies or weights of various significant terms extracted from texts, and the "topics" obtained through topic modelling are, in fact, also numeric attributes, which are defined as weighted combinations of the original attributes. The values of these composite attributes are the weights, or probabilities, of the topics for different texts. Hence, a result of topic modelling consists of two tables, or matrices: topic-term matrix, which specifies for each topic the weights (probabilities) of the terms, and document-topic matrix, which specifies for each document the weights (probabilities) of the topics. These are standard numeric data, which can be visualised using usual methods applicable to numeric data. Particularly, a natural way to visualise both matrices is a matrix display, as, for example, in Fig. 11.4 and Fig. 11.5. The rows and columns in such displays are ordered by means of special algorithms. Thus, in Fig. 11.4, the ordering reveals groups of terms related to common topics (i.e., having high weights for these topics), and in Fig. 11.5, the ordering reveals groups of topically similar documents (i.e., with the same topics having high weights). Please note that the topics in Fig. 11.5 have meaningful labels given by an analyst, differently from the automatically generated formal labels in Fig. 11.4.

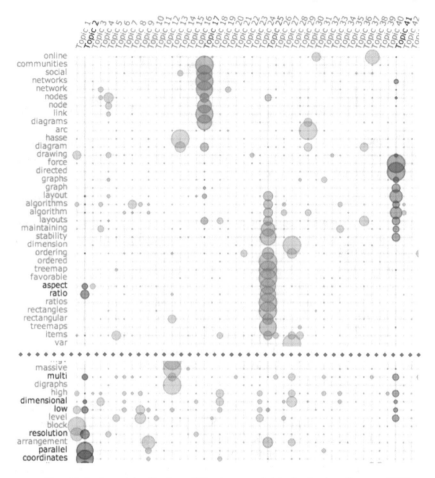

Fig. 11.4: A matrix display of the term weights for topics. Source: [42].

Results of topic modelling can also be used for data embedding and spatialisation, in which texts are represented by points in a 2D embedding space. The points are arranged according to the similarity of the combinations of the topic weights. An example is shown in Fig. 11.6. Here, an interactive display does not only show documents arranged in an embedding space by the topic weights but also allows the analyst to obtain a more detailed topic model for a selected subset of documents.

The second approach, which focuses on representing textual labels of extracted items, is not limited to generation of a single word cloud for all texts taken together. One thing that is reasonable to do is to generate word clouds from subsets of documents, for example, from positive and negative customers' reviews. In our introductory example in Chapter 1, we generated word clouds for subsets of tweets

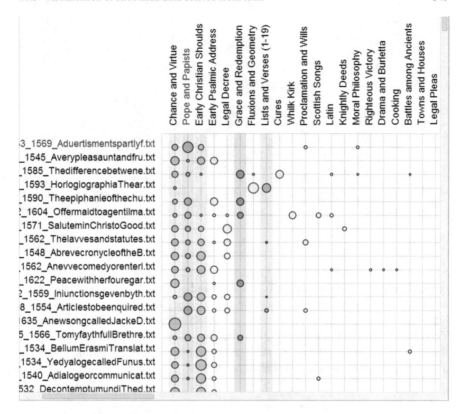

Fig. 11.5: A matrix display of the topic weights for documents. Source: [4].

selected based on their spatial and temporal references. The keywords that got high prominence in these word clouds allowed us to discover the existence of two distinct diseases and to understand the reason for the epidemic outbreak.

When you use word clouds for summarising text contents, it may also be reasonable to remove from them terms that are not relevant to analysis or occur in very many documents. For example, in customer reviews of printers, the term "printer" may appear frequently, but it is not highly informative since all reviews refer to printers. Such words can be temporarily added to the list of stop words, which makes the tool generating text clouds ignore these words and give higher prominence to other words.

Another application of the idea of text cloud is comparison of documents or collections of documents. An example is shown in Fig. 11.7, where the text clouds generated from different collections of documents are represented in the form of parallel vertical lists. The analyst can interactively select a term in any list and see in which other lists this term appears and compare its weights in these lists.

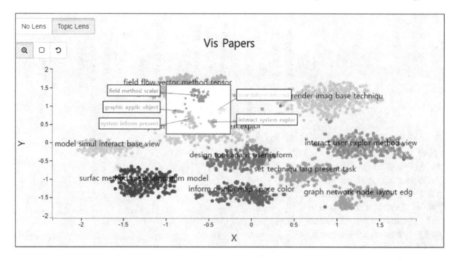

Fig. 11.6: 2D embedding has been applied to a set of documents represented by combinations of topic weights. The dots in the 2D embedding space represent the documents; the concentrations of dots correspond to groups of documents with common prevailing topics. These topics are indicated by the terms with the highest weights. The system providing this display allows the analyst to refine the topic model by selecting a region in the projection space and applying the topic modelling algorithm to the documents located in this region. Source: [82].

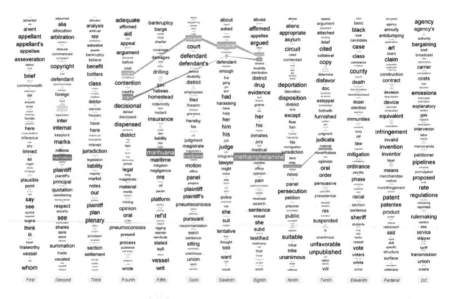

Fig. 11.7: Comparison of the contents of several collections of documents by means of parallel text clouds. Source: [45].

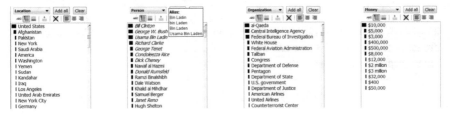

Fig. 11.8: Lists of entities of the categories Location, Person, Organisation, and Money extracted from the 9/11 Commission Report (see
`https://en.wikipedia.org/wiki/9/11_Commission_Report`).
Source: [56].

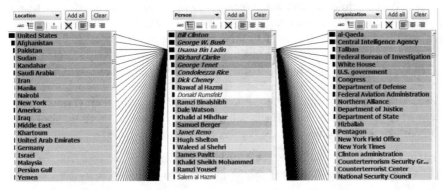

Fig. 11.9: The List View, as in Fig. 11.8, showing locations, persons, and organisations connected to Usama Bin Ladin. Source: [56].

11.5.3 Named entities and relationships

Named Entity Recognition (NER) means identification of words referring to particular people, places, or organisations. NER techniques also identify textual expressions of date and time, quantities, percentages, monetary values, and the like. Hence, an output of NER consists of lists of the entities by these categories, as shown in Fig. 11.8. The extracted entities can be dealt with similarly to other kinds of significant items with numeric measures, which may be in this case the counts of the documents referring to the entities. Thus, in Fig. 11.8, the measures are represented by the lengths of the horizontal bars drawn on the left of the labels denoting the entities. It is also possible to summarise the extracted data in word clouds.

Entities identified in texts are considered as being related if references to them appear in the same documents or in the same text fragments (such as paragraphs) when documents are long. The more such co-occurrences are detected, the stronger the relationship is. An interactive display of text-extracted entities can allow users to select particular entities and show in response which other entities are related to the selected entities. An example is shown in Fig. 11.9. Here, the saturation of the background colour represents the strength of the relationship. The general approaches to the visualisation and analysis of pairwise relationships between entities are described in Chapter 7.

11.6 Analysing word occurrences and their contexts

Apart from analysing various kinds of structured data that can be derived from texts, it can also be helpful to look at the contexts in which particular words appear in a text. A nice way to visualise this information is the Word Tree [148]. To show an example, we applied a web tool for online generation of word trees[1] to the description of the introductory example in Chapter 1. For an interactively selected word, such as "distribution" in Fig. 11.10, the display shows the phrases that either follow or precede this word in the text. The selected word appears as the root of the tree. Phases having common beginnings or ends are grouped, and their common parts appear as roots of sub-trees. These common parts are shown using the font sizes proportional to the numbers of the phrases in which they appear.

For other techniques that can be used for visualisation of text-derived data, we refer the readers to published surveys [88, 89] and online repositories, as, for example, TextVis[2].

11.7 Texts in geographic space

In investigating the epidemic outbreak in Vastopolis in Chapter 1, we dealt with text data (microblog messages) having references to spatial locations. By comparing summarised contents of the messages posted in different geographic areas, we made an important discovery of the existence of two different diseases. To analyse the spatial variation of the text contents, we selected areas in a map display and the corresponding subsets of messages by means of a spatial filtering tool and built separate word cloud displays from these subsets (Fig. 1.12). For a more convenient analysis, word clouds can be shown directly on a map, as is done in the system ScatterBlog [131]. This approach is illustrated in Fig. 11.11. The system extracts significant terms from georeferenced Twitter messages and applies a variant of density-based clustering for detection of spatio-temporal concentrations of term occurrences. These concentrations are represented on a map display by placing the respective terms approximately at the spatial locations of the cluster centres

---

[1] https://www.jasondavies.com/wordtree/

[2] http://textvis.lnu.se/

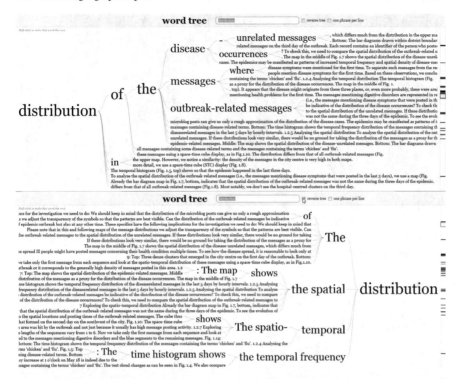

Fig. 11.10: A word tree generation tool has been applied to the text describing the introductory example of visual analysis in Chapter 1. The display shows the contexts for the word "distribution", i.e., the words and phrases following (top) and preceding (bottom) the word "distribution" in the text.

while avoiding overlapping labels. The level of detail in showing the terms depends on the map scale. The analyst can interactively select subsets of the messages, and the system will show the spatial distribution of the terms from this subset only, as demonstrated on the right of Fig. 11.11.

Another approach to representing the spatial distribution of texts on a map involves aggregation of the text data by areas in space and categories according to text semantics. For example, in analysing a set of georeferenced Twitter messages posted on the territory of Seattle (USA), a group of analysts identified 22 key topics, including home, work, education, transport, food, and others [132]. The topics were determined based on occurrence of pre-specified indicative keywords in the message texts. Then, the analysts divided the territory into areas based on the spatial distribution of the tweets and aggregated the data into per-area counts of the posted tweets by the topics. The resulting aggregated were represented on a map display by pie charts, as shown in Fig. 11.12.

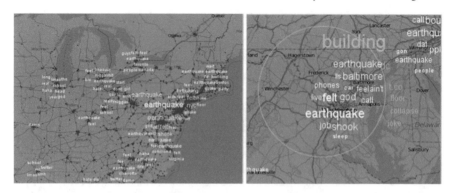

Fig. 11.11: The left image shows an overview of the spatial distribution of the most prominent terms from the tweets posted during the earthquake on 23 August 2011. The right image demonstrates the use of an interactive lens tool for a subset of the messages mentioning 'damage' and 'building'. Source: [131]

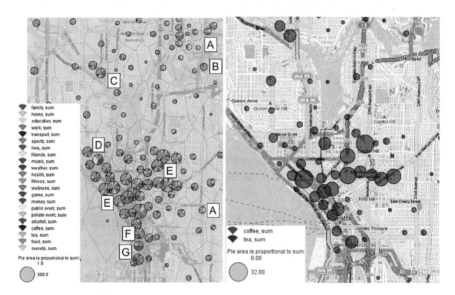

Fig. 11.12: Spatial distribution of the Twitter messages if different thematic categories in Seattle, USA. Source: [132]

On the left, the pies represent all 22 topics. The labels from A to G have been added to mark the places where particular topics were more popular than in others: A – "education" (areas of universities), B – "sports" (University of Washington sports arenas), C – "love" (an artsy and Bohemian district of Freemont), D – "music" and "public event" (Seattle Cente – the location of Bumbershoot music and arts festival, the US's largest arts festival), E – "coffee" (the most of Seattle downtown area), F – "sports" and "music" (Pioneer Square known for its lively bar and club scene),

G – "sports" and "game" (the CenturyLink Field multi-purpose stadium and the Safeco Field baseball park). The map on the right was created using the subset of the messages referring to the topics "coffee" and "tea".

In this example, which demonstrates the use of text data for understanding the spatial distribution of people's activities, the texts were used for derivation of semantically meaningful items (topics) and assigning feature vectors to the data records indicating by values 1 and 0 whether each of the items is referred to in the text. These feature vectors were spatially aggregated and visualised as standard spatially referenced numeric data.

11.8 Texts over time

Any text is created at a certain time. When you analyse a collection of texts, the times of the text creation or publication may be important to take into account. This refers, in particular, to dynamic text streams, with new texts continuously added to them. Examples are online news articles and messages in social media. Besides, a single text document may also change due to revising and editing. Hence, there may be multiple versions of a document, and the corresponding task may be to understand how the document evolved over time.

A sequence of versions of an evolving document can be treated as a time series of complex states. In Chapter 8 (Section 8.5.3), we considered two general approaches to dealing with such data. One approach is clustering of the states and visual analysis of the temporal distribution of the clusters. The other approach is spatialisation of the states and connection of the dots representing the states in the chronological order by straight or curved lines. The latter technique is known as Time Curve [28]. Thus, the upper image in Fig. 2.12 demonstrates a time curve representing the evolution of the Wikipedia article "Chocolate". The display includes a part where the curve alternates between the exact same revisions (blue halos), suggesting a so-called "edit war". The two blue halos are rather dark, suggesting a long edit war. One of the opponents finally won, and the article continued to progress.

Both clustering and spatialisation use some distance function (Section 4.2) for numeric assessment of the dissimilarity between two states. For the spatialisation in Fig. 2.12, top, the article versions were treated as sequences of symbols (words), and the dissimilarity was assessed using one of the edit distance functions (Section 4.2.6).

A result of spatialisation can tell the analyst how much the document changed from one version to another but does not tell what specifically changed in the document content. The analyst needs to look at the texts to understand this. Such a way of analysis may be daunting when changes are numerous. Particularly, dynamic text streams require a different approach, which may involve summarisation of text con-

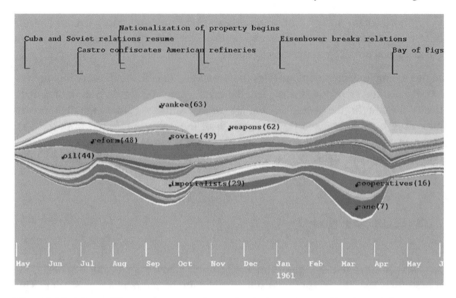

Fig. 11.13: ThemeRiver represents thematic changes in a collection of documents over time using a river metaphor. Source: [66].

tents by time steps. For each time step, the contents of the texts that appeared at this time can be summarised into a list of semantically meaningful items, such as significant keywords, topics, or entities, and their "weights", that is, values of some numeric measure representing their frequency or degree of importance. For each item, there will be a time series of the corresponding weights. Hence, the result of the summarisation will be a collection of numeric time series corresponding to different items extracted from the texts. This transformation is applicable not only to a text stream but also to a long text document, which can be divided into segments, such as sections or paragraphs. In this case, the relative ordering of the segments in the document plays the same role as the chronological sequence of time steps in a text stream.

Time series of item weights can be visualised and analysed using the standard methods suitable for numeric time series; see Chapter 8. There are also techniques proposed specifically for texts. The most popular is the ThemeRiver, which represents changes of item weights over time using the river metaphor. An example is shown in Fig. 11.13. The same information can also be shown by means of a time histogram with segmented bars corresponding to time steps. Unlike a histogram, ThemeRiver depicts time in a continuous manner. Each theme (i.e., one of semantically meaningful items extracted from the texts) is shown as a "current", which "flows" along the time. Hence, each theme maintains its integrity as a single entity throughout the graph. To create a continuous representation from discrete time steps, the data are interpolated into soft curves that look like currents in a river.

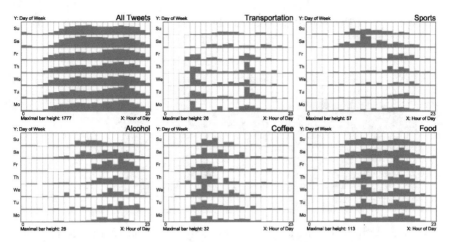

Fig. 11.14: The 2D histograms with the rows corresponding to the days of the week and columns to the hours of the day show the distribution of the Twitter messages with different thematic contents with respect to the daily and weekly time cycles. Source: [132].

The ThemeRiver, as well as a standard time series graph, shows the variation of the text amount and contents along the time line. There may be a task to understand how the text characteristics vary with respect to time cycles. A possible approach is demonstrated in Fig. 11.14. These 2D time histograms were generated for the collection of the Twitter messages posted in Seattle we discussed before (see Fig. 11.12). The rows and columns of the histogram correspond to the days of the week and hours of the day, respectively. The bars within the cells represent the counts of the posted tweets referring to all topics (top left) and to a few selected topics. The figure demonstrates that the variation of the thematic content of social media may be related to temporal cycles, according to the typical times of human activities.

This property of the social media contents needs to be taken into account when the analysis task is to detect unusual events signified by temporal or spatio-temporal bursts of thematically similar messages. To separate unusual peaks from normal patterns of social media activities, it was proposed to employ a seasonal-trend decomposition procedure [39]. For the anomaly detection task, globally and seasonally trending portions of the data need to be ignored, whereas major non-seasonal elements can be considered as potentially anomalous and, therefore, relevant. The workflow involves topic extraction (Section 4.6) and construction of time series of the daily counts of the topic-related messages. The seasonal-trend decomposition transforms each time series into a sum of three components: a trend component, a seasonal component, and a remainder. The values from the remainder are utilised for detecting anomalous outliers. The deviation of a daily remainder value from the 7-days mean by more than 2 standard deviations is considered as an anomaly. For example, Fig. 11.15 demonstrates how an abnormally high remainder value in a time series signified the Virginia earthquake on August 23rd, 2011.

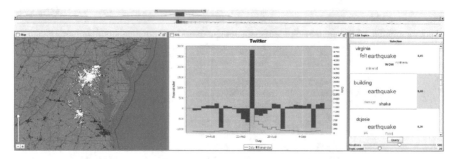

Fig. 11.15: Abnormal events are detected by applying seasonal-trend decomposition to time series of daily counts of messages related to different topics. An earthquake in Virginia on August 23rd, 2011 has been detected from an abnormally high amount of messages related to the "earthquake" topic. [39].

The examples in this section demonstrate that analysis of the distribution and variation of text amounts and contents over time involves derivation of numeric data, which are then processed, visualised, and analysed using the standard methods suitable for time-related data, i.e., events and time series.

11.9 Concluding remarks

This chapter contains examples for two classes of text analysis tasks: tasks focusing on content-unrelated text characteristics, such as amount of texts, document length, sentence complexity, as well as spatial and temporal distributions, and tasks focusing on text contents. The first class of tasks does not require any text-specific approaches to data analysis, except that calculation of some numeric measures may require text-specific processing, such as sentence parsing. After the necessary text attributes are obtained, the data are analysed in the standard ways suitable for multidimensional, spatial, temporal, or spatio-temporal data.

The second class of tasks requires such processing of texts that the resulting derived data convey some aspects of the text contents. Such aspects can be represented by extracted keywords, topics, or named entities. The analysis relies on the capability of human analysts to understand the semantics of these items. Typically, extracted semantically meaningful items are characterised by numeric attributes expressing their importance, prominence, or specificity. The analysis requires visual representations showing both the items ad their attributes. While a few text-specific techniques exist, such as word cloud and word tree, the analytical opportunities they can provide are quite limited. A more general approach is to treat the extracted items with the respective measures as attributes and analyse the data as usual multidimensional numeric data. Examples can be seen in Figs. 11.4, 11.5, and 11.6. Another possibility is data aggregation by subsets corresponding to the items, possibly, in combination with spatial and/or temporal aggregation. The resulting aggregates, on the one hand, have clear semantics, on the other hand, are standard numeric data that can be visualised and analysed in standard, text-unspecific ways. Examples have been demonstrated in Figs. 11.3, 11.12, 11.13, 11.14, and 11.15.

The chapter demonstrates that text analysis may be done for various purposes, that there exist a variety of techniques allowing derivation of task-relevant structured data, and that there are many ways to visualise and analyse these derived data. Human knowledge of text properties and understanding of the analysis tasks are vital for selection of right text processing techniques and analysis methods for the derived data.

11.10 Questions and exercises

- Consider the task of comparing election agendas of several political parties. What are the relevant aspects to compare? What kind of structured data you need to derive? What text processing methods can be used for this purpose? What visualisation of the derived data can help you to perform efficient comparison?

- Assuming that you have a collection of messages of social media users from some city (e.g., London) for a period of 3 months from November to the next year's January. How can you identify specific themes appearing in the communication before and during Christmas and New Year? How can you analyse the overall evolution of the message contents? How can you identify messages mentioning particular places in the city? Assuming that you have selected these messages, how can you summarise and represent the contents and the sentiments of the texts concerning the different places?

- Perform sentiment analysis of your emails (e.g. by applying Python code from Afinn[3]), study the dynamics of the sentiments (e.g. compare the sentiments on Friday evening and Monday morning), compare the results for different partners in communication, compare your private emails against the business emails etc.

---

[3] https://github.com/fnielsen/afinn

Chapter 12

Visual Analytics for Understanding Images and Video

Abstract Images and video recordings are commonly categorised as unstructured data, which means that they are not primarily suited for computer analysis. The contents of unstructured data cannot be adequately represented by numbers or symbols and require the power of human vision for extracting meaningful information. While images and video are well suited for human visual perception, coping with large amounts of unstructured data may be daunting for humans. Hence, humans need computer support but cannot be substituted by computers.

Computer processing of unstructured data begins with deriving some kind of structured data. One possible kind of structured data is summary statistics characterising each image or video frame as a whole. Such data can be used for arranging the images or frames by similarity to provide an overview of an image collection or a video recording and enable search for data items similar to a given sample. Another possibility is to use the existing image processing techniques for detecting particular objects represented in the visual contents and computing their characteristics, such as sizes, shapes, and positions. These characteristics can then be analysed in various ways suitable for structured data, but these analyses must be complemented with human perception and understanding of the original visual contents. This chapter includes several examples of approaches in which computers support the unique capabilities of humans. At the end, we summarise these approaches in a general scheme showing the possible operations and the types of data that are derived and analysed.

12.1 Motivating example

Imagine you are an architect that is asked to improve the lobby of your client's office building. For this, you first want to find out how people actually use this part of the

© Springer Nature Switzerland AG 2020   361
N. Andrienko et al., *Visual Analytics for Data Scientists*,
https://doi.org/10.1007/978-3-030-56146-8_12

building – how many people move through it from the elevator to the main entrance, and vice versa? Which routes do they take? When do they take these routes, are there traffic peaks when people start getting in each others' way? What about users of the staircase, and the side entrance? If people are loitering in the lobby, where do they do it?

To help you out, your client is providing you with several hours of video surveillance recordings capturing peoples' movements from typical office days. But how would you go about analysing these videos? Sure, you can watch them one after another, then spend hours rewinding and fast-forwarding to take stock of, and compare, different traffic situations. You would probably take notes and do coarse sketches of individual persons' trajectories on paper to finally get an idea of the overall movement patterns that will form the basis of your refurbishment planning. However, this manual approach will be time-consuming, tedious, and error-prone.

Luckily, video analysis methods can help with this kind of data exploration by performing a lot of the tedious and error-prone work automatically. Specifically, you delegate the detection of moving people in the videos and the extraction of their 2D trajectories from the 3D video images (i.e., lines on a floor plan of the building) to a software tool like [68]. But this would still require you to manually compare the 2D images of trajectories to find common patterns. And because there will be a lot of trajectories from many people, images with superimposed trajectories from all observations will be very cluttered and hard to read. Even worse, your analysis needs to account for further properties of the movement beyond just their geometric shape: when people where moving, and how fast for example (are loitering/slow walkers often get in the way of those in a hurry?).

Thus for an efficient analysis, you need means for **summarising** groups of trajectories, but in a flexible way. Again following Shneiderman's principle of "overview first, zoom and filter, details on demand" [124], you will want to start with a rough overview of where the majority of people walk, and which are outliers (i.e., a few unusual paths) that you want to ignore for further analysis. At this stage, just looking at the trajectories' geometry, or positions, is a good approach.

Next, you want to filter and drill down on the large flow of people coming from the elevator and going to the main entrance, and compare them to the group of people using the stairs after entering the building (Fig. 12.1). Staircase users likely do this as a form exercise, while people using the elevator may either be more convenient, or simply are in a hurry. Distinguishing between these groups for summarisation purposes works better if looking at another **facet** of movement: walking speeds. You select the speed facet and indeed find it nicely separates three major movement flows, two to the elevators (fast and slow movers), and one to the staircase.

However, you wonder if in the two groups of elevator users there might also be some people that would have rather liked to use the stairs but simply did not find them after a sensible time (they are around that large pillar and the direction sign is a little small, after all). This group of people would probably exhibit some sort of hesitant movement: walking into the lobby relatively briskly, then slowing down or mean-dering while looking for the stairs, and then making a beeline to the elevators after giving up their search. To see whether such a group exists, you select both subsets of trajectories belonging to all elevator users from the previous analysis step, and apply yet other movement facets to find and summarise new subgroups according to your hypothesis: variations of movement speeds, variations from average positions (i.e., deviations from straight-line movement), and a combinations of the two. To make sure your new summarisation really captures the assumed behaviour of "tak-ing a beeline", you interactively play with the detail settings for the measure applied to the "positional deviation" facet – specifically, you adjust and visually check how different settings for dividing trajectories into episodes affect the generated sum-marisations. You find that indeed there is a group of would-be stair users, and that using four episodes captures their behaviour best (i.e., on average the first quarter of their path represent walking in, the middle two quarters are spent searching, and the last quarter captures the beeline to the elevators).

This example adapted from [67] illustrates how a visual analysis approach is applied on top of the results of computational video content analysis: during the analysis process, the human analyst interactively selects interesting subsets based on a com-bination of movement data **facets**, lets the computer re-calculate a summarisation, and derives conclusions from its visualisation (Fig. 12.1). By further drilling down on sub-subsets the analyst interactively defines a hierarchical model that captures all relevant patterns in the examined data set with respect to the analysis task. By contrast, a non-interactive approach would be much less flexible and might not be able to account for the specifics of a given task. For example, security personnel would be looking for a different subgroup and facet structure that is better suited to distinguish "normal" from "susceptible" behaviour.

12.2 Specifics of this kind of phenomena/data

While images and videos are are very well aligned with human perception and can be a rich source of information to a human viewer, analysing large collections of images or extracting important information from many hours of surveillance video records may be too difficult, time consuming, and/or boring for humans. At the same time, such data are also poorly suited for computational analysis. That is, without additional meta data, an image or a single frame from a video is just a meaning-less grid of colour values, or 'pixel soup'. Any analysis thus typically starts with derivation of some kind of structured data from the pixels. There are two possible approaches:

- Characterise each image as a whole by various summary statistics (such as the mean, quantiles, or frequencies by value intervals) of the lightness, hue, and sat-

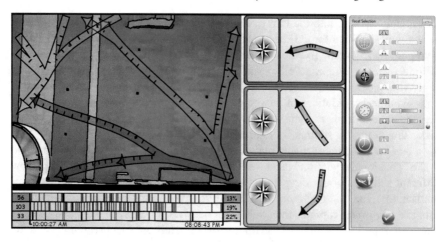

Fig. 12.1: An interactive schematic summary of trajectories of moving persons extracted from a video. The trajectories have been clustered by similarity according to different facets of movement (right, from top to bottom: position, direction, speed, time, object class); the resulting clusters are represented in a summarised form as flows (left). Example taken from [67].

uration of the pixels. For a more refined characterisation, images are divided into several parts by a coarse grid, and the summaries are computed for the parts.

- Detect specific kinds of relevant objects (e.g., human faces in photos, organs in medical images, moving entities in video records, etc.) and derive attributes characterising these objects, such as the position, size, and colour components.

There exist many computational tools and algorithms for image and video processing that can be utilised for deriving structured data. However, before doing that, it is reasonable to assess the quality of the images and take precautions against possible impacts of the quality issues, if detected. Thus, images may be noisy or contain artefacts resulting from compression. Other relevant attributes of image quality are sharpness, contrast, and presence of a moire pattern. Automated evaluation of image quality can be performed using special algorithms for no-reference image quality assessment (NR-IQA) [76], that is, algorithms measuring the perceptual quality of an image without access to a high-quality reference image. Image quality can be improved using techniques for image enhancement [73, 94].

12.3 Analysis tasks

In analysing images, typical tasks requiring visual analytics techniques are:

- Gain an overall understanding of the contents of a large collection of images.

- Organise a collection of images to facilitate finding images of interest.

- Detect and analyse changes in a temporal sequence of images.

Concerning videos, human analysts require help from computers for two major purposes:

- Efficient acquaintance with the contents of a long video recording and finding scenes of interest.

- Detection and analysis of significant events, changes, or movements of objects.

For these tasks, automated methods for image or video processing are combined with visualisations supporting human understanding, finding relevant information, and analytical reasoning.

12.4 Visual Analytics techniques

As stated in Section 12.2, computer-assisted analysis of images and video begins with derivation of structured data, which usually have the form of tables with values of multiple numeric attributes and, possibly, time references in the cases of video or temporal sequences of images. Among the attributes, there may be coordinates of detected objects. These derived data can be visually analysed in the same ways as data that have such a form originally, i.e., as usual multi-attribute data (Chapter 6), time series (Chapter 8), or trajectories of moving objects (Chapter 10). However, when visualisation is applied only to the derived structured data, the human analyst does not see the original images or video. This may be inappropriate in applications that require understanding of the semantic contents of the original data, which cannot be adequately represented by derived attributes. Therefore, it is necessary to involve visualisation and/or interaction techniques enabling human viewing and understanding of the original image or video data. Let us consider several examples of possible approaches.

12.4.1 Spatialisation of image collections or video frames

Structured data derived from images or video frames can be used for assessing the similarity between the images or frames. The analyst chooses or defines a distance function, i.e., a numeric measure of the difference between images, which is based on the differences between the values of the derived attributes; see Section 4.2. As for usual multi-attribute data, the most commonly used distance functions are the Euclidean and Manhattan distances. Once a distance function is defined, some dimensionality reduction method (Section 4.4) is applied to represent the images by points in an abstract two-dimensional space, so that similarity of images is represented by proximity of the corresponding points. Figure 12.2 shows examples of visual displays created with the use of this technique. When the image collection is relatively small, the available display space may permit showing reduced versions of the images (thumbnails), as on the left of Fig. 12.2. In this example, the images are not only placed on a plane but also connected into a tree-like structure, which is achieved using a point-placement technique called Neighbor-Joining (NJ) similarity tree [46]. When there is no enough space for thumbnails, the images are represented by dots, as on the right of Fig. 12.2. In any case, the analyst should be able to view the images corresponding to the thumbnails or dots in the display. Interactive functions allow the analyst to select images for viewing.

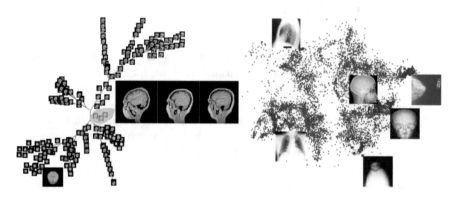

Fig. 12.2: Spatialisation of smaller and larger collections of images based on similarity of image features. Source: [50].

When spatialisation is applied to a temporal sequence of images or to video frames, it is appropriate to connect the dots representing the images by lines in the chronological order to form a time curve [28], as shown in Fig. 12.3. The temporal ordering is additionally represented by colouring the dots from light to dark. In Section 8.5.3, we discussed the application of spatialisation to time series of states of a dynamic phenomenon. This discussion fully applies to temporal sequences of images or frames of a video. The main idea is that the patterns formed by the points and lines of a time curve indicate the character of the development, which may include stagnation (minor or no changes), gradual changes, large changes, oscillation, cycles, or returns to earlier states. Please note that only seeing a curve is not enough for understanding what is going on. The analyst needs to open and view the video frames corresponding to selected dots.

Examples of time curves summarising contents of video recordings were included in Fig. 2.12. The image in the middle of the figure shows a one-minute footage from a security camera. There is a large cluster of dots that correspond to frames where only minor changes happen, such as slight changes in illumination or moving leaves. Outlier dots indicate frames where people cross the scene. The curve at the bottom of Fig. 2.12 was created from a video of an animated map showing precipitation across the United States over the course of one year (averaged 1981-2010). The curve closes itself at the end of the year, suggesting a yearly cycle. It also crosses itself, revealing that geographical precipitation patterns around October/November were the same as in March/April. Similarly, the time curve in Fig. 12.3, top, summarises a video of worldwide cloud coverage and precipitations over one year. It reveals large-scale changes across the entire year, along with small oscillations. On December, the weather does not come back to where it was on January, but instead to where it was on April.

The time curve in Fig. 12.3, bottom, represents an eight-minute animated film. Different scenes appear as clusters. The movie mostly employs scene cuts, without camera motion or transition effects. Some scenes are visited twice. The large-scale structure of the movie can be inferred from dot colours. The brightest and darkest dots are clumped together on the top left, suggesting little action at the beginning and at the end of the movie. In contrast, red dots undergo large changes, with frequent scene cuts.

These examples demonstrate that spatialisation can produce useful visual summaries of videos or collections of still images. What you need is a suitable distance function for assessing the similarity of the video frames or images. Having a suitable distance function, you can also apply clustering techniques (Section 4.5) or search for images similar to specific images of interest. Interestingly, the examples in this section do not use any sophisticated image processing. The similarity between images is assessed merely based on the distributions of the pixel values over the images. Nevertheless, even this simple approach yields quite useful visualisations, which may be sufficient in many cases. The next two sections include examples of approaches in which image processing techniques are employed to detect particular relevant objects in images and extract their characteristics.

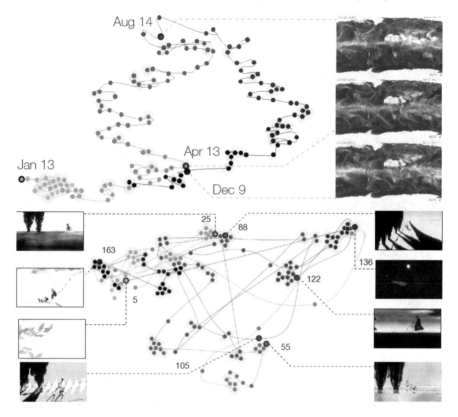

Fig. 12.3: The use of spatialisation for representing contents of video recordings in a summarised form. The Time Curve technique ([28]; Section 8.5.3) has been applied to a video of an animated map showing the worldwide cloud coverage and precipitation over one year (top) and an eight-minute movie (bottom). Source: [28].

12.4.2 Detection of relevant objects and analysis of their changes

The analysis task in the paper [65] is to observe and examine changes in the human spleen, which is the largest organ in the lymphatic system, based on series of medical images obtained by means of computer tomography (CT). The authors of the paper utilise advanced image processing techniques for detection of the spleen in each image and extraction of its characteristics, including the metric dimensions (length, width, and thickness), shape, and texture.

The overall analytical workflow is schematically represented in Fig. 12.4. Image processing techniques are used for segmenting the images, detecting the spleen, and extracting its characteristics, or features. The derived data are stored in a database

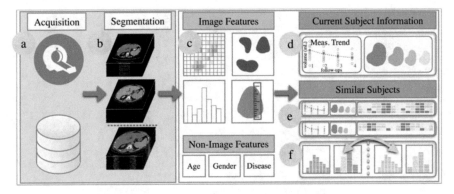

Fig. 12.4: A workflow designed for a specific application problem (source: [65]) demonstrates a general approach to image analysis with visual analytics.

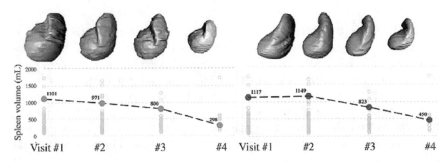

Fig. 12.5: A visual display supporting image analysis combines pictures with plots showing derived numeric data. Source: [65].

together with data about the patients. At the stage of analysis, a doctor can select the data of a specific patient and use visual representations of the spleen features and appearance to observe the temporal progression of the spleen condition of this patient. Simultaneously, the system searches in the database for other patients with similar spleen condition and temporal evolution. The spleen characteristics obtained from different CT scans are compared using a distance function suitable for multi-attribute data. The authors of the paper apply a special version of the cosine similarity function (Section 4.2) allowing attributes to have different weights. To compare the time series of the spleen conditions of different patients, they use another distance function, the Dynamic Time Warping (DTW) [112].

For a selected patient, the doctor can see plots and graphs representing the numeric characteristics of the spleen as well as 3D representations of the appearance of the spleen reconstructed from the tomography data, as shown in Fig. 12.5, left. In parallel, cases similar to the selected one are presented in the same way. Thus, the right part of Fig. 12.5 shows the most similar case to the one on the left. Both cases have the same decrease in the organ volume.

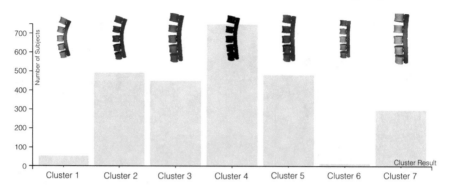

Fig. 12.6: Clustering of image-derived 3D models of the lumbar spine of different patients. Source: [83].

Another example [83] also comes from the field of medical image analysis. The images are MRI (magnetic resonance imaging) scans of the lumbar spine area. Image processing tools are used to detect the lumbar spine in the images and reconstruct a 3D model (i.e., a mesh consisting of polygons) of the spine shape from each scan. Besides the 3D mesh, the central line of the lumbar spine canal is extracted. It captures essential information about the spine shape deformation. Due to the same way of reconstruction, the corresponding points of the meshes can be easily matched, and the same applies to the central lines; therefore, differences between meshes and between lines can be straightforwardly measured based on the distances between corresponding points. Besides, it is easy to compute an "average shape" from a set of meshes or lines. These opportunities can be utilised for supporting analysis through clustering and comparative visualisation.

Figure 12.6 demonstrates application of clustering to image-derived spine data of multiple patients. The patients have been clustered according to the similarity of the central lines of the lumbar spines. The light grey bars show the sizes of the clusters (the bluish segments correspond to the female patients). The 3D images above the bars represent the average spine shapes corresponding to the clusters. The colouring encodes the local differences of the shapes with respect to a selected reference shape. In this example, the average shape of cluster 4 (in the middle of the plot) has been selected as the reference. Red denotes the differences on the X-axis, blue on the Y-axis, and green on the Z-axis.

The examples in this section demonstrate the use of image processing techniques when it is necessary to analyse particular objects reflected in images rather than the whole images. The resulting derived data can be visualised and analysed as usual multi-attribute data. However, these data usually cannot capture all information essential for analysis that is contained in the image, whereas human eyes can easily do this. Therefore, visualisations of derived numeric data need to be combined with representations of the appearance of the extracted objects. Both examples in the

Figs. 12.5 and 12.6 include visualisations of 3D models, due to the nature of the objects and the possibility to reconstruct the models from the original data. In other cases, appearances of objects can be represented by 2D excerpts from the original images, or, depending on the application, it may be even more appropriate to show the whole images, where the objects can be seen together with the surrounding context. Anyway, whatever sophisticated computational techniques are applied for image processing and analysis, the unique capabilities of the human vision are irreplaceable.

12.4.3 Analysis of object movements

A typical task in analysing video recordings is detection and tracking of moving objects. There are video processing techniques that can distinguish moving spots from non-moving background and track these moving spots from frame to frame. Quite obviously, the coordinates of some representative points of the spots (e.g., their centres) can be taken from each frame and put in a temporal order, thus forming trajectories. You should keep in mind that what you get in this way is not trajectories of real moving objects in the real world but trajectories of their images in the space of the video frame. It may be sufficient in many cases, provided that there is a human analyst who can see and interpret the scene in which the movement took place. When it is really necessary to reconstruct the real-world trajectories from the video, further and more sophisticated processing is needed. There must be an algorithm for transforming the in-frame coordinates to real-world positions. This kind of processing is applied, for example, for acquisition of positional data reflecting sport activities, such as trajectories of football or basketball players and the ball during a game [128].

There is nothing special in trajectories that were obtained by means of video recording and transformed to real-world coordinates. They are analysed like trajectories of any other origin, as described in Chapter 10. Frame-space trajectories are not very special either. From the perspective of movement analysis, the main difference of frame-space trajectories from trajectories consisting of real-world coordinates is that some information cannot be derived from the positions and time steps, namely, the speed of the movement, travelled distance, and direction, as well as distances to other moving objects. Aside of that, frame-space trajectories can be visualised in the same ways as any other trajectories, except that an image of the scene represented in the video is used instead of a background map. An example can be seen on the right of Fig. 12.7, where a trajectory is shown in a space-time cube with an image in the base. Figure 12.1 demonstrates that frame-space trajectories can be clustered by similarity and aggregated into flows.

The picture on the left of Fig. 12.7 shows an interesting and potentially quite useful way of representing a trajectory of a single moving object within the frame space.

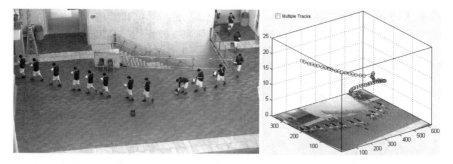

Fig. 12.7: After extraction of object movements from video, they can be analysed using space-time cube and other visualisation techniques suitable for movement data. Source: [98].

Instead of drawing a line or a sequence of dots marking the object's positions, it is possible to use the snippets of the video containing the object. These snippets need to be extracted anyway in the course of image processing prior to obtaining the coordinates of the object. The advantage of showing snippets is that they may contain valuable additional information concerning the appearance and/or actions of the moving objects, which can be seen and interpreted by a human analyst. Thus, the snippets in Fig. 12.7 represent a person who not just walked across the scene but bent at some position and put something on the floor. This way of representing trajectories may be very helpful for detection of suspicious activities.

The examples in this section demonstrate the following points:

- There are technical possibilities to extract trajectories of moving objects from video recordings.

- The usual ways of visualisation and aggregation of trajectories are applicable to trajectories consisting of in-frame coordinates.

- Trajectories extracted from video should not be analysed in separation from the original video, which provides important information concerning the context of the movements and activities of moving objects.

12.5 General scheme for visual analysis of unstructured data

As we wrote in Section 12.2, unstructured data, such as images and video, are meant to be perceived and understood by humans, whereas computers can only deal with some kinds of structured data derived from the unstructured data. Since it is not possible to obtain such structured data that comprehensively capture the human-

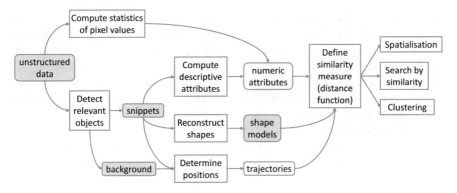

Fig. 12.8: A schematic summary of the approaches to analysing unstructured data.

oriented contents of images or videos, analysis of such data requires combining computational processing with the unique human capabilities to see, understand, and think about the things that have been seen. The examples included in this chapter demonstrate, first, different kinds of structured data that can be extracted from images or video and, second, different ways in which a human analyst can obtain relevant visual information.

The scheme in Fig. 12.8 summarises the approaches we have presented. The boxes with rounded corners represent different types of data. The grey background indicates the types of data that can convey the semantic contents of the unstructured data to a human analyst. The rectangular boxes represent computational operations used for derivation of structured data and for analysis of the derived data. The possible types of structured data that can be derived include numeric attributes, shape models, and trajectories. For any of them, it is possible to define a suitable measure of similarity, which can be employed in a range of analytical techniques, including spatialisation, search by similarity, and clustering.

To avoid over-complication of the scheme, we have not reflected in it that all kinds of data and any results of computational analysis need to be visualised and seen by a human analyst. In addition to graphs and plots presenting numeric attributes, results of spatialisations, or statistical summaries of clusters, the analyst needs to perceive the information that cannot be adequately represented by numbers. Depending on the application, properties of the original data, and the type of their contents, the analyst may need or prefer to see different things: the entire contents of selected items of the original data (images of video frames), or extracted fragments (snippets) containing relevant information, or artificially created representations incorporating parts of the original data, such as scene background (which is used for showing trajectories and aggregated movements) and snippets containing relevant objects (these can be used for showing multiple positions and/or states of the objects).

A special case is models of shapes of 3D objects reconstructed from multiple 2D images that were obtained by means of tomography. On the one hand, such models can be considered as structured data suitable for computer processing. On the other hand, it may be more convenient for a human to perceive shapes by viewing 3D displays of these models than by looking at the original sequences of tomographic images. Such a model can be considered as a surrogate of the object that was imaged using tomography. Hence, a 3D display of a model can adequately convey essential information from the original images to a human. It may be a sufficient representation of the image contents in many cases; however, the analyst may sometimes need to see the original images for observing fine details.

12.6 Questions and exercises

Consider the following examples of unstructured data:

- a collection of photographs of flower beds;
- a sequence of webcam photos of a mountain summit taken at the same time every morning during a year;
- a set of video recordings made by a motion-activated camera for wildlife observation installed in a forest.

For each of these examples, invent an exploratory task and make a plan of an approach to the data analysis. The following questions should be considered:

- What kinds of structured data would be appropriate to derive from the original data?
- What can be a suitable measure of the similarity between the data items?
- What computational analysis techniques can be applied? How can their results be used?
- What kinds of information (including the original data, derived data, and results of computational analysis) need to be seen by a human analyst to fulfil the task?
- In what form can each kind of relevant information be presented?

Chapter 13
Computational Modelling with Visual Analytics

Abstract Data scientists usually aim at building computer models. Computer-oriented modelling methods and software tools are developed in statistics, machine learning, data mining, and various specialised disciplines, such as spatial statistics, transportation research, and animal ecology. However, valid and useful computer-based models cannot be obtained by mere application of some modelling software to available data. Modelling requires understanding of the phenomenon that is modelled, the available data, and the model produced by the computer, which means that computer modelling is a process requiring involvement of a human. This chapter formulates the principles of thoughtful model building and includes examples of the use of visual analytics approaches for fulfilling these principles. The examples cover the tasks of feature engineering and selection, iterative model evaluation and progressive improvement, and comparison of different models.

13.1 Basic concepts

The data science process usually aims at creating a mathematical or computer model. A mathematical/computer model represents relationships between two or more components of a real-world phenomenon in such a way that it can be seen as a function (in mathematical sense) where some of the components play the role of "independent variables" and the others are "dependent variables". The task of a model is to estimate, or predict, the values or behaviours of the dependent variables for given values or behaviours of the independent variables. Since a model takes some inputs and produces outputs, the independent and dependent variables can also be called the input and output variables, respectively. Another frequently used term for an independent variable is a *feature*.

© Springer Nature Switzerland AG 2020

N. Andrienko et al., *Visual Analytics for Data Scientists*,

https://doi.org/10.1007/978-3-030-56146-8_13

For creating models, numerous methods and algorithms are developed in statistics, machine learning, and various domain-oriented sciences. According to the types of the input and/or output variables, there are several large classes of models:

- *Classification models*, also called *classifiers*, predict values of qualitative, or categorical, attributes, usually with a small set of possible values, called *classes*. When there are only two values, the classification is called *binary*; otherwise, it is called *multi-class* classification. The common modelling methods include Logistic Regression, Decision Tree, Random Forest, Naive Bayes, and Support Vector Machine (SVM).

- *Numeric value forecast models* estimate values of a numeric attribute. This class includes a range of *quantitative regression* models. Regression models working with a single input variable are called *univariate*, and models working with multiple input variables are called *multivariate*. Linear and polynomial regression are univariate regression models. Multivariate models include Stepwise Regression, Ridge, Lasso, Elastic Net, and Regression Tree. Besides, there exist spatial regression modelling methods, which are used when the values of the output variable are distributed in space. These methods account for the property of spatial dependence and autocorrelation (Section 9.3.1).

- *Time series models* predict the future behaviour of time-varying output variables based on their past behaviour. These models account for the specifics of temporal phenomena and data (Section 8.2), particularly, temporal dependency (autocorrelation) and possible influence of time cycles. Two commonly used forms of time series models are autoregressive models (AR) and moving-average (MA) models. These two are combined in ARMA and ARIMA (autoregressive integrated moving average) models, which are used for stationary and non-stationary time series, respectively. A time series is stationary if its statistical properties such as the mean, variance, and autocorrelation, are all constant over time and non-stationary if these properties change over time.

From another perspective, models can be categorised into *deterministic* and *probabilistic*. A deterministic model returns a single output for each given input. A probabilistic model estimates the probabilities of different possible outputs and returns a probability distribution among the possible outcomes. For example, in binary classification between classes A and B, a deterministic model returns either A or B, while an outcome of a probabilistic model may be "70% A and 30% B". Thus, most of the classification modelling methods create probabilistic models.

Modelling methods can also be distinguished based on the technology involved in model creation. Currently, a very popular technology is neural networks, which "learn" the relationship between input and output variables through training. In the context of this book, the modelling technology is irrelevant, as well as the specifics of different modelling methods. We may refer to specific methods in the examples, but our goal is to introduce the general principles of model building and demonstrate how they can be fulfilled using visual analytics approaches and techniques.

In the following discussion, we shall use a few more concepts and terms specific to modelling. First of all, what makes a model good? On the one hand, it should *accurately represent* the available data. It means that for any combination of values of the input variables that exists in the data the model returns the value of the output variable that is the same as or very close to the corresponding value in the data. The difference between the value specified in the data and the value predicted by the model is called *model error* or *residual*. It is desirable that the model errors are as low as possible. The process of minimising the model errors is often called *model fitting*.

On the other hand, the model needs to *generalise* the data in order to be applicable to new data, which have not been used for model construction. This means that the model needs to represent correctly the general character of the relationship between the input and output variables but should not represent in fine detail all individual correspondences between particular values. Typically, data contain random variations, called *noise*, due to unavoidable errors in measurement or many different factors that can affect measurements or observations. For example, the measured size of a thing may slightly vary depending on the temperature. It is rarely possible to take all such factors into account. Hence, it should always be assumed that the data used for modelling contain noise. A good model should leave the noise off. It may happen that model that has very low errors reflects the noise from the data that were used for model creation. In this case, the model reproduces well these old data but fails to give good predictions for new data. This is a phenomenon called *overfitting*, and the model is said to be *overfitted*.

Hence, in building models, people strive to achieve high accuracy (= low errors) but avoid overfitting (= capturing unessential data variations). To assess the goodness of a model, people analyse the residuals of the model. The goal is not to make the residuals not as small as possible but to ensure that *the residuals are random*. For example, when the model predicts values of a numeric attribute, the mean difference between the value in the data and the predicted value should be zero. Otherwise, the model tends to either overestimate or underestimate the value. It may happen, however, that the overall mean residual is zero, but the mean residuals for different subsets of the data, such as younger and older people or men and women, deviate from zero. Therefore, it is not sufficient to just check the mean residual for evaluating the model. It is necessary to analyse the *distributions of the residuals* that are relevant to the phenomenon being modelled. These may include the overall value frequency distribution, the distribution over subsets of entities, the distribution over space, and the distribution over time. All such distributions should not contain any patterns except a random pattern.

Other desirable characteristics of a good model are low complexity, understandable behaviour, fast performance, and low uncertainty for probabilistic models. Creating a good model certainly requires selection of a right modelling method and appropriate setting of its parameters. However, not less important is good selection of the input variables, or features, that will be used in the model. It would be absolutely wrong to think that the more features a model involves, the better. On the opposite: a crucial task in modelling is choosing a *minimal subset* of independent variables enabling sufficiently accurate prediction of the value of the dependent variable. This task is called *feature subset selection*. Involvement of redundant features entails a risk of obtaining an over-fitted and biased model. Involvement of irrelevant features can impair model performance, and it increases model complexity, slows down the performance, and hinders understanding.

There exist many methods for automatic selection of features. Different techniques often yield different results depending on the optimisation criteria used. Hence, a problem of selecting among different selections may arise. Besides, automatic methods cannot incorporate domain knowledge that would allow better generalisation but can instead select features that reflect unessential variation in the training data. These methods also cannot create meaningful and useful new features from what is available. Hence, it should be understood that "automated variable selection procedures are no substitute for careful thought" [3]. The following example demonstrates thoughtful selection of features for modelling.

13.2 Motivating example

13.2.1 Problem statement

The example we are going to discuss is taken from a paper "A Partition-Based Framework for Building and Validating Regression Models" [102]. The goal of a data scientist is to create a model that can predict the consumption of natural gas in a large city based on meteorological and other factors. Accurate prediction is important for minimising costs and guaranteeing supply. The available dataset spans in time over 5 years and includes hourly amounts of the gas consumed in the city as well as hourly measurements of weather attributes: air temperature, precipitation, wind speed, and global radiation; 42,869 data records in total. Each data record includes a time reference specifying the date and time of the measurements. The output variable of the model to be built is thus the amount of gas consumption, and the weather attributes need to be used as input variables (features). However, the data scientist understands that not only these attributes may be important but also the time, more precisely, the temporal cycles: daily, weekly, and seasonal. Therefore, the data scientist uses the time references of the records to derive attributes 'Day of Year', 'Hour' (i.e., hour of the day), and 'Weekend' (with two values 'true' and 'false'), which are added to the data and can be used as features.

The data scientist divides the available data into a training set, consisting of the records from the first three years, and a validation set consisting of the remaining

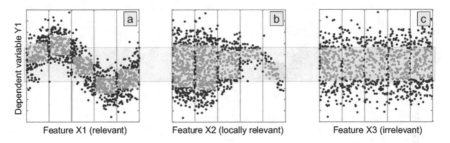

Fig. 13.1: Synthetic examples demonstrating different relationships between variables. The vertical axes in all plots correspond to a dependent variable Y1, and the horizontal axes to three different features labelled X1, X2, and X3. Source: [102].

records. The training set will be used for deriving the model and the validation set for evaluating the predictive capabilities of the model.

The amount of gas consumption, which needs to be predicted, is a quantitative attribute. The most commonly used class of modelling methods for predicting quantitative attribute values is quantitative regression. The data scientist does not rely on automatic selection of relevant features and generation of a model in a black-box manner. Instead, she prefers a more human-controlled approach by which she can incorporate her understanding of the modelled phenomenon and build a logically sound, well explainable and thus trustful model.

13.2.2 Understanding relationships among variables

In order to make a good selection of features and understand how they need to be used in the model, the data scientist needs to investigate the relationships of the different independent variables to the dependent variable. Figure 13.1 demonstrate how visual displays can help a human analyst to spot complex relationships, as in Fig. 13.1(a), or local relationships, as in Fig. 13.1(b), or determine that a feature is irrelevant to predicting the value of the dependent variable, as in Fig. 13.1(c).

In Fig. 13.1(a), the relationship between the feature X1 and the dependent variable Y1 is non-monotonic. X1 is certainly relevant to predicting the value of Y1, but the relationship between X1 and Y1 cannot be adequately represented by a single linear or polynomial regression model. Instead, a combination of several linear regression models can be appropriate. For this purpose, the data analyst needs to *partition* the range of the values of X1 into intervals so that the relationship between the values of X1 within each interval and the corresponding values of Y1 is monotonic. Being monotonic means that value of the dependent variable tends to either increase or decrease as the value of the independent variable increases.

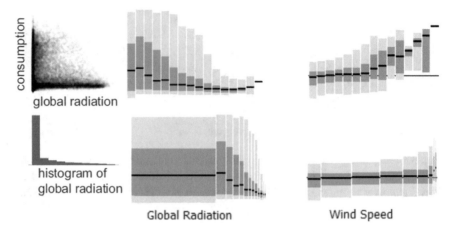

Fig. 13.2: Percentile plots show relationships between two numeric variables in cases of large data amounts. Source: [102].

In Fig. 13.1(b), the feature X2 is partly relevant to predicting the value of the output variable Y1. Specifically, for high values of X2, the corresponding ranges of the values of Y1 are quite narrow, which means that the value of Y1 can be predicted quite accurately. However, for low and medium values of X2, Y1 can take almost any value. It can be reasonable to involve X2 in the model, but use it only in the cases when the values of X2 are high.

Hence, the data scientist not only selects features but also decides whether it is reasonable to *partition* the data into subsets bases on the feature values and, if so, how the ranges of the feature values should be divided. Application of partitioning means that the data scientist creates a compound model consisting of several simple models. Each of these simple models is applied under specific conditions.

A scatterplot can clearly show how two variables are related when the data are not very numerous but does not scale to large amounts of data. Having a large amount of data, the data scientist uses another kind of visual display, called *percentile plot*. The idea is illustrated in Fig. 13.2, left. The value range of the input variable, which is represented on the X-axis of the plot, is divided into intervals. For each interval, the statistics of the corresponding values of the output variable is calculated, namely, the median, the quartiles (i.e., 25th and 75th percentiles), and the 5th and 95th percentiles. These statistics are represented by along the Y-axis. The horizontal black line represents the median, the lower and upper edges of the dark grey vertical bar correspond to the quartiles, and the lower and upper edges of the light grey bar correspond to the 5th and 95th percentiles. Such a bar is drawn for each of the value intervals of the input variable.

Figure 13.2 demonstrates two possible ways of dividing the input variable's value range into intervals. In the upper part of the figure, the intervals are of equal length.

For example, when the value range from 0 to 100 is divided into 20 equal-length intervals, the resulting intervals are from 0 up to 5, from 5 up to 10, and so on; the length of each interval is 5. The problem with such division is that the intervals may greatly differ in the amounts of data records that include values from these intervals. Thus, the scatterplot and the histogram on the left of Fig. 13.2 show that most of the data contain very low values of the input variable. The upper percentile plot does not show this information. This may lead to wrong understanding of the relationship. For example, the plot on the top right of Fig. 13.2 induces an impression of a strong dependency of the output variable (gas consumption) from the input variable (wind speed). However, high values of the input variable occur very rarely; therefore, their correspondences to high values of the output variable may be occasional.

To represent the relationship in a more truthful way, the value range of the input variable is divided into intervals differing in the length but containing approximately equal amounts of data records. These are called equal-frequency intervals. The percentile plots in the lower part of Fig. 13.2 have been built based on the equal-frequency division. The vertical bars differ in their widths, which are proportional to the lengths of the corresponding intervals of the values of the input variables. These plots show that the global radiation and the wind speed are irrelevant to predicting the gas consumption, because the values of the output variable (represented by the statistical summaries) do not substantially change along the ranges of the input variables.

13.2.3 Iterative construction of a model

So, the data scientist has an instrument to explore the relationships between the variables, namely, the percentile plots, which can be used for finding the features that are the most relevant to predicting the gas consumption. It is reasonable to begin with constructing a simple univariate regression model based on the most relevant feature. The percentile plots in Fig. 13.3 suggest that the most relevant feature is Temperature. The data scientist selects this feature and creates a univariate regression model M1. Specifically, M1 is a third degree (i.e., cubic) polynomial function.

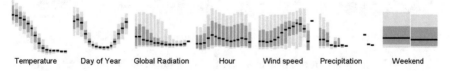

Temperature   Day of Year   Global Radiation   Hour   Wind speed   Precipitation   Weekend

Fig. 13.3: Selection of the most relevant feature (Temperature) for the initial model. Source: [102].

To evaluate the accuracy of the model, the data scientist computes the residuals, or errors, which are simply the differences between the values in the data and the values predicted by the model. A frequently used measure of the model error is RMSE (Root Mean Squared Error), which is calculated as the square root of the sum of the squared errors. The problem with this measure is that it is hard to judge whether the value is too high or OK. Thus, the RMSE of M1 is 24853; is it good or bad? Such a global numeric measure can only be useful when two or more models are compared; the model with a smaller total error is more accurate. So, the data scientist needs to build another model and check if the RMSE of the new model is significantly smaller than the RMSE of M1. However, it is pointless to construct just any model that somehow differs from M1, but it makes sense to build a model that can be expected to perform better than M1.

It may seem that a reasonable approach is to create a bivariate regression model involving two most relevant features, which are Temperature and Day of Year, according to Fig. 13.3. However, this is a bad idea for at least two reasons. First, not necessarily the combination of the two features with the highest relevance will give the most accurate prediction among all possible pairs. It is possible that the combination of Temperature with another feature may yield better accuracy than Temperature and Day of Year. However, trying all possible pairs would take too much effort and time. Second, as we explained in Section 13.1, the absolute values of the model residuals are not as important as the absence of non-random patterns in their distributions over different components of the data.

In our example, it is necessary to investigate the distributions of the residuals over the value ranges of the available variables. This can be done using percentile plots. Instead of the distributions of the values of the output variable, as in Fig. 13.3, they can show the distributions of the residuals of a model, as in Fig. 13.4. The plots not only reveal notable patterns in the distributions of the residuals but also suggest the data scientist which feature can be the best to use in combination with Temperature for refining the model M1. The most prominent pattern is observed for the feature Hour rather than for Day of Year. For Hour, not only the medians of the residuals notably differ in the night and day hours but also the percentile ranges are located on different sides of the zero line. For Day of year, the percentile ranges are wide and stretch on both sides of the zero line.

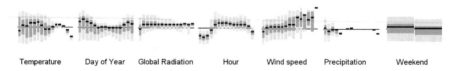

Temperature   Day of Year   Global Radiation   Hour   Wind speed   Precipitation   Weekend

Fig. 13.4: Exploration of the distribution of the residuals of the model M1 over the ranges of feature values. Source: [102].

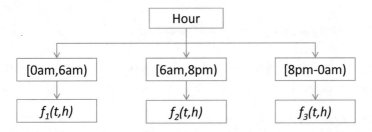

Fig. 13.5: A schematic representation of the structure of the model M2. It includes three sub-models, which are functions of the input variables Temperature (t) and Hour (h).

Hence, the data scientist decides to improve the model using the feature Hour. As it is seen from the percentile plot, the relationship between Hour and the M1 residuals is non-monotonous: the values almost do not change during the night, then increase in the morning, then tend to keep nearly stable over the day, and then decrease in the evening. It is very difficult to represent this relationship by a single function that would be sufficiently simple and easy to understand. Therefore, the data scientist decides to proceed by partitioning the range of Hour and creating a combination of several partial models rather than a single model. She divides the range of Hour into three sub-ranges, [0am-6am), [6am-8pm), and [8pm-0am). In these sub-ranges, the relationships can be represented by polynomials of the second degree, i.e., quadratic. So, the data scientist creates the next model M2 as a combination of three sub-models. Each sub-model comprises the terms from the model M1 involving the variable Temperature and two additional terms, linear and squared, involving the variable Hour. The structure of the model M2 is schematically represented as a tree in Fig. 13.5. The RMSE of the new model is 14385, which is a substantial reduction with respect to M1 (24853).

Now the computation and analysis of the residuals is repeated for M2 in the same way as it was done for M1. The percentile plots (Fig. 13.6) show that the effect of Hour is captured well as there is no pattern in the distribution of the residuals over the range of Hour anymore. However, there are patterns in the distributions over Day of Year and Weekend. At the first glance, it seems that there is also a strong pattern (increasing trend) in the distribution with respect to Wind speed, but the apparent pattern almost disappears when the division of the value ranges of the variables is changed from equal-interval to equal-frequency and the percentile plots are re-built; see the lower row of plots in Fig. 13.6. At the same time, the remaining patterns for the features Day of Year and Weekend indicate that these features should be included in the model. In other words, the model needs to account to the seasonal and weekly variations of the gas consumption.

Again, the percentile plot for Day of Year exhibits a non-monotonous relationship to the output variable. To capture this relationship, the data scientist again applies

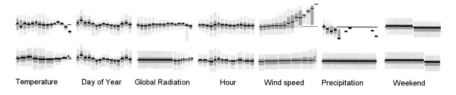

Temperature   Day of Year   Global Radiation   Hour   Wind speed   Precipitation   Weekend

Fig. 13.6: Exploration of the distribution of the residuals of the model M1 over the ranges of feature values. The plots in the upper row are built based on intervals of equal length and in the lower row on intervals of equal frequency. Source: [102].

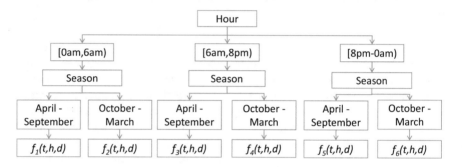

Fig. 13.7: A schematic representation of the structure of the model M3. It includes six sub-models, which are functions of the input variables Temperature (t), Hour (h), and Day (d).

partitioning. She divides the year into two periods: from the beginning of April till the end of September and from the beginning of October till the end of March. On this basis, she transforms model M2 into model M3 containing 6 sub-models (Fig. 13.7).

The RMSE of M3 is 10832, which is not as dramatic improvement with respect to M2 (14385) as it was with M2 compared to M1 (24853). More importantly, the refinement of M2 to M3 removes the previously present pattern from the residual distribution with respect to Day of Year. However, the pattern with respect to Weekend still remains, indicating that this variable needs to be involved in the model. Since this is a categorical variable with two values, the data scientist again refines the model through partitioning, separating the weekends from the weekdays. In this way, she creates model M4 containing 12 sub-models (Fig. 13.8). Its RMSE is 10251.7.

While the pattern of the distribution of the model residuals with respect to Wind Speed can be judged as insignificant, the data scientists, based on her domain knowledge, suspects that the effect of the wind speed may depend on the temperature. To check this, the data scientist uses a 2D error plot with one dimension corresponding to Temperature and the other to Wind Speed. The plot area is divided into cells,

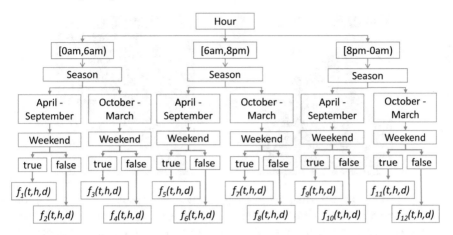

Fig. 13.8: A schematic representation of the structure of the model M4. It includes 12 sub-models, which are functions of the input variables Temperature (t), Hour (h), and Day (d).

which are painted in shades of two colours (Fig. 13.9, left). Yellow corresponds to positive residuals, when the values in the data are higher than the predicted values, which means that the model underestimates the values. Blue corresponds to negative residuals, which mean that the model overestimates the values of the output variable. The yellow area in the plot in Fig. 13.9, left, shows that model M4 underestimates the gas consumption for combinations of high wind speeds with lower temperatures. To improve the prediction, the data scientist refines M4 by involving the variable Wind Speed. The resulting model M5 has the same hierarchical structure as M4 (Fig. 13.8) but the 12 polynomial functions at the bottom level include additional terms with the variable Wind Speed. The RMSE of M5 is 10059.8, which is a modest improvement with respect to M4.

Since M5 is more complex than M4, the data scientist wants to compare the accuracy of M4 and M5 in more detail. She again uses 2D error plots, but the colouring of the cells shows this time the differences between the absolute values of the residuals of two models. On the right of Fig. 13.9, yellow indicates superiority (i.e., lower errors) of model M5 and blue means lower errors of M4. For all but one pairs of input variables, the shades of yellow clearly dominate over the entire plot areas whereas shades of blue occur rarely, and the occurrences are randomly scattered. However, in the plot of Day of Year against Temperature, which is shown in Fig. 13.9, right, there is a relatively large area where blue shades prevail; the area is marked in the figure. It means that M4 gives more accurate predictions than M5 for certain temperatures during spring and summer. Nevertheless, the data scientist is more satisfied with the accuracy of M5 and takes it as the final result of the model building process.

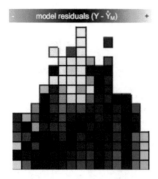 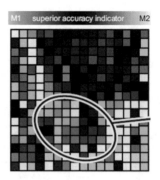

Fig. 13.9: Left: Investigating the interplay of the wind speed (vertical dimension) and the temperature (horizontal dimension). Right: Comparison of the distributions of the residuals of two models with respect to two input variables, Temperature (vertical dimension) and Day of Year (horizontal dimension). Source: [102].

13.2.4 Main takeaways from the example

This example is intended to demonstrate several important things:

- Model building is not a mere application of a modelling tool to available data. It requires involvement of human knowledge, judgement, and reasoning.

- Model building is an iterative process consisting of repeated evaluation and subsequent refinement of the current version of the model until suitable quality is achieved.

- Global indicators of model quality (RMSE and others) are insufficient and uninformative. More detailed analysis of model performance, such as investigation of distributions of model errors, can show where and why the model performs poorly and indicate directions for improvement.

- A model can be refined not only by using more complex functions involving more variables but also by partitioning the value domains of the input variables and creating a structured model that includes a specific sub-model for each partition. Meaningful partitioning makes the resulting structured model easy to understand whereas the accuracy may be much higher than can be achieved with a complicated monolithic model.

- A gain in model accuracy is often achieved at the cost of increased complexity. The model builder usually needs to find an appropriate trade-off between the accuracy and complexity.

13.3 General tasks in model building

The example we have considered involved several activities that are often performed in the course of model building. We shall call such activities *general tasks*. These include:

- *Data preparation:* Data collection when needed, examination of the properties of the data, checking the suitability for the modelling, handling problems if detected.

- *Feature engineering:* Creation of potentially relevant input variables from available data components.

- *Feature selection:* Finding a good combination of features to involve in the model.

- *Method selection:* Choosing an appropriate modelling method.

- *Data division:* Dividing the available data into parts one of which will be used for model creation (this part is usually called *training set*) and another for model evaluation (this part is called *test set*). Some modelling methods additionally require a *validation set*, which may be used for setting model parameters or for choosing one of multiple alternative models.

- *Method application:* Creation of a model by applying the chosen method to the data, which typically involves setting and tuning method parameters.

- *Model evaluation:* Analysis of model prediction quality and checking for possible biases. Evaluation also includes testing model sensitivity to small changes in input data and/or parameter settings.

- *Model refinement:* Actions to improve the prediction quality, eliminate biases, and increase robustness.

- *Model comparison:* Comparing the performance and complexity of two or more alternative models.

This list of tasks should not be treated as a one-way pipeline that needs to be followed once. These tasks (not necessarily all) are performed as steps of an iterative process with multiple returns to previous steps. Thus, in our motivating example (Section 13.2), the data scientist repeatedly returned to the task of feature selection and repeatedly performed method application, model evaluation, and model refinement. The example included feature engineering: it was creation of the attributes 'Day of Year', 'Hour', and 'Weekend' from the timestamps available in the raw data. The data scientist divided the available data into parts used for model building (the first three years) and for evaluation (the following two years). The modelling method (regression) was chosen according to the type of the output variable. The modelling process included comparison between the models M4 and M5 for deciding if the gain in the accuracy and reduced bias justify the increase of the complexity.

An example of modelling was also discussed in Chapter 11 (Section 11.1). It was building of a model for assessing text readability. The input data were in a form that is not suitable for direct use in modelling. The first task was to engineer a set of potentially relevant features. The analysts generated a large number of variables but did not know which of them were really relevant. To select truly relevant features among all, they compared the distributions of the feature values in two contrasting datasets consisting of complex and simple texts and chose the features whose distributions significantly differed. Then the analysts grouped the chosen features based on their correlations and chose a single most distinctive feature from each group.

The chosen modelling method in that example was very simple, just the mean of several numeric scores, but it was appropriate for the purpose. The example shows that not a sophisticated method makes a model accurate and useful but thoughtful creation and selection of relevant features. Moreover, a sufficiently good model built with a simple method but meaningful features is preferable to a more complex model, even if the latter performs a bit better. A simpler model is easier to understand, to check, and to use correctly.

13.4 Doing modelling tasks with visual analytics

Our motivating example (Section 13.2) demonstrates how visual analysis helps in fulfilling several tasks in the model building process. Now we shall generalise and extend what was demonstrated.

Feature construction is often done based on domain knowledge and may not need a help from visual analytics. However, useful features may also be constructed based on patterns observed in data. In this case, visualisation plays a crucial role as a means enabling the analyst to observe patterns. Let us take the example of the epidemic outbreak in Vastopolis (Section 1.2). If we had a goal to build a model predicting if a person gets sick and, if so, the kind of the disease, we would construct features reflecting the observed patterns in the spatio-temporal distribution of the outbreak-related microblog posts. The features would express whether a person had been in the area affected by air pollution in the next few days after the truck accident and whether a person had been close to the river in the southwestern part of the city on the third day after the accident.

Visual analytics is definitely useful for feature selection. In the example in Section 13.2, relevant features were selected based on analysis and comparison of frequency distributions of the feature values and model residuals. In the text readability assessment example from Section 11.1, the analysts explored a visual representation of the correlations between the features in the form of correlation matrix as shown in Fig. 4.5 [106]. Special ordering of the rows and columns of the matrix by means of a hierarchical clustering algorithm helped the analysts to detect groups of correlated features and select the most expressive representatives from the groups. Hence, we can conclude that visualisation supports feature selection by showing distributions and correlations.

Method selection does not always require visualisation, but there are cases when an appropriate method needs to be chosen according to patterns observed in the data, which makes visualisation essential. The most common case is modelling of time series. The choice of the method depends on the presence of cyclic variation patterns. Moreover, when the variation is related to two or more cycles, such as weekly and daily, the analyst's choice may be to decompose the variation into components and create a structured model where each component is captured by a special sub-model. An example can be found in paper [18]. The decomposition of the modelling problem and definition of the structure of the future model are parts of the method selection task.

For data division into training and test sets and, when needed, a validation set, the key requirement is that all sets have the same distribution patterns. Therefore, visual analytics is certainly relevant as a tool to explore and compare distributions and uncover patterns. Thus, in the example in Section 13.2, the data scientist could use percentile plots, as in Fig. 13.3, to compare the distributions of the values of the output variable over the features in the subsets of the data covering the first three years and the remaining two years, although it is not said explicitly in the paper as the data division step is not described in detail [102].

In application of modelling methods to data, model builders often use visual displays for judging how well the prediction matches the training data. The most commonly used display is a scatterplot showing data as points and the relationship extracted by the modelling method as a line. Another common case is to use a time series plot when modelling temporal variation of a numeric variable. Depending on the nature and structure of the data, other visualisations may be helpful. For example, mosaic plots can be used for seeing how a classification model separates classes based on ordinal or categorical variables [99]. The model builder uses such displays to make an approximate judgement of the model fitness for deciding whether the parameter settings of the modelling method need to be changed (for example, the order of the polynomial regression model) or, possibly, another method needs to be chosen. This judgement is not yet a full-fledged evaluation of the model but a preliminary estimation of its potential appropriateness. When it is clearly seen that a model does not fit, it makes little sense to do a thorough evaluation.

In evaluating a model, it is insufficient to use just statistical indicators of model quality, which do not provide any hint of what is wrong and how the quality can be improved; see our arguments concerning RMSE in Section 13.2.3. In our motivating example, the data scientist used percentile plots (Figs. 13.4 and 13.4) and 2D error plots (Fig. 13.9) for a detailed inspection of the distributions of the model errors expressed as numeric values. When the model output is not numeric, it is anyway important to inspect the distributions of the correct and wrong model results. In the following section, we shall show examples of doing this for classification models.

Model refinement can be done by including additional features or changing model structure, in particular, through partitioning, as demonstrated in Section 13.2.3. By no doubt, visual analytics techniques can help the model builder to understand what features should be added or how the model should be structured to reflect patterns observed in the data. The visual displays used for this purpose need to be sufficiently detailed and expressive to enable seeing patterns. Thus, in our example, the level of data aggregation applied in the percentile plots was appropriate for exhibiting patterns. When a plot was not expressive enough due to a skewed distribution of the feature values, the data scientist increased the expressiveness by changing the division of the feature value range from equal-length to equal-frequency intervals. Model refinement can also be done in other ways, for example, by providing additional labelled data instances when models are obtained using machine learning techniques. Also in such cases visual representations of current model results can help the model builder to understand where and what kind of refinement is needed and thus find or create appropriate instances to be used for the training.

Comparison can be applied to models of the same kind having similar quality, as models M4 and M5 in our motivating example (Section 13.2), or to models obtained using different modelling methods, as will be shown in one of the further examples. Like in other tasks, it is insufficient to compare just statistical indicators of model performance but it is necessary to compare the distributions of model errors, if they are possible to compute, or the distributions of the right and wrong predictions, or the degrees of the certainty or uncertainty in the predictions. This information gives a ground for an informed judgement of the relative merits and disadvantages of the models with respect to each other.

To show how some of these tasks can be done for models of a different kind than in Section 13.2, we shall discuss a couple of examples dealing with classification models. Such models are also called shortly *classifiers*.

13.5 Further examples: Evaluation and refinement of classification models

Before describing the examples, we shall discuss the usual practices in assessing the quality of classification models.

	Actual Fraud = 'yes'	Actual Fraud = 'no'
Predicted Fraud = 'yes'	True positive 57	False positive 14
Predicted Fraud = 'no'	False negative 23	True negative 171

	Actual Dog	Actual Cat	Actual Rabbit
Classified Dog	23	12	7
Classified Cat	11	29	13
Classified Rabbit	4	10	24

Fig. 13.10: Confusion matrices for classifiers with two (left) and three (right) classes.

13.5.1 Assessment of the quality of a classifier

For a classification model, good quality means correct assignment of items to their classes. In practice, a model can rarely achieve absolute correctness. For binary classifiers, which assign items to two possible classes 'yes' and 'no', or 'true' and 'false', there are special terms for correct and wrong assignments. "True positive" means a correct assignment to the class 'yes' and "false positive" means a wrong assignment to this class. Similarly, "true negative" and "false negative" mean correct and wrong assignments, respectively, to the class "no". For multi-class classifiers, these concepts are applied individually to each class. Thus, a false positive for class X is an item assigned to class X but actually belonging to another class, whereas a false negative for class X is an item belonging to class X but assigned to another class by the classifier. The numbers or percentages of correct and wrong assignments to each class can be represented in a *confusion matrix*, as demonstrated in Fig. 13.10.

Commonly used statistical metrics of classification model quality are:

- *accuracy*: the ratio of the number of correct classifications (i.e., true positives and true negatives) to the total number of classifications; in other words, the sum of the diagonal cells of the confusion matrix divided by the sum of all cells;

- *true positive rate*, also called *recall* or *sensitivity*: the ratio of the number of true positives to the size of the 'yes' class, that is, to the sum of the numbers of true positives and false negatives;

- *precision*: the ratio of the number of true positives to the total number of the items assigned to the 'yes' class, that is, the sum of the numbers of true and false positives;

- *true negative rate*, also called *specificity*: the ratio of the number of true negatives to the size of the 'no' class, that is, the sum of the numbers of true negatives and false positives;

- *false positive rate*: the ratio of the number of false positives to the size of the 'no' class, that is, the sum of the numbers of true negatives and false positives.

As we said in Section 13.2.3 regarding regression models, such overall quality metrics are not very helpful, because they do not tell the model builder what and how can be improved.

There are many methods that create probabilistic classifiers whose results are not crisp assignments of items to singular classes but probabilities for the items to belong to each of possible classes. These probabilities are also called *prediction scores*. The sum of all probabilities is 1; therefore, for a binary classifier, it is sufficient to return a single value between 0 and 1, where 0 means 100% 'no' and 1 means 100% 'yes'. To assign items to either the 'yes' or 'no' class, some threshold value between 0 and 1 is chosen; it is often called *prediction threshold*. An item is then assigned to the class 'yes' if the probability is above the prediction threshold. When using a multi-class classifier, a possible approach is to assign an item to the class having the highest probability. Another approach is to apply a threshold, as for a binary classifier. With this approach, it may happen that the probabilities of all classes are below the threshold, and some items may get no class assignment.

Multiclass classifiers can also be built by training multiple binary classifiers and then combining their outputs to make predictions for individual items. The one-vs-rest method (also known as one-vs-all) means that each binary classifier is trained to discriminate one of the classes from the remaining classes. The class of each item is determined by the classifier that produces the highest score. With the one-vs-one (or all-vs-all) method, binary classifiers are created for every pair of classes, and majority voting is used to select the winning class prediction for each item.

To asses the performance of a classifier and choose a suitable value of the prediction threshold, model builders plot a so-called *ROC (Receiver Operating Characteristic) Curve*, as shown in Fig. 13.11. The X-axis corresponds to the false positive rate and the Y-axis to the true positive rate. The points on the curve correspond to the values of these rates obtained with different threshold settings. The threshold value is the lowest at the right end of the curve and the highest at the left end. With a very low threshold value (close to 0), all items will be assigned to the given class (in a binary case, to the 'yes' class) irrespective of their actual class membership. This means that the true positive rate will be 1, but the false positive rate will also be 1, because all "negative" items will be classified as "positive". As the threshold increases, fewer and fewer items are assigned to the given class. When fewer "negative" items are assigned to the class, the false positive rate decreases. When fewer "positive" items are assigned to the class, the true positive rate decreases. In an ideal case, only the false positive rate decreases, but the true positive rate does not decrease. This means that the plot has the shape of a horizontal line approaching the Y-axis at the level $Y = 1$ (the green line in Fig. 13.11). In reality, both rates decrease as the threshold increases. It is good if the true positive rate decreases much slower than the false positive rate. The slower the true positive rate decreases, the closer the curve approaches the upper horizontal line. The faster the decrease of the true positive

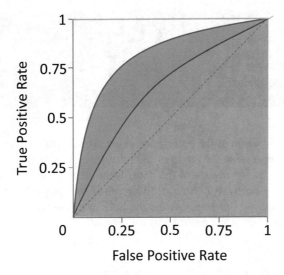

Fig. 13.11: Receiver Operating Characteristic (ROC) Curve.

rate is, the closer the curve approaches the diagonal line, where both rates have equal values. Thus, in Fig. 13.11, the classifier represented by the blue ROC curve is better than the classifier represented by the red curve, because the blue curve deviates from the diagonal more and approaches the level $Y = 1$ closer than the red curve.

The quality of a classifier can be expressed numerically by the *Area Under the Curve (AUC)*: the larger this number is, the better. Thus, the area under the blue curve in Fig. 13.11 is larger than the area under the red curve.

The ROC curve shows the overall quality of a probabilistic classifier and the impact of the prediction threshold on the results. However, this information does not suggest how the quality can be improved. The model builder needs to see what is wrong and understand the reason for it in order to make targeted improvements. This often requires direct examination of individual prediction errors of a model.

The most common reasons for classification errors include

- mislabelled data,

- inadequate features to distinguish between classes, and

- insufficient data for generalising from existing examples.

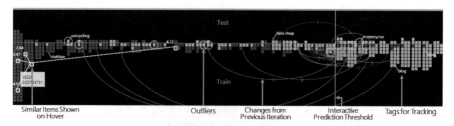

Fig. 13.12: Visualisation of the performance of a probabilistic binary classifier. The horizontal axis represents the range of the probability values returned by the classifier. The small squares represent the items from the train and test set (below and above the axis, respectively). A square is green if the respective item belongs to the 'yes' class and red otherwise, while the horizontal position corresponds to the probability of the 'yes' class returned by the classifier.

13.5.2 Example: improving a binary classifier

In this example [6], a binary classifier needs to determine whether a text document (more specifically, a web page) concerns a certain subject, such as cycling. The data scientist applies a modelling method (specifically, logistic regression) that creates a probabilistic model. Two sets of labelled documents are used for model training and testing. To understand how the model performs and how this can be improved, the data scientist needs to see in detail the model predictions for the labelled documents. For this purpose, she uses the display shown in Fig. 13.12.

This is a variant of the general visualisation technique called *stacked dot plot*, where data items are represented by dots positioned along a horizontal axis according to values of a numeric attribute. When several items have the same or very close values, the corresponding dots have the same horizontal position and are arranged in a stack in the vertical dimension of the display. In Fig. 13.12, the dots have square shapes. Each square represents a document. It is drawn below the horizontal axis if the document belongs to the train set and above the axis if the document is from the test set. The green and red colouring of the squares corresponds to the labels 'yes' and 'no', respectively. The positions along the horizontal axis correspond to the scores, that is, the probabilities of the 'yes' class, returned by the classifier. The display includes a vertical line that shows the current prediction threshold.

For an ideal classifier, all red squares would be at the left end and all green squares at the right end of the display. In a real case, the red and green squares are mixed. The data scientist wants to increase the separation between the classes, to be able to find such a position for the threshold line that all or almost all squares on the left of it are red and all or almost all squares on the right are green. To find out how to achieve this, the data scientist needs to understand the reasons for the model errors, that is, low probabilities for the items labelled 'yes' and high probabilities for the items labelled 'no'. This requires examination of particular documents. The interactive display facilitates such examination by providing an opportunity to select a dot and obtain a hyperlink to the corresponding document.

Let us see how the visualisation can help the data scientist. One of possible reasons for classification errors is mislabelled data items. Assuming that wrong class labels occur rarely in the train set, correct data will have higher influence on the classifier's training. Therefore, the classifier's predictions for the mislabelled data items are likely to correspond to their true classes. In our case, documents wrongly labelled as belonging to the 'yes' class will receive low scores and will appear in the display as green squares positioned close to the left edge of the plot. On the opposite, documents wrongly labelled as representatives of the 'no' class will receive high scores and will appear as red squares drawn close to the right edge of the plot. When the data scientist sees such squares, she can open the corresponding documents and check if they have correct class labels. Having encountered mislabelled documents, the data scientists can modify the labels and re-train the classifier. After each re-training, the display is updated.

Another possible reason for poor class separation is inadequacy of the features that are used for the classification. When this is the case, items belonging to different classes may have very similar values of the features. In our example, the features are presence or absence of specific keywords. The set of keywords may be insufficient for distinguishing relevant documents (i.e., talking about the subject of interest) from irrelevant. Thus, when the subject is "cycling", the model may give a low score to a document about unicycling, which is relevant and has the class label 'yes', and a high score to a document about motorcycling, which is irrelevant and has the class label 'no'. The relevant document will be represented by a green rectangle drawn on the left of the plot, and the irrelevant document will appear as a red rectangle positioned on the right.

To find out whether the reason is feature inadequacy, the data scientist should be able to see which other documents are similar to the documents whose scores do not correspond to the class labels. Using the interactive tool shown in Fig. 13.12, the data scientist can select any square, and the tool will link this square with squares representing similar documents by lines. Another possibility would be to use an additional projection plot built with the use of some dimensionality reduction method (Section 4.4), where the dots representing the documents are positioned according to the similarity of their features. In such a plot, items with similar features will be represented by groups (clusters) of close dots. When the feature set is insufficiently distinctive, green dots will appear in or close to groups of red dots, and the other way around. When such situations are detected, the data scientist can inspect the contents of the documents and extend the set of features with additional keywords, such as "motorcycling" and "unicycling".

As mentioned at the end of Section 13.5.1, the data set used for model training may be insufficient for generalising. When this happens, the model may classify the examples from the train set well enough but perform poorly for the test set.

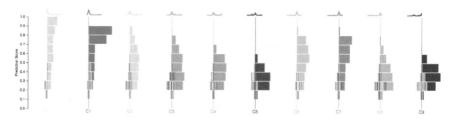

Fig. 13.13: Visualisation of the performance of a probabilistic classifier with 10 classes. For each class, there is a frequency histogram of the item scores (class probabilities) given by the classifier.

Another possible indication of such a problem is presence of outliers, that is, data items that are very dissimilar to all others according to the features used for the classification. In the tool shown in Fig. 13.12, outliers are specially marked, but it would be even more convenient to use an additional projection plot, as we suggested in the previous paragraph. In such a plot, outliers would appear as isolated dots positioned far from all others. An obvious remedy for this problem is to extend the train set with additional examples that are similar to the outliers or to the items from the test set that were poorly classified.

Hence, appropriate visualisation of model performance allows model builders to detect and inspect errors, understand their reasons, and choose appropriate actions for targeted improvement. To enable error detection, the visualisation must show model results for individual data items. Detailed inspection is enabled by interactive operations exhibiting selected data items. For understanding reasons for the errors, model builders need to compare the features of the data items. This can be enabled by interactive linking of similar items or by an additional projection plot where similarities between data items are represented by distances between corresponding visual marks.

13.5.3 Example: analysing and comparing performances of multi-class classifiers

In this example [115], the data scientist analyses the performance of a probabilistic multi-class classifier. The task is to recognise hand-written digits from 0 to 9; hence, there are 10 possible classes. The data scientist creates and trains a classification model, which yields the accuracy 0.87. Can this be improved?

To understand this, the data scientist creates a visualisation with 10 histograms showing the distribution of the item scores for each class (Fig. 13.13). As we explained in Section 13.5.1, each item receives some probability value, or score, for each class. The histogram of a class shows the frequency distribution of the probabilities of this class received by all items. The histograms in Fig. 13.13 are drawn so that the value axes are oriented vertically; the lower ends correspond to the probability 0 and the upper ends to the probability 1. The bars are oriented horizontally and show the value frequencies by intervals of the length 0.1. A distinctive colour is chosen for each class. These colours are used for showing correct and wrong class assignments. Some bars in the histograms are painted in uniform solid colours. This means that all items represented by these bars were correctly recognised as members of the respective classes. Other bars contain segments with textured painting. These segments represent groups of items that were classified wrongly. A textured segment drawn on the right of the vertical axis represents items that were assigned to the class corresponding to this histogram but belong to another class, which is indicated by the colour of the segment paining. A textured segment drawn on the left of the axis represents items that actually belong to the class of this histogram but were assigned to another class indicated by the colour of the segment. Hence, the histograms show the confusions between the classes and the scores received by the correctly and wrongly classified items.

The overall shapes of the histograms are also highly informative. They show that, except for the classes C0 and C1, the classifier tends to give quite low class probabilities. In many cases, the probabilities of the winning classes were less than 0.3. This means that the probabilities of the remaining classes did not differ much from the highest class probability. This, in turn, means that, even when the classes are identified correctly, the certainty of the classification results is quite low. A very small difference in the input data may change the class prediction.

Additional information about the between-class confusions is represented by the small plots drawn above the histograms. One of these plots is enlarged in Fig. 13.14. Each plot contains multiple polylines (polygonal lines) corresponding to all true members of the respective class and showing the scores the members of this class received for each of the ten classes. A plot that has a single high peak and a flat remainder indicates that all or almost all members of this class received high probabilities of this class, which is very good. The presence of two or more peaks means that some class members received relatively high probabilities of classes they do not belong to. Thus, the enlarged plot in Fig. 13.14 indicates that many of the members of the class C5 received high probabilities of the class C3. For some of them, the probabilities of C3 were higher than the probabilities of C5, and they were wrongly assigned to C3. The red segments in the histogram of the class C3 correspond to these items. Reciprocally, many members of class C3 received relatively high probabilities of C5.

From the histograms and plots in Fig. 13.13, the data scientist gains an overall idea about the performance of the classifier, but she needs to see details for individual items for understanding the reasons for the poor performance and finding ways to improve it. In particular, she needs to look at several members of the classes C3 and C5 in order to understand the reasons for the confusions between these two classes.

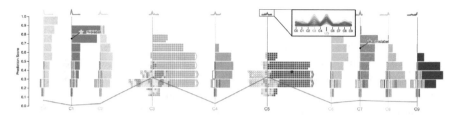

Fig. 13.14: Some histogram bars are transformed into arrays of dots representing individual data items.

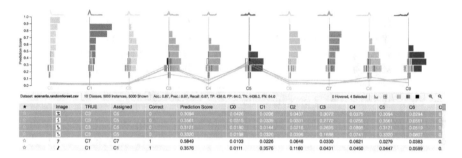

Fig. 13.15: The class probabilities for several selected items are shown by lines connecting corresponding positions on the histogram axes of the different classes.

In Figs. 13.14 and 13.15, selected items are represented by polylines connecting the positions on the vertical axes corresponding to the probabilities of the classes these axes correspond to. In Fig. 13.14, the data scientist has selected one member of the class C5 that received equal probabilities of the classes C5 and C3. In Fig. 13.15, the data scientist has selected four members of the class C3 that were wrongly classified as C5. Since the items being classified are images (of handwritten digits), the data scientist looks not only at the scores of the selected items but also at the images.

Please note that detailed information for selected items does not need to be represented by polylines on top of the histogram display, as in Figs. 13.14 and 13.15. It is possible to use a separate display, or even a table, as at the bottom of Fig. 13.15, can be appropriate. What is really important is selection and detailed inspection of representative problematic items for understanding the reasons for the problems. Such items can be selected using query tools.

In our example, the data scientist finds that some handwritten variants of the digits 3 and 5 may be hard to distinguish due to low resolution of the input images (7x7 pixels). This is a special case of using inadequate features. Increasing the image resolution to 14x14 pixels greatly improves the recognition of the digits.

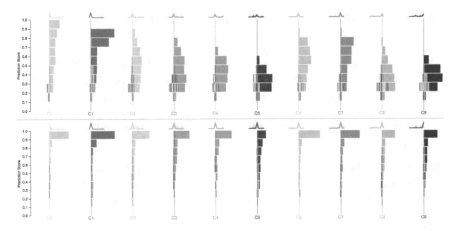

Fig. 13.16: Comparison of the performance of two classification models.

Like with a binary classifier (Section 13.5.2), there may be mislabelled examples. Such an example is likely to receive a high probability of its true class rather than the class specified by the label. In the histogram display, mislabelled examples will be manifested as segments located high in some of the class histograms and coloured differently from the colours of the respective classes.

Quite often, people create several classifiers using different modelling methods in order to choose the one that performs the best. It may not be enough to compare the statistical indicators of each model's quality. When there is no obvious winner in terms of these indicators, it is appropriate to compare the behaviours of the classifiers, particularly, the distributions of the class probabilities. In Fig. 13.16, the classification model that we have discussed so far is compared with another model created for the same classification problem using the same input data but a different modelling method. Both models have the same accuracy 0.87, but the distributions of the class probabilities are very different. While the first model (Random Forest) tends to give low scores that do not differ much between the classes, the second model (SVM - support vector machine) has high frequencies of very high scores (close to 1). The long bars at the tops of the histograms indicate that there are many items for which one class has a very high probability whereas the probabilities of the remaining classes are close to 0. Hence, the predictions made by the second model have much higher certainty than the predictions of the first model. With the first model, slight variations or noise in the input data may easily flip the prediction from correct to wrong or vice versa. The second model will be more robust and therefore should be preferred.

13.5.4 General notes concerning classification models

Both examples, with the binary and multiclass classifiers, demonstrate the useful-ness of visualising the distributions of the prediction scores with indication of the correct and wrong class assignments. Apart from enabling understanding of the model behaviour, the visualisation suggests what items should be inspected indi-vidually for understanding the reasons of wrong classifications and finding ways to improve the model.

Additionally to the visualisations proposed by the authors of the discussed papers [6, 115], we suggest using a projection plot where the items from the train and test sets are arranged on a plane according to the similarity of their features and coloured according to the true classes. When items from different classes happen to be close in the projection plot, it means that they have similar features, which, it turn, may mean that the set of features currently used is inadequate for distinguishing members of different classes.

13.6 Visual analytics in modelling time series

In modelling time series, visualisation of the data to be modelled is essential for understanding the character of the temporal variation and, on this basis, choos-ing an appropriate modelling method, often called *model class*. In particular, it is important to identify the presence of variations related to temporal cycles and the existence of a long-term trend. After choosing a modelling method and fitting the model to the data by adjusting the parameters, the quality of the model is analysed very similarly to numeric value forecast models, as in the example considered in Section 13.2. The similarity is not occasional, because time series models forecast numeric values; hence, it is possible to compute the residuals by subtracting the pre-dicted values from the values given in the input data. Like for any numeric forecast model, the distribution of the model residuals needs to be random. In the example in Section 13.2.3, the data scientist analysed the distribution of the residuals over different components of the data. In time series modelling, it is necessary to analyse the distribution of the residuals over time.

Figure 13.17 demonstrates several complementary ways of visually checking the distribution of time-related values (particularly, the values of the residuals of a time series model) for randomness. The image is a part of an illustration from a pa-per [36], and the labels of the image sections come from the source paper. In the section 4a, standardised residuals (i.e., residuals divided by the standard deviation) are plotted over time. This plot can reveal episodes where the variation does not look like random fluctuation. In 4b, the values of the autocorrelation function[1] (ACF) are

---
[1] https://en.wikipedia.org/wiki/Autocorrelation

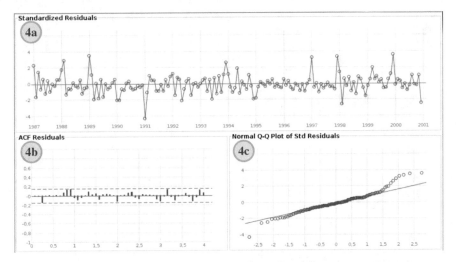

Fig. 13.17: Visualisation of the temporal and statistical distributions of the standardised residuals of a time series model (source: [36]). 4a: the residuals plotted over time; 4b: the values of the ACF (autocorrelation function) plotted over different time lags; 4c: the Q-Q (quantile-quantile) plot of the quantiles of the residuals against the quantiles of an "ideal" normal distribution.

plotted over a range of temporal lags. The ACF is the correlation between any two values in a time series with a specific time shift, called lag.

4c is a Q-Q plot, or quantile-quantile plot. Such a plot is used for comparison of the statistical distributions of two sets of numeric values. Quantiles (e.g., 0.01, 0.02, ..., or 1%, 2%, and so on) of one set are plotted against the corresponding quantiles of the other set. An $x\%$ quantile of a set of numeric values is a particular value v_x such that $x\%$ of all values are less than v_x and the remaining $(100-x)\%$ are above v_x. When two sets have the same statistical distribution, all dots in the Q-Q plot fall on the 45° diagonal. Deviations from the diagonal indicate how different the distributions are.

A fully random set of numeric values has a normal, or Gaussian, statistical distribution. Therefore, the randomness of the distribution of model residuals can be checked by comparing it with the "ideal" (theoretical rather than real) normal distribution. Hence, the plot labelled 4c is the Q-Q plot of the quantiles of the set of the residuals against the quantiles of the theoretical normal distribution.

As for any kind of model, building of a time series model is an iterative process, in which the model builder repeatedly evaluates the current version of the model and performs some actions for its improvement until getting satisfied with the model quality. This process requires comparing of each new version with the previous version. Figure 13.18 demonstrates a possible way of comparing residuals of two models.

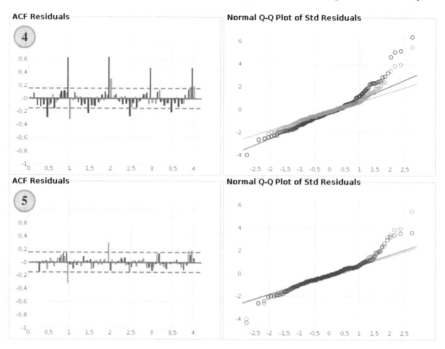

Fig. 13.18: Comparison of residuals of two models in two selected steps (4 and 5) of an iterative model building process. Source: [36].

The images labelled as 4 and 5 correspond to two consecutive steps, 4 and 5, of a process of building and refining a time series model. The ACF residual plot and the Q-Q plot include residuals of two models distinguished by colours. In step 4, the residuals of the previous model version are shown in bright pink and the residuals of the new model version obtained in this step are shown in green. We can see in the ACF plot that the autocorrelation of the residuals is much lower for the new model than for the previous model. In the Q-Q plot, the arrangement of the green dots deviates less from the diagonal than the arrangement of the pink dots, which means higher similarity to the normal distribution. In step 5, the "green" model from the previous step is compared with a new model represented in orange. The randomness of the residuals is further improved.

The examples in this section demonstrate the importance of human critical judgements for creation of good computer models and the usefulness of visual analytics techniques in the model building process.

13.7 Explaining model behaviour

Being able to understand why a model makes a certain prediction is extremely important for the users to gain trust and use the model confidently. Lack of understanding may preclude model use in many applications, such as health care or law enforcement. Understanding of model behaviour also provides insight into how a model may be improved and supports understanding of the phenomenon being modelled. However, in response to the challenges of big data, there is a growing trend of developing and using methods of machine learning (ML) that create models whose operation is not comprehensible to people, so-called "black boxes". To address the problem of model non-transparency, even a special research field called eXplainable Artificial Intelligence (XAI) has emerged [63]. Artificial intelligence (AI) is the overarching discipline that includes ML as well as other areas focusing on making machines behave smart, such as robotics and natural language processing.

Although a large part of the AI algorithms and the resulting models cannot be directly explained (for instance, deep learning models), XAI methods aim to create human-understandable explanations of some aspects of the behaviour of these models. Thus, there are methods that support checking whether a model makes use of features that a human expert deems important. Other methods explain a black box model by creating an understandable surrogate model (such as a set of rules or a decision tree) that is supposed to replicate most of the behaviour of the primary model. In fact, any explanation of a model's prediction can be viewed as a model itself. Based on this view, Lundberg and Lee have proposed a unified framework for interpreting predictions called SHAP, which assigns each input variable an importance value for a particular prediction [92]. There are also methods revealing the internal structure of a model and/or flows of data. Such methods may be useful for model developers. A comprehensive survey [63] can be recommended for understanding the main concepts and the variety of approaches to explaining different aspects of black box models. A brief overview of the major approaches and representative methods can be found in paper [126].

Since explanations are meant for humans, they often involve visualisations. A good example is visualisation of the contributions of different input variables into model predictions[23] based on the SHAP framework [92].

An increasing amount of research in the field of visual analytics is focused on supporting the process of building deep learning models [70]. Model developers use visual representations of the network architecture, filters, or neuron activations in response to given input data. We shall not describe these representations, since specific technical knowledge would be required for understanding them. What is im-

---

2   https://towardsdatascience.com/explain-your-model-with-the-
shap-values-bc36aac4de3d

3   https://towardsdatascience.com/explain-any-models-with-the-
shap-values-use-the-kernelexplainer-79de9464897a

portant to note is that visual analytics can and should be involved in building AI models, as with other kinds of models.

Nevertheless, before rushing into creating a human-incomprehensible AI model, it is appropriate to think about a possibility to create instead an understandable model that could serve the intended purpose sufficiently well. We highly recommend reading the paper with the expressive title "Stop explaining black box machine learning models for high stakes decisions and use interpretable models instead" [119]. The author Cynthia Rudin convincingly argues that XAI methods provide "explanations" that cannot have perfect fidelity with respect to what the original models actually do. This entails the danger that the representation of the behaviour of the target model can be inaccurate for some of possible inputs, which limits the trust in the explanations and in the model itself.

Here is what the author says: "An explainable model that has a 90% agreement with the original model indeed explains the original model most of the time. However, an explanation model that is correct 90% of the time is wrong 10% of the time. If a tenth of the explanations are incorrect, one cannot trust the explanations, and thus one cannot trust the original black box. If we cannot know for certain whether our explanation is correct, we cannot know whether to trust either the explanation or the original model" [119, p.207].

The paper contains a number of other serious arguments against the use of black box models, even when they are supplied with "explanations" (which is a misleading term, because these are not real explanations). The author also refutes the widespread beliefs that more complex models are more accurate and that AI methods are always superior to the "traditional" modelling techniques. "In data science problems, where structured data with meaningful features are constructed as part of the data science process, there tends to be little difference between algorithms, assuming that the data scientist follows a standard process for knowledge discovery" [119, p.207].

Another belief that prompts some people to use AI techniques creating black box models is that these techniques are capable to capture hidden patterns in the data the users are not aware of. The author objects that a transparent model may be able to uncover the same patterns, if they are important enough to be leveraged for obtaining better predictions.

It is also worth noting that the existing spectrum of XAI methods does not cover all types of data. Thus, the authors of the survey [63] note that they did not find any works addressing the interpretability of models for data different from images, texts, and tabular data.

Together with Cynthia Rudin, we call for responsible use of ML and AI techniques and avoidance of creating black box models if they are intended to support important decisions. For solving more serious problems than classification of photos of cats and dogs, efforts should be put into creation of interpretable models. However, even when there are good reasons for creating a black box model, the model developer should not understand the term "Artificial Intelligence" literally and believe that a machine can be intelligent enough to create a good model without human control and involvement of human knowledge and intellect. Any kind of model requires thoughtful preparation of data and making a number of selections and settings, and any kind of model needs to be carefully verified, checked for sensitivity to the settings made and to variation of input data, and compared with possible alternatives. Visual analytics approaches are particularly relevant to these activities.

13.8 General principles of thoughtful model building with visual analytics

The main and most general principle to obey in model building is that all tasks involved in the model building process need to be done thoughtfully and responsibly. We have listed these tasks in Section 13.3 and discussed in Section 13.4 how visual analytics can help a thoughtful modeller to fulfil these tasks. Here we would like to attract your special attention to several activities that are not always considered by model builders.

One of the activities that may not come to the model builder's mind is *decomposition of the overall modelling problem into sub-problems*. A combination of several partial models may perform much better than a single global model; moreover, several partial models may be simpler and easier to understand than a single model intended to cover everything. Very often appropriate partitioning can be suggested by domain knowledge or even common sense. Thus, it is well known that in the life and activities of people weekdays differ from weekends and summers from winters. The example in Section 13.2 demonstrated how a modelling problem could be aptly decomposed by taking these differences into account. Essential differences that can motivate creation of several sub-models exist also in the geographic space: cities differ from the countryside, coastal areas from inland, and mountains from plains. Among people, there may be differing subgroups that cannot be adequately represented in a single universal model. Many other examples can be added to this list.

Essential differences between parts of the subject that is modelled can be known in advance or expected, or they can be discovered in the process of data analysis prior to model building. In any case, the model builder needs to have a careful look at the data to decide whether partitioning is appropriate and, if so, to define the sub-models to be built and the portions of the data to be used for building and testing them.

While the task of model testing and evaluation is not likely to be omitted, modellers very often rely too much on the statistical metrics of model quality and may not investigate the model behaviour in sufficient detail. As we noted several times, statistical metrics do not tell what and where is wrong and do not give the model builder any hint of how the model can be improved. We demonstrated by examples how useful is *visualisation of the distribution of the model results and/or model errors* over the set of available data and over its components, such as domains of different attributes and time (the same refers to space). The reason is that a model may perform differently for different parts of the data. Thus, in predicting a numeric value, a model can underestimate it for one part of data and overestimate for another, but the overall statistics may look OK; particularly, the statistical distribution of the residuals may appear close to normal. For classification models, visualisation of the distribution of the model results may reveal errors in labelling, insufficient distinctiveness of the features, insufficiently representative set of examples for training, or low confidence in assigning items to classes. Information provided by visual displays of model results or errors distributions can be enlightening and suggestive of suitable ways towards model improvement. One of possible ways may be problem decomposition, as discussed above.

Comparison of models is also a task that can hardly be neglected: since a model needs to be iteratively evaluated and improved, it is necessary at least to compare the next, supposedly improved, version of the model with the previous one. It is also often reasonable to create several variants of models using different methods or different parameter settings. Such variants also need to be compared. What we said earlier about model evaluation also applies to model comparison. Comparing only statistical indicators of model quality is, generally, not sufficient. Even statistical graphs, such as ROC (Fig. 13.11) or Q-Q plots (Fig. 13.18, right), that seem to provide more information than just numbers, do not tell how the models perform on different subsets of inputs or how much uncertainty is in their results. More useful information can be gained from visual comparison of two distributions, which can be juxtaposed, as in Fig. 13.16, or superposed in the same display, as in Fig. 13.18, left. Visualisation of the distribution of the differences between the model results or residuals, as in Fig. 13.9, right, may be very helpful.

These considerations can be briefly summarised in the following principles:

- Do not expect that the machine is more intelligent than you; apply your knowledge and reasoning using visualisations that provide you relevant food for thought.

- Do not rely on overall indicators and statistical summaries; see and inspect details (not only the devil may be there but also the angel suggesting you what to do).

13.9 Questions

- Why does a computer model need to be built in an iterative process?
- What are the general tasks in model building?
- Why aren't statistical summaries and numeric metrics of model performance sufficient for evaluating and improving a model?
- What needs to be visualised for supporting the different tasks along the model building process?
- What approaches exist for supporting model comparison?

Chapter 14
Conclusion

Abstract This chapter very briefly summarises the main ideas and principles of visual analytics, while the main goal is to show by example how to devise new visual analytics approaches and workflows using general techniques of visual analytics: abstraction, decomposition, selection, arrangement, and visual comparison. We take an example of an analysis scenario where the standard approaches presented earlier in this book do not work, and we demonstrate the possibility to construct a suitable new approach starting with abstract operations and then inventively elaborating these abstractions according to the specifics of the data and analysis tasks.

14.1 What you have learned about visual analytics

In the context of data science, visual analytics is used for (1) data exploration and assessment of their fitness to purpose; (2) gaining understanding of a piece of real world reflected in data; (3) conscious derivation of good and trustable computational models. Visualisation is the most effective way for conveying information to human mind and promoting cognition, but it has its principles that must be followed to avoid being mislead. Visualisation methods need to be carefully chosen depending on (a) the structure and properties of data and (b) the goals of analysis. Interaction techniques complement visualisation enabling seeing data from different perspectives, digging into detail, or focusing on relevant portions. Visual analytics approaches combine visualisation and interaction with the power of computational processing thereby enabling effective division of labour and synergistic cooperation of the human and the computer for appropriate data analysis and problem solving, in which each partner can employ its unique capabilities.

N. Andrienko et al., *Visual Analytics for Data Scientists*,
https://doi.org/10.1007/978-3-030-56146-8_14

14.2 Visual analytics way of thinking

We summarise the principles of the visual analytics way of thinking:

- You (as a human analyst) have the leading role in analysis and the responsibility for its result. It is you, not the computer, who gains understanding and constructs new knowledge.

- The computer is your assistant, who needs to be oriented, steered, and helped with your knowledge and experience. Whatever comes from the computer must be checked, as well as the assumptions of the analysis methods run by the computer.

- You look carefully at different aspects of data, understand their properties, and gain awareness of existing quality issues before trying to apply any processing.

- You are mindful of intrinsic incompleteness of data with regard to the reality they reflect. You cautiously apply abstraction, generalisation, and extrapolation being aware of the initial uncertainties and their propagation and proliferation over the process of analysis.

- You do not allow yourself to be fooled with apparently obvious patterns. You always question your first impressions and try to look at the same thing differently.

- Whenever any parameter settings are involved in the analysis, you check how modification of the settings affects the outcomes.

Visualisation and interaction are your instruments with which you can fulfil these principles.

14.3 Examples in this book

In the first part of our book, we introduced several general approaches and classes of methods that are most commonly used in analytical workflows where human reasoning plays the leading role. Throughout the book, we presented numerous examples of the use of these approaches and methods. Some examples focus only on application of particular techniques. Other examples describe analytical workflows where techniques are used in combination. In such examples, we tried to present and explain the reasoning of the analysts and their decisions concerning further steps in the analysis.

How can you use these examples of analysis processes in your data science practice? A trivial way of use would be to just reproduce the presented workflows when you have very similar data and analysis tasks. However, such situations are not likely to happen frequently. Therefore, what you need to learn from the examples is how to choose suitable methods and plan your own workflows based on critical assessment of what you have (data) and what you need to do (tasks) and matching these to the capabilities and applicability conditions of the approaches and methods known to you. As one of the main differences of humans from computers is inventiveness and capability to act reasonably in new situations, you need to involve these features for creating new workflows from generic building blocks.

At the most general level, the main visual analytics techniques are **abstraction** (e.g., by grouping and aggregating), **decomposition** (e.g., by partitioning or taking complementary perspectives), **arrangement** (e.g., by spatialisation, re-ordering, or transformation of coordinates), **selection** (querying and filtering), and **visual comparison**. (e.g., by juxtaposition of several displays, superposing several information layers in one display, or by explicit visualisation of differences). Each of these basic techniques has multiple realisations in various kinds of methods, which are chosen based on the specifics of particular data and tasks. Seeing it from the opposite side, each generic technique is an abstraction of multiple specific methods used for common purposes. When you devise an approach to handling your data and achieving your analysis goals, you can begin with considering these generic techniques: how they can help you, for what sub-goals you can use them, in what sequence, how their outcomes can be used or accounted for in applying the other techniques, and so on. In this way, you create an abstract plan of your analysis. After elaborating this plan according to the particulars of your data and tasks, you will instantiate it by choosing specific methods suitable for your data and producing the types of results you need.

In the following section, we shall present one more example of analysis, where we shall emphasise the process of planning the analysis and choosing suitable techniques.

14.4 Example: devising an analytical workflow for understanding team tactics in football

The example, which is based on the paper [9], is taken from the domain of sport analytics, specifically, analysis of spatio-temporal data describing movements and events occurring in a football (a.k.a. soccer) game. The analysis workflow has been designed in a close collaboration with domain experts who defined analysis tasks, provided data sets, discussed the approach and methods, and validated findings of the study. As it is usual for specialists in different domains and, more generally, customers of data analysis, the football experts prefer simple and easily understandable methods and workflows dealing with data in transparent and reasoned ways. Visualisations play an important role as facilitators of understanding of the analysis process and results.

14.4.1 Data description and problem statement

Football[1] is one of the most popular spatio-temporal phenomena that attracts attention of billions of people. Professional football requires not only well-trained brilliant players and wise coaches but also understanding of opponent's tactics, strengths and weaknesses, as well as assessment of team's own behaviour. Not surprisingly, many professional clubs nowadays hire data scientists for analysing data. In almost all important games, the movements of the players and the ball are tracked, resulting in trajectories with high temporal and spatial resolution. In addition to automatically acquired trajectories, data about game events, such as changes of ball possession, passes, tackles, shots and goals, are collected semi-automatically or manually. An overview of methods for acquisition of football game data can be found in paper [128].

Typically, positions of football players and the ball are recorded at 25Hz frequency, i.e. 25 frames per second. For 22 players moving over 2 x 45 minutes (135,000 frames), this produces about 3,000,000 time-stamped position records. This is quite a large amount of data for analysis and finding relevant patterns. Moreover, the kinds of potentially interesting patterns are very complex. The major interest of analysis is collective behaviours of the teams and the contributions of the individual players into these collective behaviours. The behaviours include two aspects: cooperation within the teams and competition between the teams. The behaviours are not just combinations of spontaneous actions but implementations of certain intended tactics, which need to be revealed in analysis. However, it is hardly possible to describe teams' behaviours and tactics using the kinds of patterns that are typically looked for in spatial, temporal, and spatio-temporal data and that we had in our examples in Chapters 8, 9, and 10. Thus, such patterns as spatio-temporal clusters of events, groups of trajectories following similar routes, or occurrence of similar spatial situations are not relevant or not sufficient for characterising team behaviours and tactics.

14.4.2 Devising the approach

The data consist of mere positions (x- and y-coordinates) of game participants at many different time steps plus a few hundreds of records specifying the times, positions, types, and participants of elementary events. Gaining an overall picture of the behaviour of a team and understanding of the tactics definitely requires high degree of **abstraction** above the elementary level of the raw data. The abstraction can be enabled by aggregation of the data. However, it would not be appropriate to aggregate data from the whole game or to do this by equal time steps, as is typi-

---

[1] https://en.wikipedia.org/wiki/Association_football

cally done in aggregating (spatial) events or trajectories into (spatial) time series. The team's behaviour depends on the current circumstances: which team possesses the ball, where on the pitch the ball and the team players are, how many opponents are around, how they are distributed, what the current score is, how much time is left till the game end, and others. It is therefore necessary to consider separately the data from the time intervals characterised by different circumstances, which means that we need to apply **decomposition** to the data (i.e., partition the game into intervals with distinct characteristics) and to the analysis task (i.e., investigate the team behaviour in different classes of situations). We use the term 'situation' in the sense of a combination of circumstances in which players behave, and we shall use the term 'episode' to refer to all movements, actions, and events that occur during a continuous time interval. A class of situations consists of situations with similar characteristics.

We need to take into account that the total time of any football game includes quite many intervals of different duration when the game is stopped by the referee. Such situations, which are called "out of play", need to be disregarded. For this purpose, **selection** techniques can be used. Furthermore, since it may be very difficult to consider simultaneously the behaviours in all possible classes of situations, selection is also needed, as a complement to partitioning, to enable focusing on particular parts of the data one after another.

The tactics of a team is partly reflected in the team formation[2], which is the spatial arrangement of the players in the team characterised by the relative positions of the players with respect to each other. A formation is often represented (particularly, in mass media) by showing the average positions of the players on the pitch computed from all positions they had during a game. However, team tactics also involves certain changes of the formation depending on the situation. For example, a team may extend in width when possessing the ball and condense when defending. It is thus reasonable to reconstruct the formations separately for different classes of situations. Furthermore, while the players of a team are constantly moving on the pitch, they strive to keep their relative positions. Therefore, it makes sense to consider, in addition to the positions on the pitch, such an **arrangement** of the data that would be independent of the movements on the pitch and represent only the relative positions of the players with respect to their teammates, as in Fig. 9.6 and Fog. 9.7.

Naturally, since we are going to apply decomposition and consider team behaviours in different situation classes, it is necessary to apply **visual comparison** techniques to compare these behaviours and understand how the team responds to situation changes.

Hence, at the high level of abstraction, our approach to analysis will involve the following components:

---

[2] https://en.wikipedia.org/wiki/Formation_(association_football)

- **Decomposition**: division of the game into situations and episodes with distinct characteristics.

- **Selection**: disregarding irrelevant episodes and focusing on groups of episodes with particular characteristics.

- **Abstraction**: aggregation of data from groups of episodes and generation of visual summaries.

- **Arrangement**: creation of representations showing the relative players' positions irrespective of their movements on the pitch.

- **Visual comparison**: creation of visualisations showing differences between behaviours in different groups of episodes.

Our analysis plan consists of iterative selection of different groups of episodes, examination and interpretation of visual summaries of these groups of episodes in terms of the players' positions on the pitch and their arrangements within the teams, and comparison of the summaries representing different groups of episodes. By means of these operations, we hope to understand how the teams' behaviours depend on the circumstances and ultimately gain an insight into the intended tactics of the teams.

Let us now elaborate the components of our approach.

Decomposition. To do the intended decomposition, we need to define and construct the relevant attributes (features) by which situations and episodes can be characterised and distinguished. This means that our workflow needs to include *feature engineering*. Relevant features need to be derived from what we have: positions of the players and the ball and elementary events.

Selection. We need a query tool allowing selection of time intervals based on combinations of attribute values and/or occurrences of certain types of events. In terms of Section 8.4, we need a tool for conditions-based temporal filtering. Our specific requirement to this tool is to be able to ignore momentary changes of conditions, which often happen in such a highly dynamic game as football. For example, during an attack of one team, the ball may be seized for a short moment by the other team but quickly re-gained by the attacking team. Hence, we need a tool where we can set the minimal meaningful duration of an episode, so that the tool skips shorter intervals of fulfilment of query conditions and unites shorter intervals of non-fulfilment with the preceding and following intervals of query fulfilment.

Abstraction. Since we need to see the contribution of the players in the team's behaviour and implementation of the tactics, we cannot use such aggregation methods where the positions or movements of all players are aggregated all together. Thus, density fields (Fig. 10.27) and flow maps (Fig. 10.25) will not be helpful to us. A suitable approach is computation and representation of the players' average positions. Since the mean of a set of positions may be affected by outliers, it makes sense to look at the median positions. To see also the variation of the positions around the average, we can build polygons (convex hulls) enclosing certain percentages of the positions (e.g., 50% and/or 75%) taken in the order of increasing distances from the average.

Arrangement. The idea of "team space" (Fig. 9.6) is suitable for our purposes. The aggregated players' positions can be computed for the pitch space and for the team space.

Visual comparison. We need to compare arrangements (configurations made by average players' positions) in different subsets of episodes. Besides, we need to see changes in the arrangements during the episode development, i.e., along a sequence of relative time steps with respect to the episode starts or ends. Simple juxtaposition of multiple images may not be very effective for detection and assessment of the differences and changes. It is preferable to have the differences or changes shown explicitly.

With the elaborated specification of the techniques we want to use, we proceed to the selection of specific methods and tools.

14.4.3 Choosing methods and tools

To enable **partitioning**, we need to have a number of relevant features for characterising situations. Only a few important features are available in the original data, namely, the attributes specifying whether the ball is in or out of game and which team is in possession, and also the event occurrences. We thus compute for each frame additional attributes, such as the distances of the ball and the team centres to each of the goals, teams' widths and depths, counts of team players in different zones on the pitch, etc.

To enable **selection**, we need a tool for condition-based temporal querying or filtering. We shall use the interactive temporal filtering tool called *Time Mask* [11, 9] that provides the required functionality. This is not the only possible way to implement these functions; similar operations could also be done in other ways. What is convenient with the Time Mask is that includes not only interactive controls for making queries but also a temporal display of the features that can be used for setting query conditions, and it shows the selected time intervals in this display. Figure 14.1 demonstrates the appearance and the use of the Time Mask interface. In the first step, we exclude the "out-of-play" situations by interacting with the upper section of the temporal display, which represents the game status (the blue colour encodes "in play" and grey "out of play"). In the second step, we interact with the second from top section of the display containing a bar with red and yellow segments corresponding to ball possession by two teams. We select the time steps (frames) with the

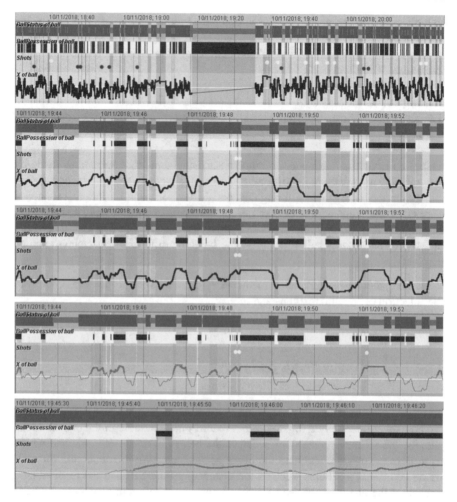

Fig. 14.1: An incremental process of query construction in Time Mask. The horizontal dimension represents time. In the vertical dimension, the display is divided into sections. Each section shows the variation of values of an attribute or a sequence of events. Categorical attributes are represented by segmented bars, the values being encoded in segment colours. Numeric attributes are represented by line plots. Events are represented by dots coloured according to event categories. The yellow vertical stripes mark the time intervals satisfying query conditions. The blue vertical stripes mark the time selected after applying a duration threshold and/or extension (or reduction, or shifting) of the time intervals.

ball possession by one of the teams. Next, we set a duration threshold to disregard episodes shorter than 1 second. Then, we add a new query condition that the ball must be in the one third of the pitch adjacent to the goal of the defending team. Finally, we extend the intervals selected by the query by adding 1 second before each of them. In the lower image, zooming in the temporal dimension has been applied to show in more detail how the selected intervals are marked in the display. The semi-transparent painting in yellow colour marks all intervals that satisfy query conditions in terms of features. The semi-transparent painting in light blue marks the selected time intervals after applying the duration threshold and time extension.

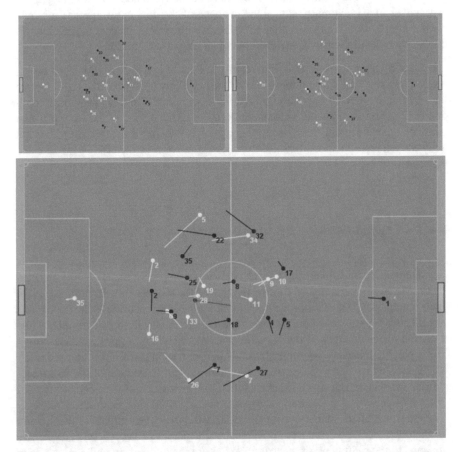

Fig. 14.2: Average positions of the players and the ball (in blue) under the ball possession of the red team (top left) and yellow team (top right). The lower image shows the changes of the average positions due to the changes of the ball possession.

As we discussed previously, the technique for **abstraction** is aggregation of players' positions from selected episodes. The aggregates, including the average (median) positions and, optionally, polygons around them indicating the position variation, as discussed previously, are shown on maps of the pitch and the team space, which implements the general technique of **arrangement**. For the sake of simplicity, we do not include the polygons in the following illustrations. The **visual comparison** of two or more combinations of players' position aggregate is supported by superposing the combinations in the same visual display and connecting the average positions of each player by lines. Figure 14.2 demonstrates this idea by example of differences due to changes of the ball possession (after excluding the "out-of-play" intervals). The upper two images in Fig. 14.2 show the average positions of the players and the ball under the possession of the red team (left) and yellow team (right). The lower image explicitly shows the changes of the average positions from one subset of episodes to another. Note that the total number of players in each team is higher than the obviously expected value of 11 due to substitutions.

The idea of superposing and linking combinations of average positions can be extended to more than two combinations. This will allow us to trace the development of a group of selected episodes by creating a sequence of aggregates corresponding to consecutive time steps along the duration of the episodes, for example, first second, second second, third second, and so on. We shall thus be able to see not only the average positions but also the average movement patterns in the selected episodes.

14.4.4 Implementing the analysis plan

To demonstrate an implementation of our analysis plan by means of the selected techniques, we shall describe an investigation of the red team's behaviour related to their long passes that advance the pitch for at least 20 meters towards the opponent's goal. The 20 metres threshold has been selected after considering the frequency distribution of the changes of the X-coordinates of the ball in all passes made during the game. By means of condition-based temporal filtering, we select all time intervals including the long passes of the red team. Visual inspection of the ball traces during the passes informs us that they belong to two different categories: passes from the vicinity of the team's own goal, mostly originating from the goalkeeper or central defenders, and attacking passes directed from the midfield to the vicinity of the opponents' goal. Interestingly, most of the passes of the second category are directed to the right wing of the team. We discard the intervals with the passes from the first category and focus on the attacking long balls. Next, we make a series of temporal queries for considering the development of the pass episodes. Specifically, we select consecutive time steps of 2 seconds length relative to the beginning of the pass episodes: the initial 2 seconds of the pass, the following 2 seconds, and so on. For each step, the average positions of the players and the ball are computed and visualised. The consecutive positions are connected by lines, as shown in Fig. 14.3. These lines resemble trajectories and can be called pseudo-trajectories. The time references in the pseudo-trajectories are relative times with respect to the episode starts, and the positions are abstractions (aggregates) of corresponding positions

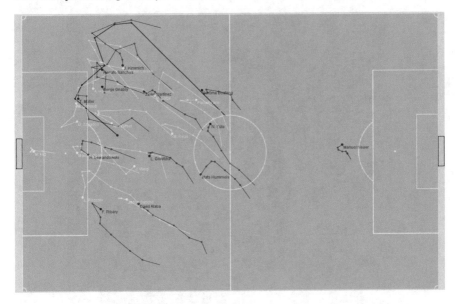

Fig. 14.3: Coloured lines depict pseudo-trajectories made of sequences of average positions of the players of two teams and the ball during seven two-seconds time steps starting from the beginnings of the game episodes containing long forward passes.

from multiple episodes. Hence, the pseudo-trajectories give us an abstraction of the players' and ball movements in multiple selected episodes.

We see that the pseudo-trajectories of the key players who were positioned close to the ball are very similar in their starting phases but then diverge as the ball turns and moves closer to the goal after the pass is completed. Obviously, the players spread themselves in front of the opponents' goal taking suitable positions for developing the attacks.

Now we want to consider the net displacements of the players during the times of the passes only. We adjust the Time Mask filter accordingly and visualise the vectors connecting the original and final abstracted positions of the players and the ball (Fig. 14.4). The visualisation shows us good coordination between the back-side players of the red team who consistently moved to the right wing and forward. The front-side players did not move in exactly the same direction but slightly diverged to be better distributed in front of the opponents' goal. The rightmost player moved almost twice as far as the others, aiming at receiving the passes. The other front-side players, evidently, were trying to position themselves suitably for receiving the following passes from this player. As to the players of the yellow team, we see that they moved faster than the players of the red team (the yellow lines are longer) attempting to intercept or hinder the expected following passes.

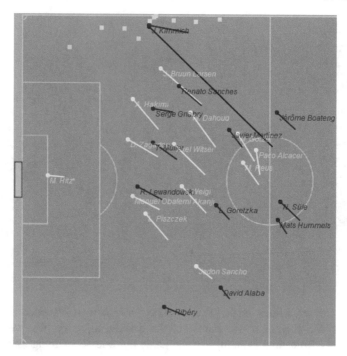

Fig. 14.4: Abstracted vectors of the displacements of the players and the ball from the start to the end moments of the long forward passes of the red team. Like in the previous figures, there are more than 11 lines in each team due to substitutions. The destinations of the are marked by cyan dots.

To see the relative displacements of the players of the two teams, we also look at the similar summaries of the long-ball episodes presented in the relative spaces of the teams. Figure 14.5 shows the displacement vectors of the players and the ball in the space of the defending yellow team. The vertical axis corresponds to the direction from the goal of the yellow team (bottom) to the goal of the red team (top). This representation shows us an interesting pattern that all long balls of the red team were directed into a small area in the space of the yellow team. A possible reason was that the red team was aware of a certain weakness of the left wing of the yellow team's defence. It seems to be a part of the red team's tactics to send repeatedly long balls in the direction where they expected to have more chances for successful reception of the pass and further development of the attack. Perhaps, to implement this tactics, the red team's coach assigned one of the fastest runners of the team to play on the right wing. The team space display also shows us that the remaining players of the red team were effectively increasing the area covered by their team and spreading themselves among the opponents.

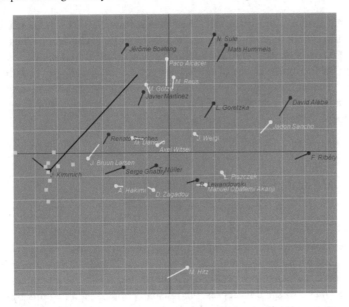

Fig. 14.5: The abstracted displacement vectors, as in Fig. 14.4, are shown in the team space of the defending yellow team.

Here we have described the investigation into the team behaviours in tactics in one class of episodes. Examples of analyses focusing on other classes of episodes can be seen in the paper [9] from which the example has been taken.

14.4.5 Conclusion

The analytical workflow in this example did not include any advanced computational methods but only easily understandable and implementable techniques. Like many other examples in this book, it demonstrates that human reasoning is the main instrument of analysis. This statement remains valid also when more sophisticated methods of computational processing are used. However, the use of simpler methods, when possible (i.e., does not increase substantially the human's workload), gives several advantages. You can better understand what is done and what is obtained, and it will be easier for you to explain this to others. You can easier validate the results and convince others of their validity.

With this example, we primarily aimed to demonstrate not the analysis process as such but the process of finding an approach to the analysis. We intentionally took an analysis scenario in which we could not re-apply any of the analytical workflows described earlier in the book. Moreover, the common methods that are usually applied to this kind of data (i.e., trajectories; see Section 10.3.3) were not suitable for this scenario. In these settings, we wanted to show that new approaches to analysis can be devised by inventive application of general ideas and principles of visual analytics. Inventiveness is one of the key aspects of the superiority of humans over machines. The philosophy of visual analytics assumes that humans use their advantages and do interesting part of the work requiring creative thinking, while the routine, tedious processing is left to computers.

14.5 Final remarks

Certainly, this book does not present visual analytics science in all its width and depth. We introduced the fundamental concepts and ideas, and we made a selection of the methods that we deemed sufficiently general and widely applicable, easy to understand, and easy to reproduce. We hope that this book convinces you how exciting is can be to see data represented visually and to gain insights by means of the great power of the human brain. We hope that you will love visual analytics, try to learn more from it, and keep trace of new developments in this area.

Glossary

Base of a distribution is a component (e.g., space, time, set of entities) or a composition of components (e.g., space+time) viewed as a container or carrier for elements of other components. See section 2.3.1.

Centroid of a set of data items consisting of values of multiple numeric attributes is the combination of the means of the values of the attributes. It is common to use centroids of *clusters* as representing the characteristics of the cluster members.

Cluster A group of objects that have similar properties or close positions in space and/or time.

Clustering is a class of computational analysis methods aiming to produce sets of similar items where the items in the same group share common characteristics that make them distinguishable from items in the other sets. See section 4.5.

Computer model : In the context of this book, this term refers to any kind of model (e.g., statistical model, simulation model, machine learning model, etc.) that is meant to be executed by computers, typically for the purpose of prediction.

Distance function A method that, for two given items, expresses the degree of their relatedness or similarity by a numeric value such that a smaller value means higher similarity or relatedness. Distance functions are used for clustering, projection, search by similarity, detection of frequent patterns, etc. Other names are **similarity function** and **dissimilarity function**. See section 4.2.

Distribution is a set of correspondences between elements of two or more components of a subject, such that one component or a composition of components is treated as a **base** (i.e., a container or a carrier) for elements of other components, which are called **overlays** over the base. For example, space and time can be seen

© Springer Nature Switzerland AG 2020
N. Andrienko et al., *Visual Analytics for Data Scientists*,
https://doi.org/10.1007/978-3-030-56146-8

as bases (i.e., containers) of entities and values of attributes, and a set of entities can be seen as a base (i.e., a carrier) of attribute values. In this view, the elements of the base play the role of positions that can be filled by elements of the overlay. Then, the *distribution* of a component O over (or in) another component B viewed as a base is the relationship between the elements of O and the positions in B, that is, which positions in B are occupied by which elements of O. See section 2.3.1.

Heatmap : A display, such as a matrix or a map, where the display space is divided into uniform compartments, in which values of a numeric attribute are represented by colours or degrees of lightness.

Interrelation between two or more components distributed over the same **base** means a tendency of particular elements of different components to co-occur at the same or close positions in the base. Correlation (in the statistical sense) is a special case of interrelation. See section 2.3.4.

Medoid of a set or *cluster* of data items of any nature is the data item having the smallest sum of the distances (in terms of a certain *distance function*) to all other data items.

Outlier is an item that cannot be put in a group together with any other items because of being very different from everything else. See section 2.3.2.

Overlay of a distribution is the component (e.g., a set of entities or an attribute) or a combination of components (e.g., multiple attributes) whose elements (entities, attribute values, or combinations of values) are distributed over the **base of the distribution**. See section 2.3.1.

Pattern is a group of items of any kind that can be perceived or represented as an integrated whole having its specific properties that are not mere compositions of properties of the constituent items. See section 2.3.2.

Space-time cube (STC) A visual display providing a perspective view of a 3D scene where the horizontal plane represents the geographic space (or another kind of 2D space) and the vertical dimension represents time. Spatio-temporal data are represented by symbols or lines positioned in the cube according to the spatial and temporal references of the data.

Spatialisation refers to arranging visual objects within the display space in such a way that the distances between them reflect the degree of similarity or relatedness between the data items they represent. Spatialisation exploits the innate capability of humans to perceive multiple things located closely as being united in a larger shape or structure, that is, in some pattern. It also exploits the intuitive perception of spatially close things to be more related than distant things. See section 2.3.5.

Subject of analysis A thing or phenomenon that needs to be studied and modelled. A subject can be seen as a system consisting of **components**, or **aspects**, linked by **relationships**. See section 2.1.

Variation pattern in a distribution is a category of patterns consisting of the similarity-difference relationships between the elements of the overlay corresponding to different elements of the base. See Section 2.3.3.

Visual variables are aspects of a display that can be used for visual representation of information. These include x-position, y-position, width, height, size, hue, lightness, saturation, orientation, etc. See section 3.2.2.

References

1. IEEE VAST Challenge 2011. `http://hcil.cs.umd.edu/localphp/hcil/vast11/`, last accessed 03/22/2018 (2011)
2. Abello, J., Hadlak, S., Schumann, H., Schulz, H.: A modular degree-of-interest specification for the visual analysis of large dynamic networks. IEEE Transactions on Visualization and Computer Graphics **20**(3), 337–350 (2014). DOI 10.1109/TVCG.2013.109
3. Agresti, A., Finlay, B.: Statistical Methods for the Social Sciences. Pearson Prentice Hall (2007)
4. Alexander, E., Kohlmann, J., Valenza, R., Witmore, M., Gleicher, M.: Serendip: Topic model-driven visual exploration of text corpora. In: 2014 IEEE Conference on Visual Analytics Science and Technology (VAST), pp. 173–182. IEEE (2014)
5. Alsallakh, B., Aigner, W., Miksch, S., Gröller, M.E.: Reinventing the contingency wheel: Scalable visual analytics of large categorical data. IEEE Transactions on Visualization and Computer Graphics **18**(12), 2849–2858 (2012). DOI 10.1109/TVCG.2012.254
6. Amershi, S., Chickering, D.M., Drucker, S.M., Lee, B., Simard, P.Y., Suh, J.: Modeltracker: Redesigning performance analysis tools for machine learning. In: Proceedings of the 33rd Annual ACM Conference on Human Factors in Computing Systems, CHI '15, pp. 337–346. ACM, New York, NY, USA (2015). DOI 10.1145/2702123.2702509
7. Andrienko, G., Andrienko, N.: Interactive maps for visual data exploration. International Journal of Geographical Information Science **13**(4), 355–374 (1999). DOI 10.1080/136588199241247
8. Andrienko, G., Andrienko, N.: Exploring spatial data with dominant attribute map and parallel coordinates. Computers, Environment and Urban Systems **25**(1), 5 – 15 (2001). DOI https://doi.org/10.1016/S0198-9715(00)00037-5. GISRUK 2000
9. Andrienko, G., Andrienko, N., Anzer, G., Bauer, P., Budziak, G., Fuchs, G., Hecker, D., Weber, H., Wrobel, S.: Constructing spaces and times for tactical analysis in football. IEEE Transactions on Visualization and Computer Graphics pp. 1–1 (2019). DOI 10.1109/TVCG.2019.2952129
10. Andrienko, G., Andrienko, N., Bak, P., Keim, D., Wrobel, S.: Visual Analytics of Movement. Springer-Verlag Berlin Heidelberg (2013)
11. Andrienko, G., Andrienko, N., Budziak, G., Dykes, J., Fuchs, G., von Landesberger, T., Weber, H.: Visual analysis of pressure in football. Data Mining and Knowledge Discovery **31**(6), 1793–1839 (2017). DOI 10.1007/s10618-017-0513-2
12. Andrienko, G., Andrienko, N., Fuchs, G.: Understanding movement data quality. Journal of Location Based Services **10**(1), 31–46 (2016). DOI 10.1080/17489725.2016.1169322
13. Andrienko, G., Andrienko, N., Fuchs, G., Wood, J.: Revealing patterns and trends of mass mobility through spatial and temporal abstraction of origin-destination movement data. IEEE Transactions on Visualization and Computer Graphics **23**(9), 2120–2136 (2017). DOI 10.1109/TVCG.2016.2616404
14. Andrienko, G., Andrienko, N., Hurter, C., Rinzivillo, S., Wrobel, S.: Scalable analysis of movement data for extracting and exploring significant places. IEEE Transactions on Visualization and Computer Graphics **19**(7), 1078–1094 (2013). DOI 10.1109/TVCG.2012.311
15. Andrienko, G., Andrienko, N., Rinzivillo, S., Nanni, M., Pedreschi, D., Giannotti, F.: Interactive visual clustering of large collections of trajectories. In: 2009 IEEE Symposium on Visual Analytics Science and Technology, pp. 3–10 (2009). DOI 10.1109/VAST.2009.5332584
16. Andrienko, N., Andrienko, G.: Informed spatial decisions through coordinated views. Information Visualization **2**(4), 270–285 (2003). DOI 10.1057/palgrave.ivs.9500058
17. Andrienko, N., Andrienko, G.: Exploratory analysis of spatial and temporal data: a systematic approach. Springer Science & Business Media (2006)
18. Andrienko, N., Andrienko, G.: A visual analytics framework for spatio-temporal analysis and modelling. Data Mining and Knowledge Discovery **27**(1), 55–83 (2013). DOI 10.1007/s10618-012-0285-7

19. Andrienko, N., Andrienko, G.: State transition graphs for semantic analysis of movement behaviours. Information Visualization **17**(1), 41–65 (2018). DOI 10.1177/1473871617692841
20. Andrienko, N., Andrienko, G., Barrett, L., Dostie, M., Henzi, P.: Space transformation for understanding group movement. IEEE Transactions on Visualization and Computer Graphics **19**(12), 2169–2178 (2013). DOI 10.1109/TVCG.2013.193
21. Andrienko, N., Andrienko, G., Fuchs, G., Jankowski, P.: Scalable and privacy-respectful interactive discovery of place semantics from human mobility traces. Information Visualization **15**(2), 117–153 (2016). DOI 10.1177/1473871615581216
22. Andrienko, N., Andrienko, G., Fuchs, G., Rinzivillo, S., Betz, H.: Detection, tracking, and visualization of spatial event clusters for real time monitoring. In: 2015 IEEE International Conference on Data Science and Advanced Analytics (DSAA), pp. 1–10 (2015). DOI 10.1109/DSAA.2015.7344880
23. Andrienko, N., Andrienko, G., Garcia, J.M.C., Scarlatti, D.: Analysis of flight variability: a systematic approach. IEEE Transactions on Visualization and Computer Graphics **25**(1), 54–64 (2019). DOI 10.1109/TVCG.2018.2864811
24. Andrienko, N., Andrienko, G., Stange, H., Liebig, T., Hecker, D.: Visual analytics for understanding spatial situations from episodic movement data. KI - Künstliche Intelligenz **26**(3), 241–251 (2012). DOI 10.1007/s13218-012-0177-4
25. Angelini, M., Santucci, G., Schumann, H., Schulz, H.J.: A review and characterization of progressive visual analytics. Informatics **5**(3) (2018). DOI 10.3390/informatics5030031
26. Arnheim, R.: Visual Thinking. London: Faber (1969)
27. Arun, R., Suresh, V., Veni Madhavan, C.E., Narasimha Murthy, M.N.: On finding the natural number of topics with latent dirichlet allocation: Some observations. In: M.J. Zaki, J.X. Yu, B. Ravindran, V. Pudi (eds.) Advances in Knowledge Discovery and Data Mining, pp. 391–402. Springer Berlin Heidelberg, Berlin, Heidelberg (2010)
28. Bach, B., Shi, C., Heulot, N., Madhyastha, T., Grabowski, T., Dragicevic, P.: Time curves: Folding time to visualize patterns of temporal evolution in data. IEEE Transactions on Visualization and Computer Graphics **22**(1), 559–568 (2016). DOI 10.1109/TVCG.2015.2467851
29. Bazán, E., Dokládal, P., Dokládalová, E.: Quantitative Analysis of Similarity Measures of Distributions. In: British Machine Vision Conference (BMVC). Cardiff, United Kingdom (2019). URL https://hal-upec-upem.archives-ouvertes.fr/hal-02299826
30. Behrisch, M., Bach, B., Henry Riche, N., Schreck, T., Fekete, J.D.: Matrix reordering methods for table and network visualization. Computer Graphics Forum **35**(3), 693–716 (2016). DOI 10.1111/cgf.12935
31. Bertin, J.: Semiology of Graphics: Diagrams,Networks, Maps (1983)
32. Blei, D.M.: Probabilistic topic models. Communications of the ACM **55**(4), 77–84 (2012). DOI 10.1145/2133806.2133826
33. Blei, D.M., Ng, A.Y., Jordan, M.I.: Latent dirichlet allocation. Journal of machine Learning research **3**(Jan), 993–1022 (2003)
34. Blokker, E., Furnass, W.R., Machell, J., Mounce, S.R., Schaap, P.G., Boxall, J.B.: Relating water quality and age in drinking water distribution systems using self-organising maps. Environments **3**(2), 10 (2016)
35. Blondel, V.D., Esch, M., Chan, C., Clérot, F., Deville, P., Huens, E., Morlot, F., Smoreda, Z., Ziemlicki, C.: Data for development: the d4d challenge on mobile phone data. ArXiv **abs/1210.0137** (2012)
36. Bögl, M., Aigner, W., Filzmoser, P., Lammarsch, T., Miksch, S., Rind, A.: Visual analytics for model selection in time series analysis. IEEE transactions on visualization and computer graphics **19**, 2237–46 (2013). DOI 10.1109/TVCG.2013.222
37. Boriah, S., Chandola, V., Kumar, V.: Similarity measures for categorical data: A comparative evaluation. In: Society for Industrial and Applied Mathematics - 8th SIAM International Conference on Data Mining 2008, Proceedings in Applied Mathematics 130, Society for Industrial and Applied Mathematics - 8th SIAM International Conference on Data Mining 2008, Proceedings in Applied Mathematics 130, pp. 243–254 (2008)

38. Brandes, U., Erlebach, T.: Network analysis: methodological foundations, vol. 3418. Springer Science & Business Media (2005)
39. Chae, J., Thom, D., Bosch, H., Jang, Y., Maciejewski, R., Ebert, D.S., Ertl, T.: Spatiotemporal social media analytics for abnormal event detection and examination using seasonal-trend decomposition. In: 2012 IEEE Conference on Visual Analytics Science and Technology (VAST), pp. 143–152 (2012). DOI 10.1109/VAST.2012.6400557
40. Chen, S., Andrienko, N., Andrienko, G., Adilova, L., Barlet, J., Kindermann, J., Nguyen, P.H., Thonnard, O., Turkay, C.: Lda ensembles for interactive exploration and categorization of behaviors. IEEE Transactions on Visualization and Computer Graphics pp. 1–1 (2019). DOI 10.1109/TVCG.2019.2904069
41. Chu, D., Sheets, D.A., Zhao, Y., Wu, Y., Yang, J., Zheng, M., Chen, G.: Visualizing hidden themes of taxi movement with semantic transformation. In: 2014 IEEE Pacific Visualization Symposium, pp. 137–144 (2014). DOI 10.1109/PacificVis.2014.50
42. Chuang, J., Manning, C.D., Heer, J.: Termite: Visualization techniques for assessing textual topic models. In: Proceedings of the International Working Conference on Advanced Visual Interfaces, pp. 74–77. ACM (2012)
43. Chuang, J., Roberts, M.E., Stewart, B.M., Weiss, R., Tingley, D., Grimmer, J., Heer, J.: Topiccheck: Interactive alignment for assessing topic model stability. In: Proceedings of the 2015 Conference of the North American Chapter of the Association for Computational Linguistics: Human Language Technologies, pp. 175–184 (2015)
44. Collins, C., Andrienko, N., Schreck, T., Yang, J., Choo, J., Engelke, U., Jena, A., Dwyer, T.: Guidance in the human–machine analytics process. Visual Informatics 2(3), 166–180 (2018)
45. Collins, C., Viegas, F., Wattenberg, M.: Parallel tag clouds to explore and analyze faceted text corpora. pp. 91 – 98 (2009). DOI 10.1109/VAST.2009.5333443
46. Cuadros, A.M., Paulovich, F.V., Minghim, R., Telles, G.P.: Point placement by phylogenetic trees and its application to visual analysis of document collections. In: Proceedings of the 2007 IEEE Symposium on Visual Analytics Science and Technology, VAST '07, p. 99–106. IEEE Computer Society, USA (2007). DOI 10.1109/VAST.2007.4389002
47. Dang, T.N., Wilkinson, L.: Scagexplorer: Exploring scatterplots by their scagnostics. In: Visualization Symposium (PacificVis), 2014 IEEE Pacific, pp. 73–80. IEEE (2014)
48. Deerwester, S., Dumais, S.T., Furnas, G.W., Landauer, T.K., Harshman, R.: Indexing by latent semantic analysis. Journal of the American society for information science 41(6), 391–407 (1990)
49. Dykes, J.A.: Exploring spatial data representation with dynamic graphics. Computers & Geosciences 23(4), 345 – 370 (1997). DOI https://doi.org/10.1016/S0098-3004(97)00009-5. Exploratory Cartograpic Visualisation
50. Eler, D., Nakazaki, M., Paulovich, F., Santos, D., Andery, G., Oliveira, M.C., Neto, J., Minghim, R.: Visual analysis of image collections. The Visual Computer 25, 923–937 (2009). DOI 10.1007/s00371-009-0368-7
51. Endert, A., Ribarsky, W., Turkay, C., Wong, B.W., Nabney, I., Blanco, I.D., Rossi, F.: The state of the art in integrating machine learning into visual analytics. In: Computer Graphics Forum, vol. 36, pp. 458–486 (2017)
52. Fisher, D.: Animation for Visualization: Opportunities and Drawbacks, beautiful visualization edn. O'Reilly Media (2010). Complete book avaialble at http://oreilly.com/catalog/0636920000617/
53. Gibson, H., Faith, J., Vickers, P.: A survey of two-dimensional graph layout techniques for information visualisation. Information Visualization 12(3-4), 324–357 (2013). DOI 10.1177/1473871612455749
54. Gleicher, M.: Explainers: Expert explorations with crafted projections. IEEE Transactions on Visualization & Computer Graphics (12), 2042–2051 (2013)
55. Gleicher, M., Albers, D., Walker, R., Jusufi, I., Hansen, C.D., Roberts, J.C.: Visual comparison for information visualization. Information Visualization 10(4), 289–309 (2011)
56. Görg, C., Kang, Y., Liu, Z., Stasko, J.: Visual analytics support for intelligence analysis. Computer 46(7), 30–38 (2013). DOI 10.1109/MC.2013.76

57. Gou, L., Zhang, X., Luo, A., Anderson, P.F.: Socialnetsense: Supporting sensemaking of social and structural features in networks with interactive visualization. In: 2012 IEEE Conference on Visual Analytics Science and Technology (VAST), pp. 133–142 (2012)

58. Gould, P.: Letting the data speak for themselves. Annals of the Association of American Geographers **71**(2), 166–176 (1981)

59. Gower, J.C.: A general coefficient of similarity and some of its properties. Biometrics **27**(4), 857–871 (1971)

60. Green, M.: Toward a perceptual science of multidimensional data visualization: Bertin and beyond. ERGO/GERO Human Factors Science **8**, 1–30 (1998)

61. Gschwandtner, T., Erhart, O.: Know your enemy: Identifying quality problems of time series data. In: 2018 IEEE Pacific Visualization Symposium (PacificVis), pp. 205–214 (2018). DOI 10.1109/PacificVis.2018.00034

62. Gschwandtner, T., Gärtner, J., Aigner, W., Miksch, S.: A taxonomy of dirty time-oriented data. In: G. Quirchmayr, J. Basl, I. You, L. Xu, E. Weippl (eds.) Multidisciplinary Research and Practice for Information Systems, pp. 58–72. Springer Berlin Heidelberg, Berlin, Heidelberg (2012)

63. Guidotti, R., Monreale, A., Ruggieri, S., Turini, F., Giannotti, F., Pedreschi, D.: A survey of methods for explaining black box models. ACM Comput. Surv. **51**(5), 93:1–93:42 (2018). DOI 10.1145/3236009

64. Guo, D., Gahegan, M., MacEachren, A.M., Zhou, B.: Multivariate analysis and geovisualization with an integrated geographic knowledge discovery approach. Cartography and geographic information science **32**(2), 113–132 (2005)

65. Gutenko, I., Dmitriev, K., Kaufman, A.E., Barish, M.A.: Anafe: Visual analytics of image-derived temporal features—focusing on the spleen. IEEE Transactions on Visualization and Computer Graphics **23**(1), 171–180 (2017). DOI 10.1109/TVCG.2016.2598463

66. Havre, S., Hetzler, B., Nowell, L.: Themeriver: visualizing theme changes over time. In: IEEE Symposium on Information Visualization 2000. INFOVIS 2000. Proceedings, pp. 115–123 (2000). DOI 10.1109/INFVIS.2000.885098

67. Höferlin, M., Höferlin, B., Heidemann, G., Weiskopf, D.: Interactive schematic summaries for faceted exploration of surveillance video. IEEE Transactions on Multimedia **15**(4), 908–920 (2013). DOI 10.1109/TMM.2013.2238521

68. Höferlin, M., Höferlin, B., Weiskopf, D., Heidemann, G.: Uncertainty-aware video visual analytics of tracked moving objects. Journal of Spatial Information Science **2011**(2), 87–117 (2011). DOI 10.5311/JOSIS.2010.2.1

69. Hofmann, T.: Probabilistic latent semantic analysis. In: Proceedings of the Fifteenth Conference on Uncertainty in Artificial Intelligence, UAI'99, pp. 289–296. Morgan Kaufmann Publishers Inc., San Francisco, CA, USA (1999)

70. Hohman, F., Kahng, M., Pienta, R., Chau, D.H.: Visual analytics in deep learning: An interrogative survey for the next frontiers. IEEE Transactions on Visualization and Computer Graphics **25**(8), 2674–2693 (2019). DOI 10.1109/TVCG.2018.2843369

71. Holten, D., Van Wijk, J.J.: Force-directed edge bundling for graph visualization. Computer Graphics Forum **28**(3), 983–990 (2009). DOI 10.1111/j.1467-8659.2009.01450.x

72. Huff, D.: How to lie with statistics. WW Norton & Company (1993)

73. Iwasokun, G.: Image enhancement methods: A review. British Journal of Mathematics & Computer Science **4**, 2251–2277 (2014). DOI 10.9734/BJMCS/2014/10332

74. Iwueze, I., Nwogu, E., Nlebedim, V., Nwosu, U., Chinyem, U.: Comparison of methods of estimating missing values in time series. Open Journal of Statistics (8), 390–399 (2018). DOI 10.4236/ojs.2018.82025

75. Jankowski, P., Andrienko, N., Andrienko, G.: Map-centred exploratory approach to multiple criteria spatial decision making. International Journal of Geographical Information Science **15**(2), 101–127 (2001)

76. Kamble, V., Bhurchandi, K.: No-reference image quality assessment algorithms: A survey. Optik **126**(11), 1090 – 1097 (2015). DOI https://doi.org/10.1016/j.ijleo.2015.02.093

77. Kandel, S., Paepcke, A., Hellerstein, J.M., Heer, J.: Enterprise data analysis and visualization: An interview study. IEEE Transactions on Visualization and Computer Graphics **18**(12), 2917–2926 (2012)
78. Keim, D., Andrienko, G., Fekete, J.D., Görg, C., Kohlhammer, J., Melançon, G.: Visual analytics: Definition, process, and challenges. In: A. Kerren, J.T. Stasko, J.D. Fekete, C. North (eds.) Information Visualization: Human-Centered Issues and Perspectives, pp. 154–175. Springer, Berlin (2008)
79. Keim, D., Kohlhammer, J., Ellis, G., Mansmann, F. (eds.): Mastering the Information Age : Solving Problems with Visual Analytics. Goslar : Eurographics Association (2010). DOI 10.2312/14803
80. Keim, D.A., Nietzschmann, T., Schelwies, N., Schneidewind, J., Schreck, T., Ziegler, H.: A spectral visualization system for analyzing financial time series data. In: EuroVis 2006 – Eurographics /IEEE VGTC Symposium on Visualization, pp. 195–202. Eurographics Association (2006). DOI 10.2312/VisSym/EuroVis06/195-202
81. Keogh, E., Kasetty, S.: On the need for time series data mining benchmarks: A survey and empirical demonstration. Data Min. Knowl. Discov. **7**(4), 349–371 (2003). DOI 10.1023/A: 1024988512476
82. Kim, M., Kang, K., Park, D., Choo, J., Elmqvist, N.: Topiclens: Efficient multi-level visual topic exploration of large-scale document collections. IEEE Transactions on Visualization and Computer Graphics **23**(1), 151–160 (2017). DOI 10.1109/TVCG.2016.2598445
83. Klemm, P., Oeltze-Jafra, S., Lawonn, K., Hegenscheid, K., Völzke, H., Preim, B.: Interactive visual analysis of image-centric cohort study data. IEEE Transactions on Visualization and Computer Graphics **20** (2014). DOI 10.1109/TVCG.2014.2346591
84. Kohonen, T., Schroeder, M.R., Huang, T.S.: Self-Organizing Maps, 3rd edn. Springer-Verlag, Berlin, Heidelberg (2001)
85. Krause, J., Dasgupta, A., Fekete, J.D., Bertini, E.: Seekaview: an intelligent dimensionality reduction strategy for navigating high-dimensional data spaces. In: Large Data Analysis and Visualization (LDAV), 2016 IEEE 6th Symposium on, pp. 11–19. IEEE (2016)
86. Kriegel, H.P., Kröger, P., Zimek, A.: Subspace clustering. WIREs Data Mining and Knowledge Discovery **2**(4), 351–364 (2012). DOI 10.1002/widm.1057
87. Kruskal, J.: Multidimensional scaling by optimizing goodness of fit to a nonmetric hypothesis. Psychometrika **29**(1), 1–27 (1964)
88. Kucher, K., Kerren, A.: Text visualization techniques: Taxonomy, visual survey, and community insights. In: Visualization Symposium (PacificVis), 2015 IEEE Pacific, pp. 117–121. IEEE (2015)
89. Kucher, K., Paradis, C., Kerren, A.: The state of the art in sentiment visualization. Computer Graphics Forum **37**(1), 71–96 (2018). DOI 10.1111/cgf.13217
90. von Landesberger, T., Kuijper, A., Schreck, T., Kohlhammer, J., van Wijk, J., Fekete, J.D., Fellner, D.: Visual analysis of large graphs: State-of-the-art and future research challenges. Computer Graphics Forum **30**(6), 1719–1749 (2011). DOI 10.1111/j.1467-8659.2011. 01898.x
91. Liu, S., Andrienko, G., Wu, Y., Cao, N., Jiang, L., Shi, C., Wang, Y.S., Hong, S.: Steering data quality with visual analytics: The complexity challenge. Visual Informatics **2**(4), 191 – 197 (2018). DOI https://doi.org/10.1016/j.visinf.2018.12.001
92. Lundberg, S.M., Lee, S.I.: A unified approach to interpreting model predictions. In: Proceedings of the 31st International Conference on Neural Information Processing Systems, NIPS'17, p. 4768–4777. Curran Associates Inc., Red Hook, NY, USA (2017)
93. MacEachren, A.M.: How maps work: representation, visualization, and design. Guilford Press (1995)
94. Maini, R., Aggarwal, H.: A comprehensive review of image enhancement techniques. Journal of Computing **2**(3) (2010)
95. Malczewski, J.: GIS and multicriteria decision analysis. John Wiley & Sons (1999)
96. Marey, É.J.: La méthode graphique dans les sciences expérimentales et principalement en physiologie et en médecine. G. Masson (1885)

97. Matejka, J., Fitzmaurice, G.: Same stats, different graphs: generating datasets with varied appearance and identical statistics through simulated annealing. In: Proceedings of the 2017 CHI Conference on Human Factors in Computing Systems, pp. 1290–1294. ACM (2017)

98. Meghdadi, A.H., Irani, P.: Interactive exploration of surveillance video through action shot summarization and trajectory visualization. IEEE Transactions on Visualization and Computer Graphics 19(12), 2119–2128 (2013). DOI 10.1109/TVCG.2013.168

99. Migut, M., Worring, M.: Visual exploration of classification models for risk assessment. pp. 11–18 (2010). DOI 10.1109/VAST.2010.5652398

100. Monmonier, M.: How to lie with maps. University of Chicago Press (1996)

101. Monroe, M., Lan, R., Lee, H., Plaisant, C., Shneiderman, B.: Temporal event sequence simplification. IEEE Transactions on Visualization and Computer Graphics 19(12), 2227–2236 (2013). DOI 10.1109/TVCG.2013.200

102. Mühlbacher, T., Piringer, H.: A partition-based framework for building and validating regression models. IEEE Transactions on Visualization and Computer Graphics 19, 1962–1971 (2013)

103. Munzner, T.: Visualization analysis and design. CRC press (2014)

104. Nam, J.E., Mueller, K.: Tripadvisor^{ND}: A tourism-inspired high-dimensional space exploration framework with overview and detail. IEEE transactions on visualization and computer graphics 19(2), 291–305 (2013)

105. Oelke, D., Hao, M., Rohrdantz, C., Keim, D.A., Dayal, U., Haug, L., Janetzko, H.: Visual opinion analysis of customer feedback data. In: 2009 IEEE Symposium on Visual Analytics Science and Technology, pp. 187–194 (2009). DOI 10.1109/VAST.2009.5333919

106. Oelke, D., Spretke, D., Stoffel, A., Keim, D.A.: Visual readability analysis: How to make your writings easier to read. IEEE Transactions on Visualization and Computer Graphics 18(5), 662–674 (2012). DOI 10.1109/TVCG.2011.266

107. Openshaw, S.: Ecological fallacies and the analysis of areal census data. Environment and planning A 16(1), 17–31 (1984)

108. Paatero, P., Tapper, U.: Positive matrix factorization: A non-negative factor model with optimal utilization of error estimates of data values. Environmetrics 5(2), 111–126 (1994). DOI 10.1002/env.3170050203

109. Pelekis, N., Andrienko, G., Andrienko, N., Kopanakis, I., Marketos, G., Theodoridis, Y.: Visually exploring movement data via similarity-based analysis. Journal of Intelligent Information Systems 38(2), 343–391 (2012). DOI 10.1007/s10844-011-0159-2

110. Pelleg, D., Moore, A.: X-means: Extending k-means with efficient estimation of the number of clusters. In: Proceedings of the 17th International Conference on Machine Learning, pp. 727–734. Morgan Kaufmann (2000)

111. Radoš, S., Splechtna, R., Matković, K., Đuras, M., Gröller, E., Hauser, H.: Towards quantitative visual analytics with structured brushing and linked statistics. Computer Graphics Forum 35(3), 251–260 (2016). DOI 10.1111/cgf.12901

112. Ralanamahatana, C.A., Lin, J., Gunopulos, D., Keogh, E., Vlachos, M., Das, G.: Mining Time Series Data, pp. 1069–1103. Springer US, Boston, MA (2005). DOI 10.1007/0-387-25465-X_51

113. Ratanamahatana, C., Keogh, E.: Everything you know about dynamic time warping is wrong. In: Third Workshop on Mining Temporal and Sequential Data (2004)

114. Ray, C., Dreo, R., Camossi, E., Jousselme, A.L.: Heterogeneous Integrated Dataset for Maritime Intelligence, Surveillance, and Reconnaissance (2018). DOI 10.5281/zenodo.1167595

115. Ren, D., Amershi, S., Lee, B., Suh, J., Williams, J.D.: Squares: Supporting interactive performance analysis for multiclass classifiers. IEEE Transactions on Visualization and Computer Graphics 23(1), 61–70 (2017). DOI 10.1109/TVCG.2016.2598828

116. Rinzivillo, S., Pedreschi, D., Nanni, M., Giannotti, F., Andrienko, N., Andrienko, G.: Visually driven analysis of movement data by progressive clustering. Information Visualization 7(3-4), 225–239 (2008). DOI 10.1057/PALGRAVE.IVS.9500183

117. Rosenberg, D., Grafton, A.: Cartographies of time: A history of the timeline. Princeton Architectural Press (2013)

118. Rousseeuw, P.J.: Silhouettes: A graphical aid to the interpretation and validation of cluster analysis. Journal of Computational and Applied Mathematics **20**, 53 – 65 (1987). DOI https://doi.org/10.1016/0377-0427(87)90125-7

119. Rudin, C.: Stop explaining black box machine learning models for high stakes decisions and use interpretable models instead. Nature Machine Intelligence **1**, 206–215 (2019). DOI 10.1038/s42256-019-0048-x

120. Sammon, J.W.: A nonlinear mapping for data structure analysis. IEEE Transactions on Computers **18**(5), 401–409 (1969). DOI 10.1109/T-C.1969.222678

121. Sankararaman, S., Agarwal, P.K., Mølhave, T., Pan, J., Boedihardjo, A.P.: Model-driven matching and segmentation of trajectories. In: Proceedings of the 21st ACM SIGSPA-TIAL International Conference on Advances in Geographic Information Systems, SIGSPA-TIAL'13, p. 234–243. Association for Computing Machinery, New York, NY, USA (2013). DOI 10.1145/2525314.2525360

122. Seo, J., Shneiderman, B.: A rank-by-feature framework for unsupervised multidimensional data exploration using low dimensional projections. In: Information Visualization, 2004. INFOVIS 2004. IEEE Symposium on, pp. 65–72. IEEE (2004)

123. Shearer, C.: The CRISP-DM Model: The new blueprint for data mining. Journal of Data Warehousing **5**(4), 13–22 (2000)

124. Shneiderman, B.: The Eyes Have It: A Task by Data Type Taxonomy for Information Visualization. In: Proceedings of the IEEE Symposium on Visual Languages, pp. 336–343 (1996). DOI 10.1109/VL.1996.545307

125. Sips, M., Köthur, P., Unger, A., Hege, H., Dransch, D.: A visual analytics approach to multiscale exploration of environmental time series. IEEE Transactions on Visualization and Computer Graphics **18**(12), 2899–2907 (2012). DOI 10.1109/TVCG.2012.191

126. Spinner, T., Schlegel, U., Schafer, H., El-Assady, M.: explAIner: A visual analytics framework for interactive and explainable machine learning. IEEE Transactions on Visualization and Computer Graphics pp. 1–1 (2019). DOI 10.1109/tvcg.2019.2934629

127. Stahnke, J., Dörk, M., Müller, B., Thom, A.: Probing projections: Interaction techniques for interpreting arrangements and errors of dimensionality reductions. IEEE transactions on visualization and computer graphics **22**(1), 629–638 (2016)

128. Stein, M., Janetzko, H., Lamprecht, A., Breitkreutz, T., Zimmermann, P., Goldlücke, B., Schreck, T., Andrienko, G., Grossniklaus, M., Keim, D.A.: Bring it to the pitch: Combining video and movement data to enhance team sport analysis. IEEE Transactions on Visualization and Computer Graphics **24**(1), 13–22 (2018)

129. Stolper, C.D., Perer, A., Gotz, D.: Progressive visual analytics: User-driven visual exploration of in-progress analytics. IEEE Transactions on Visualization and Computer Graphics **20**(12), 1653–1662 (2014). DOI 10.1109/TVCG.2014.2346574

130. Tague, N.R.: The Quality Toolbox, Second Edition. ASQ Quality Press (2005)

131. Thom, D., Bosch, H., Koch, S., Wörner, M., Ertl, T.: Spatiotemporal anomaly detection through visual analysis of geolocated twitter messages. In: 2012 IEEE Pacific Visualization Symposium, pp. 41–48 (2012). DOI 10.1109/PacificVis.2012.6183572

132. Thom, D., Jankowski, P., Fuchs, G., Ertl, T., Bosch, H., Andrienko, N., Andrienko, G.: Thematic patterns in georeferenced tweets through space-time visual analytics. Computing in Science & Engineering **15**(03), 72–82 (2013). DOI 10.1109/MCSE.2013.70

133. Thomas, J., Cook, K.: Illuminating the Path: The Research and Development Agenda for Visual Analytics. IEEE (2005)

134. Tobler, W.R.: A Computer Movie Simulating Urban Growth in the Detroit Region. Economic Geography **46**, 234–240 (1970)

135. Tominski, C., Schumann, H.: Interactive Visual Data Analysis. AK Peters Visualization Series. CRC Press (2020). DOI 10.1201/9781315152707

136. Torkamani, S., Lohweg, V.: Survey on time series motif discovery. Wiley Interdiscip. Rev. Data Min. Knowl. Discov. **7** (2017)

137. Tufte, E.R.: The Visual Display of Quantitative Information. Graphics Press, Cheshire, CT, USA (1986)

138. Tukey, J.W.: Exploratory Data Analysis. Addison-Wesley (1977)
139. Tukey, J.W., Tukey, P.A.: Computer graphics and exploratory data analysis: An introduction. The Collected Works of John W. Tukey: Graphics: 1965-1985 **5**, 419 (1988)
140. Turkay, C., Lundervold, A., Lundervold, A.J., Hauser, H.: Hypothesis generation by interactive visual exploration of heterogeneous medical data. In: Human-Computer Interaction and Knowledge Discovery in Complex, Unstructured, Big Data, pp. 1–12. Springer (2013)
141. van den Elzen, S., Holten, D., Blaas, J., van Wijk, J.J.: Reducing snapshots to points: A visual analytics approach to dynamic network exploration. IEEE Transactions on Visualization and Computer Graphics **22**(1), 1–10 (2016). DOI 10.1109/TVCG.2015.2468078
142. van der Maaten, L., Hinton, G.: Visualizing high-dimensional data using t-sne. Journal of Machine Learning Research **9**(nov), 2579–2605 (2008)
143. von Landesberger, T., Brodkorb, F., Roskosch, P., Andrienko, N., Andrienko, G., Kerren, A.: Mobilitygraphs: Visual analysis of mass mobility dynamics via spatio-temporal graphs and clustering. IEEE Transactions on Visualization and Computer Graphics **22**(1), 11–20 (2016). DOI 10.1109/TVCG.2015.2468111
144. von Landesberger, T., Gorner, M., Schreck, T.: Visual analysis of graphs with multiple connected components. In: 2009 IEEE Symposium on Visual Analytics Science and Technology, pp. 155–162 (2009). DOI 10.1109/VAST.2009.5333893
145. Walker, J.S., Borgo, R., Jones, M.W.: Timenotes: A study on effective chart visualization and interaction techniques for time-series data. IEEE Trans. Vis. Comput. Graph. **22**(1), 549–558 (2016)
146. Wang, Y., Xue, M., Wang, Y., Yan, X., Chen, B., Fu, C., Hurter, C.: Interactive structure-aware blending of diverse edge bundling visualizations. IEEE Transactions on Visualization and Computer Graphics **26**(1), 687–696 (2020). DOI 10.1109/TVCG.2019.2934805
147. Wattenberg, M.: Arc diagrams: visualizing structure in strings. In: IEEE Symposium on Information Visualization, 2002. INFOVIS 2002., pp. 110–116 (2002). DOI 10.1109/INFVIS.2002.1173155
148. Wattenberg, M., Viégas, F.B.: The word tree, an interactive visual concordance. IEEE Transactions on Visualization and Computer Graphics **14**(6), 1221–1228 (2008). DOI 10.1109/TVCG.2008.172
149. Weaver, C.: Building highly-coordinated visualizations in improvise. In: IEEE Symposium on Information Visualization, pp. 159–166 (2004). DOI 10.1109/INFVIS.2004.12
150. Weng, J., Lim, E.P., Jiang, J., He, Q.: Twitterrank: finding topic-sensitive influential twitterers. In: Proceedings of the third ACM international conference on Web search and data mining, pp. 261–270 (2010)
151. van Wijk, J.J., van Selow, E.R.: Cluster and calendar based visualization of time series data. In: Proc. IEEE Symposium on Information Visualization (InfoVis), pp. 4–9 (1999)
152. Wilkinson, L., Anand, A., Grossman, R.: Graph-theoretic scagnostics. In: Proceedings of the Proceedings of the 2005 IEEE Symposium on Information Visualization, p. 21. IEEE Computer Society (2005)
153. Xu, P., Mei, H., Ren, L., Chen, W.: Vidx: Visual diagnostics of assembly line performance in smart factories. IEEE Transactions on Visualization and Computer Graphics **23**(1), 291–300 (2017). DOI 10.1109/TVCG.2016.2598664
154. Yamauchi, T., Xiao, K., Bowman, C., Mueen, A.: Dynamic time warping: A single dry electrode eeg study in a self-paced learning task. In: Proceedings of the 2015 International Conference on Affective Computing and Intelligent Interaction (ACII), ACII '15, p. 56–62. IEEE Computer Society, USA (2015). DOI 10.1109/ACII.2015.7344551
155. Yan, X., Guo, J., Lan, Y., Cheng, X.: A biterm topic model for short texts. In: Proceedings of the 22nd international conference on World Wide Web, pp. 1445–1456. ACM (2013)
156. Yang, L.: Data embedding techniques and applications. In: Proceedings of the 2nd International Workshop on Computer Vision Meets Databases, CVDB '05, p. 29–33. Association for Computing Machinery, New York, NY, USA (2005). DOI 10.1145/1160939.1160948
157. Zhang, Z., McDonnell, K.T., Zadok, E., Mueller, K.: Visual correlation analysis of numerical and categorical data on the correlation map. IEEE Transactions on Visualization and Computer Graphics **21**(2), 289–303 (2015)

Index

© Springer Nature Switzerland AG 2020
N. Andrienko et al., *Visual Analytics for Data Scientists*,
https://doi.org/10.1007/978-3-030-56146-8

Index